Thermodynamics Demystified

Demystified Series

Accounting Demystified
Advanced Calculus Demystified
Advanced Physics Demystified
Advanced Statistics Demystified
Algebra Demystified
Alternative Energy Demystified
Anatomy Demystified
Astronomy Demystified
Audio Demystified
Biochemistry Demystified
Biology Demystified
Biotechnology Demystified
Business Calculus Demystified
Business Math Demystified
Business Statistics Demystified
C++ Demystified
Calculus Demystified
Chemistry Demystified
Circuit Analysis Demystified
College Algebra Demystified
Complex Variables Demystified
Corporate Finance Demystified
Databases Demystified
Diabetes Demystified
Differential Equations Demystified
Digital Electronics Demystified
Discrete Mathematics Demystified
Dosage Calculations Demystified
Earth Science Demystified
Electricity Demystified
Electronics Demystified
Engineering Statistics Demystified
Environmental Science Demystified
Everyday Math Demystified
Fertility Demystified
Financial Planning Demystified
Fluid Mechanics Demystified
Forensics Demystified
French Demystified
Genetics Demystified
Geometry Demystified
German Demystified
Global Warming and Climate Change Demystified
Hedge Funds Demystified
Investing Demystified
Italian Demystified
Java Demystified
JavaScript Demystified
Lean Six Sigma Demystified
Linear Algebra Demystified
Macroeconomics Demystified
Management Accounting Demystified
Math Proofs Demystified
Math Word Problems Demystified
Mathematica Demystified
MATLAB® Demystified
Medical Billing and Coding Demystified
Medical Charting Demystified
Medical-Surgical Nursing Demystified
Medical Terminology Demystified
Meteorology Demystified
Microbiology Demystified
Microeconomics Demystified
Nanotechnology Demystified
Nurse Management Demystified
OOP Demystified
Options Demystified
Organic Chemistry Demystified
Pharmacology Demystified
Physics Demystified
Physiology Demystified
Pre-Algebra Demystified
Precalculus Demystified
Probability Demystified
Project Management Demystified
Psychology Demystified
Quantum Field Theory Demystified
Quantum Mechanics Demystified
Real Estate Math Demystified
Relativity Demystified
Robotics Demystified
Sales Management Demystified
Signals and Systems Demystified
Six Sigma Demystified
Spanish Demystified
SQL Demystified
Statics and Dynamics Demystified
Statistics Demystified
String Theory Demystified
Technical Analysis Demystified
Technical Math Demystified
Thermodynamics Demystified
Trigonometry Demystified
Vitamins and Minerals Demystified

Thermodynamics Demystified

Merle C. Potter, Ph.D.

New York Chicago San Francisco Lisbon London
Madrid Mexico City Milan New Delhi San Juan
Seoul Singapore Sydney Toronto

The McGraw-Hill Companies

Library of Congress Cataloging-in-Publication Data

Potter, Merle C.
Thermodynamics demystified / Merle C. Potter.
p. cm.
Includes index.
ISBN 978-0-07-160599-1 (alk. paper)
1. Thermodynamics. I. Title.
QC311.15.P68 2009
536′.7—dc22 2009000165

1 2 3 4 5 6 7 8 9 0 DOC/DOC 0 1 5 4 3 2 1 0 9

ISBN 978-0-07-160599-1
MHID 0-07-160599-1

Sponsoring Editor
Judy Bass

Editing Supervisor
Stephen M. Smith

Production Supervisor
Pamela A. Pelton

Project Manager
Preeti Longia Sinha, International Typesetting and Composition

Copy Editor
Bhavna Gupta, International Typesetting and Composition

Proofreader
Upendra Prasad, International Typesetting and Composition

Art Director, Cover
Jeff Weeks

Composition
International Typesetting and Composition

Printed and bound by RR Donnelley.

ABOUT THE AUTHOR

Merle C. Potter, Ph.D., has engineering degrees from Michigan Technological University and the University of Michigan. He has coauthored *Fluid Mechanics*, *Mechanics of Fluids*, *Thermodynamics for Engineers*, *Thermal Sciences*, *Differential Equations*, *Advanced Engineering Mathematics*, and *Jump Start the HP-48G* in addition to numerous exam review books. His research involved fluid flow stability and energy-related topics. The American Society of Mechanical Engineers awarded him the 2008 James Harry Potter Gold Medal. He is Professor Emeritus of Mechanical Engineering at Michigan State University and continues to write and golf.

CONTENTS

PREFACE

This book is intended to accompany a text used in the first course in thermodynamics that is required in all mechanical engineering departments, as well as several other departments. It provides a succinct presentation of the material so that the students more easily understand the more difficult concepts. Many thermodynamics texts are over 900 pages long and it is often difficult to ferret out the essentials due to the excessive verbiage. This book presents those essentials.

The basic principles upon which a study of thermodynamics is based are illustrated with numerous examples and practice exams, with solutions, that allow students to develop their problem-solving skills. All examples and problems are presented using SI metric units. English-unit equivalents are given in App. A.

The mathematics required to solve the problems is that used in several other engineering courses. The more-advanced mathematics is typically not used in an introductory course in thermodynamics. Calculus is more than sufficient.

The quizzes at the end of each chapter contain four-part, multiple-choice problems similar in format to those found in national exams, such as the Fundamentals of Engineering exam (the first of two exams required in the engineering registration process), the Graduate Record Exam (required when applying for most graduate schools), and the LSAT and MCAT exams. Engineering courses do not, in general, utilize multiple-choice exams but it is quite important that students gain experience in taking such exams. This book allows that experience. If one correctly answers 50 percent or more of multiple-choice questions correctly, that is quite good.

If you have comments, suggestions, or corrections or simply want to opine, please email me at MerleCP@sbcglobal.net. It is impossible to write a book free of errors, but if I'm made aware of them, I can have them corrected in future printings.

Merle C. Potter, Ph.D.

Thermodynamics Demystified

CHAPTER 1

Basic Principles

Thermodynamics involves the storage, transformation, and transfer of energy. Energy is *stored* as internal energy (due to temperature), kinetic energy (due to motion), potential energy (due to elevation), and chemical energy (due to chemical composition); it is *transformed* from one of these forms to another; and it is *transferred* across a boundary as either heat or work. We will present equations that relate the transformations and transfers of energy to properties such as temperature, pressure, and density. The properties of materials thus become very important. Many equations will be based on experimental observations that have been presented as mathematical statements, or *laws*: primarily the first and second laws of thermodynamics.

The mechanical engineer's objective in studying thermodynamics is most often the analysis of a rather complicated device, such as an air conditioner, an engine, or a power plant. As the fluid flows through such a device, it is assumed to be a continuum in which there are measurable quantities such as pressure, temperature, and velocity. This book, then, will be restricted to *macroscopic* or *engineering thermodynamics*. If the behavior of individual molecules is important, *statistical thermodynamics* must be consulted.

1.1 The System and Control Volume

A thermodynamic *system* is a fixed quantity of matter upon which attention is focused. The system surface is one like that surrounding the gas in the cylinder of Fig. 1.1; it may also be an imagined boundary like the deforming boundary of a certain amount of water as it flows through a pump. In Fig. 1.1 the system is the compressed gas, the *working fluid,* and the dashed line shows the *system boundary.*

All matter and space external to a system is its *surroundings.* Thermodynamics is concerned with the interactions of a system and its surroundings, or one system interacting with another. A system interacts with its surroundings by transferring energy across its boundary. No material crosses the boundary of a system. If the system does not exchange energy with the surroundings, it is an *isolated system.*

An analysis can often be simplified if attention is focused on a particular volume in space into which, and/or from which, a substance flows. Such a volume is a *control volume.* A pump and a deflating balloon are examples of control volumes. The surface that completely surrounds the control volume is called a *control surface.* An example is sketched in Fig. 1.2.

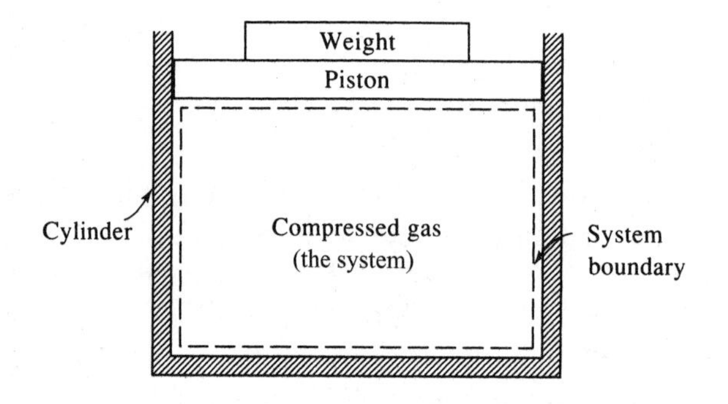

Figure 1.1 A system.

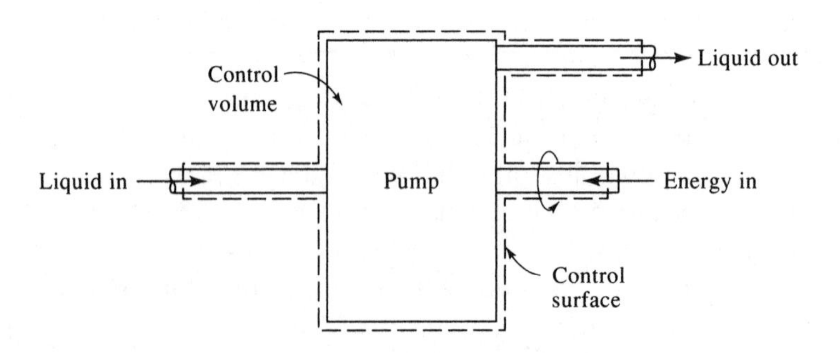

Figure 1.2 A control volume.

In a particular problem we must decide whether a system is to be considered or whether a control volume is more useful. If there is mass flux across a boundary, then a control volume is usually selected; otherwise, a system is identified. First, systems will be considered followed by the analysis of control volumes.

1.2 Macroscopic Description

In engineering thermodynamics we postulate that the material in our system or control volume is a *continuum;* that is, it is continuously distributed throughout the region of interest. Such a postulate allows us to describe a system or control volume using only a few measurable properties.

Consider the definition of *density* given by

$$\rho = \lim_{\Delta V \to 0} \frac{\Delta m}{\Delta V} \tag{1.1}$$

where Δm is the mass contained in the volume ΔV, shown in Fig. 1.3. Physically, ΔV cannot be allowed to shrink to zero since, if ΔV became extremely small, Δm would vary discontinuously, depending on the number of molecules in ΔV.

There are, however, situations where the continuum assumption is not valid; for example, the re-entry of satellites. At an elevation of 100 km the *mean free path,* the average distance a molecule travels before it collides with another molecule, is about 30 mm; the macroscopic approach with its continuum assumption is already questionable. At 150 km the mean free path exceeds 3 m, which is comparable to the dimensions of the satellite! Under these conditions, statistical methods based on molecular activity must be used.

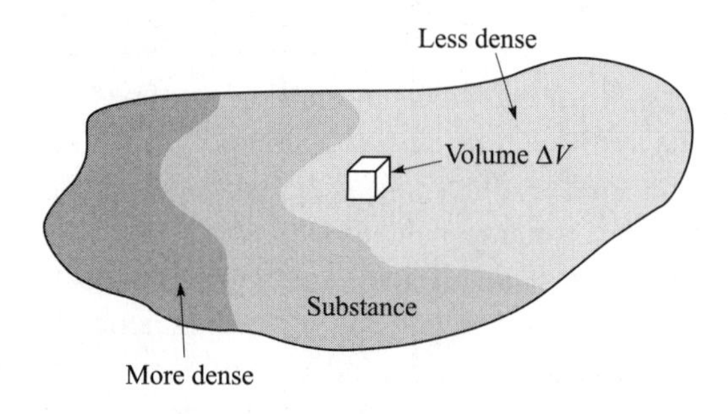

Figure 1.3 Mass as a continuum.

1.3 Properties and State of a System

The matter in a system may exist in several phases: a solid, a liquid, or a gas. A *phase* is a quantity of matter that has the same chemical composition throughout; that is, it is *homogeneous*. It is all solid, all liquid, or all gas. Phase boundaries separate the phases in what, when taken as a whole, is called a *mixture*. Gases can be mixed in any ratio to form a single phase. Two liquids that are miscible form a mixture when mixed; but liquids that are not miscible, such as water and oil, form two phases.

A *pure substance* is uniform in chemical composition. It may exist in more than one phase, such as ice, liquid water, and vapor, in which each phase would have the same composition. A uniform mixture of gases is a pure substance as long as it does not react chemically (as in combustion) or liquefy in which case the composition would change.

A *property* is any quantity that serves to describe a system. The *state* of a system is its condition as described by giving values to its properties at a particular instant. The common properties are pressure, temperature, volume, velocity, and position; others must occasionally be considered. Shape is important when surface effects are significant.

The essential feature of a property is that it has a unique value when a system is in a particular state, and this value does not depend on the previous states that the system passed through; that is, it is not a *path function*. Since a property is not dependent on the path, any change depends only on the initial and final states of the system. Using the symbol ϕ to represent a property, the mathematical statement is

$$\int_{\phi_2}^{\phi_1} d\phi = \phi_2 - \phi_1 \tag{1.2}$$

This requires that $d\phi$ be an *exact differential*; $\phi_2 - \phi_1$ represents the change in the property as the system changes from state 1 to state 2. There are several quantities that we will encounter, such as work, that are path functions for which an exact differential does not exist.

A relatively small number of *independent properties* suffice to fix all other properties and thus the state of the system. If the system is composed of a single phase, free from magnetic, electrical, and surface effects, the state is fixed when any two properties are fixed; this *simple system* receives most attention in engineering thermodynamics.

Thermodynamic properties are divided into two general types, intensive and extensive. An *intensive property* is one that does not depend on the mass of the system. Temperature, pressure, density, and velocity are examples since they are the same for the entire system, or for parts of the system. If we bring two systems together, intensive properties are not summed.

An *extensive property* is one that does depend on the mass of the system; mass, volume, momentum, and kinetic energy are examples. If two systems are brought together the extensive property of the new system is the sum of the extensive properties of the original two systems.

If we divide an extensive property by the mass, a *specific property* results. The *specific volume* is thus defined to be

$$v = \frac{V}{m} \tag{1.3}$$

We will generally use an uppercase letter to represent an extensive property (exception: m for mass) and a lowercase letter to denote the associated intensive property.

1.4 Equilibrium, Processes, and Cycles

When the temperature of a system is referred to, it is assumed that all points of the system have the same, or approximately the same, temperature. When the properties are constant from point to point and when there is no tendency for change with time, a condition of *thermodynamic equilibrium* exists. If the temperature, for example, is suddenly increased at some part of the system boundary, spontaneous redistribution is assumed to occur until all parts of the system are at the same increased temperature.

If a system would undergo a large change in its properties when subjected to some small disturbance, it is said to be in *metastable equilibrium*. A mixture of gasoline and air, and a bowling ball on top of a pyramid are examples.

When a system changes from one equilibrium state to another, the path of successive states through which the system passes is called a *process*. If, in the passing from one state to the next, the deviation from equilibrium is small, and thus negligible, a *quasiequilibrium* process occurs; in this case, each state in the process can be idealized as an equilibrium state. Quasiequilibrium processes can approximate many processes, such as the compression and expansion of gases in an internal combustion engine, with acceptable accuracy. If a system undergoes a quasiequilibrium process (such as the compression of air in a cylinder of an engine) it may be sketched on appropriate coordinates by using a solid line, as shown between states 1 and 2 in Fig. 1.4*a*. If the system, however, goes from one equilibrium state to another through a series of nonequilibrium states (as in combustion) a *nonequilibrium process* occurs. In Fig. 1.4*b* the dashed curve represents a nonequilibrium process between (V_1, P_1) and (V_2, P_2); properties are not uniform throughout the system and thus the state of the system is not known at each state between the two end states.

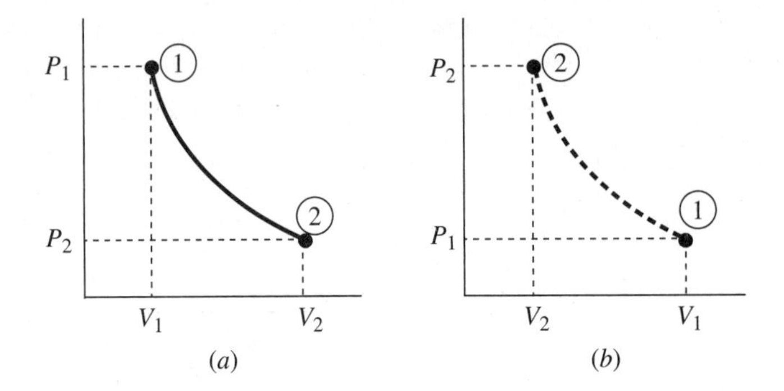

Figure 1.4 A process. (*a*) Quasiequilibrium. (*b*) Nonequilibrium.

Whether a particular process may be considered quasiequilibrium or nonequilibrium depends on how the process is carried out. Let us add the weight W to the piston of Fig. 1.5 and explain how W can be added in a nonequilibrium manner or in an equilibrium manner. If the weight is added suddenly as one large weight, as in Fig. 1.5*a,* a nonequilibrium process will occur in the gas. If we divide the weight into a large number of small weights and add them one at a time, as in Fig. 1.5*b,* a quasiequilibrium process will occur.

Note that the surroundings play no part in the notion of equilibrium. It is possible that the surroundings do work on the system via friction; for quasiequilibrium it is only required that the properties of the system be uniform at any instant during a process.

When a system in a given initial state experiences a series of quasiequilibrium processes and returns to the initial state, the system undergoes a *cycle.* At the end of the cycle the properties of the system have the same values they had at the beginning.

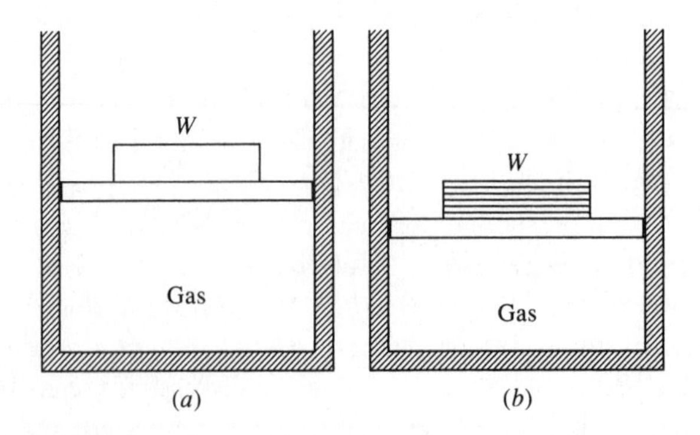

Figure 1.5 (*a*) Equilibrium and (*b*) nonequilibrium additions of weight.

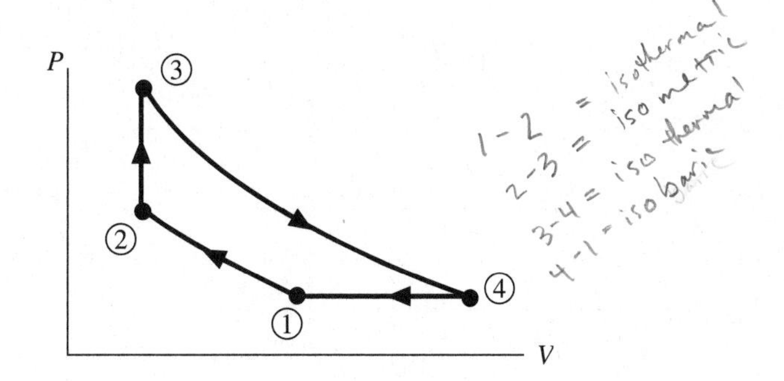

Figure 1.6 Four processes that make up a cycle.

The prefix *iso-* is attached to the name of any property that remains unchanged in a process. An *isothermal* process is one in which the temperature is held constant; in an *isobaric* process, the pressure remains constant; an *isometric* process is a constant-volume process. Note the isobaric and the isometric legs in Fig. 1.6 (the lines between states 4 and 1 and between 2 and 3, respectively).

1.5 Units

While the student is undoubtedly comfortable using SI units, much of the data gathered and available for use in the United States is in English units. Table 1.1 lists units and conversions for many thermodynamic quantities. Observe the use of V for both volume and velocity. Appendix A presents the conversions for numerous additional quantities.

When expressing a quantity in SI units, certain letter prefixes shown in Table 1.2 may be used to represent multiplication by a power of 10. So, rather than writing 30 000 W (commas are not used in the SI system) or 30×10^3 W, we may simply write 30 kW.

The units of various quantities are interrelated via the physical laws obeyed by the quantities. It follows that, no matter the system used, all units may be expressed as algebraic combinations of a selected set of *base units.* There are seven base units in the SI system: m, kg, s, K, mol (mole), A (ampere), cd (candela). The last one is rarely encountered in engineering thermodynamics. Note that N (newton) is not listed as a base unit. It is related to the other units by Newton's second law,

$$F = ma \tag{1.4}$$

If we measure F in newtons, m in kg, and a in m/s^2, we see that N = kg · m/s^2. So, the newton is expressed in terms of the base units.

Table 1.1 Conversion Factors

Quantity	Symbol	SI Units	English Units	To Convert from English to SI Units Multiply by
Length	L	m	ft	0.3048
Mass	m	kg	lbm	0.4536
Time	t	s	sec	1
Area	A	m^2	ft^2	0.09290
Volume	V	m^3	ft^3	0.02832
Velocity	V	m/s	ft/sec	0.3048
Acceleration	a	m/s^2	ft/sec^2	0.3048
Angular velocity	ω	rad/s	rad/sec	1
Force, Weight	F, W	N	lbf	4.448
Density	ρ	kg/m^3	lbm/ft^3	16.02
Specific weight	γ	N/m^3	lbf/ft^3	157.1
Pressure	P	kPa	psi	6.895
Work, Energy	W, E, U	J	ft · lbf	1.356
Heat transfer	Q	J	Btu	1055
Power	$\dot{W}$	W	ft · lbf/sec	1.356
		W	hp	746
Heat flux	$\dot{Q}$	J/s	Btu/sec	1055
Mass flux	$\dot{m}$	kg/s	lbm/sec	0.4536
Flow rate	$\dot{V}$	m^3/s	ft^3/sec	0.02832
Specific heat	C	kJ/kg · K	Btu/lbm · °R	4.187
Specific enthalpy	h	kJ/kg	Btu/lbm	2.326
Specific entropy	s	kJ/kg · K	Btu/lbm · °R	4.187
Specific volume	v	m^3/kg	ft^3/lbm	0.06242

Weight is the force of gravity; by Newton's second law,

$$W = mg \tag{1.5}$$

Since mass remains constant, the variation of W is due to the change in the acceleration of gravity g (from about 9.77 m/s^2 on the highest mountain to 9.83 m/s^2 in the deepest ocean trench, only about a 0.3% variation from 9.80 m/s^2). We will use the standard sea-level value of 9.81 m/s^2 (32.2 ft/sec^2), unless otherwise stated.

Table 1.2 Prefixes for SI Units

Multiplication Factor	Prefix	Symbol
10^{12}	tera	T
10^{9}	giga	G
10^{6}	mega	M
10^{3}	kilo	k
10^{-2}	centi*	c
10^{-3}	mili	m
10^{-6}	micro	μ
10^{-9}	nano	n
10^{-12}	pico	p

*Discouraged except in cm, cm², or cm³.

EXAMPLE 1.1

Express the energy unit J (joule) in terms of the base units.

Solution

The units are related by recalling that energy is force times distance:

$$[F \times d] = \text{N} \cdot \text{m} = \frac{\text{kg} \cdot \text{m}}{\text{s}^2} \cdot \text{m} = \text{kg} \cdot \text{m}^2/\text{s}^2$$

We used Newton's second law to relate newtons to the base units.

EXAMPLE 1.2

Express the kinetic energy $mV^2/2$ in acceptable terms if $m = 10$ kg and $V = 5$ m/s.

Solution

Using the SI system we have

$$\frac{1}{2}mV^2 = \frac{1}{2} \times 10 \times 5^2 = 125\ \text{kg} \cdot \text{m}^2/\text{s}^2 = 125\ \frac{\text{N} \cdot \text{s}^2}{\text{m}} \frac{\text{m}^2}{\text{s}^2} = 125\ \text{N} \cdot \text{m}$$

1.6 Density, Specific Volume, and Specific Weight

By Eq. (1.1), density is mass per unit volume; by Eq. (1.3), specific volume is volume per unit mass. By comparing their definitions, we see that the two properties are related by

$$v = \frac{1}{\rho} \tag{1.6}$$

Associated with (mass) density is *weight density*, or *specific weight* γ:

$$\gamma = \frac{W}{V} \tag{1.7}$$

with units N/m^3 (lbf/ft^3). (Note that γ is volume-specific, not mass-specific.) Specific weight is related to density through $W = mg$:

$$\gamma = \frac{mg}{mv} = \rho g \tag{1.8}$$

For water, nominal values of ρ and γ are, respectively, 1000 kg/m^3 and 9810 N/m^3. For air at standard conditions, the nominal values are 1.21 kg/m^3 and 11.86 N/m^3.

EXAMPLE 1.3

The mass of air in a room 3 m × 5 m × 20 m is known to be 350 kg. Determine the density, specific volume, and specific weight of the air.

Solution

Equations (1.1), (1.6), and (1.8) are used:

$$\rho = \frac{m}{V} = \frac{350}{3 \times 5 \times 20} = 1.167 \text{ kg/m}^3$$

$$v = \frac{1}{\rho} = \frac{1}{1.167} = 0.857 \text{ m}^3/\text{kg}$$

$$\gamma = \rho g = 1.167 \times 9.81 = 11.45 \text{ N/m}^3$$

1.7 Pressure

In gases and liquids, the effect of a normal force acting on an area is the *pressure.* If a force ΔF acts at an angle to an area ΔA (Fig. 1.7), only the normal component ΔF_n enters into the definition of pressure:

$$P = \lim_{\Delta A \to 0} \frac{\Delta F_n}{\Delta A} \tag{1.9}$$

The SI unit of pressure is the pascal (Pa), where 1 Pa = 1 N/m^2. The pascal is a relatively small unit so pressure is usually measured in kPa. By considering the pressure forces acting on a triangular fluid element of constant depth we can show that the pressure at a point in a fluid in equilibrium is the same in all directions; it is a scalar quantity. For gases and liquids in relative motion, the pressure may vary from point to point, even at the same elevation; but it does not vary with direction at a given point.

PRESSURE VARIATION WITH ELEVATION

In the atmosphere, pressure varies with elevation. This variation can be expressed mathematically by summing vertical forces acting on an infinitesimal element of air. The force PA on the bottom of the element and $(P + dP)A$ on the top balance the weight $\rho gAdz$ to provide

$$dP = -\rho g dz \tag{1.10}$$

If ρ is a known function of z, the above equation can be integrated to give $P(z)$

$$P(z) - P_0 = -\int_0^z \gamma \, dz \tag{1.11}$$

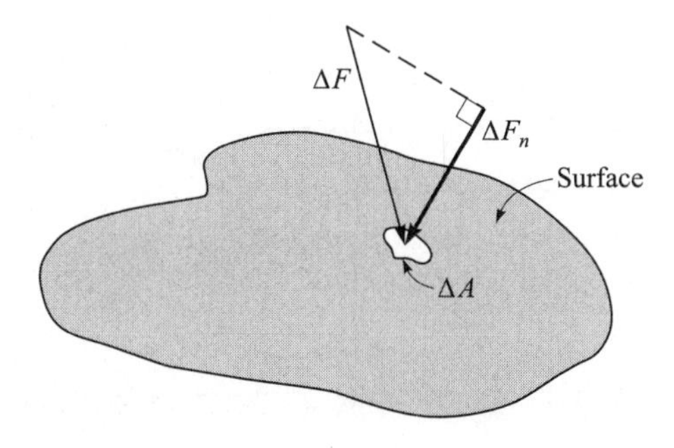

Figure 1.7 The normal component of a force.

where we used $\rho g = \gamma$. For a liquid, γ is constant. If we write Eq. (1.10) using $dh = -dz$, we have

$$dP = \gamma dh \tag{1.12}$$

where h is measured positive downward. Integrating this equation, starting at a liquid surface where usually $P = 0$, results in

$$P = \gamma h \tag{1.13}$$

This equation can be used to convert a pressure to pascals when that pressure is measured in meters of water or millimeters of mercury.

In many relations, *absolute pressure* must be used. Absolute pressure is gage pressure plus the local atmospheric pressure:

$$P_{\text{abs}} = P_{\text{gage}} + P_{\text{atm}} \tag{1.14}$$

A negative gage pressure is often called a *vacuum,* and gages capable of reading negative pressures are *vacuum gages.* A gage pressure of −50 kPa would be referred to as a vacuum of 50 kPa (the sign is omitted).

Figure 1.8 shows the relationships between absolute and gage pressure at two different points.

The word "gage" is generally specified in statements of gage pressure, e.g., $p =$ 200 kPa gage. If "gage" is not mentioned, the pressure will, in general, be an absolute pressure. Atmospheric pressure is an absolute pressure, and will be taken as 100 kPa (at sea level), unless otherwise stated. It is more accurately 101.3 kPa at standard conditions. It should be noted that atmospheric pressure is highly dependent

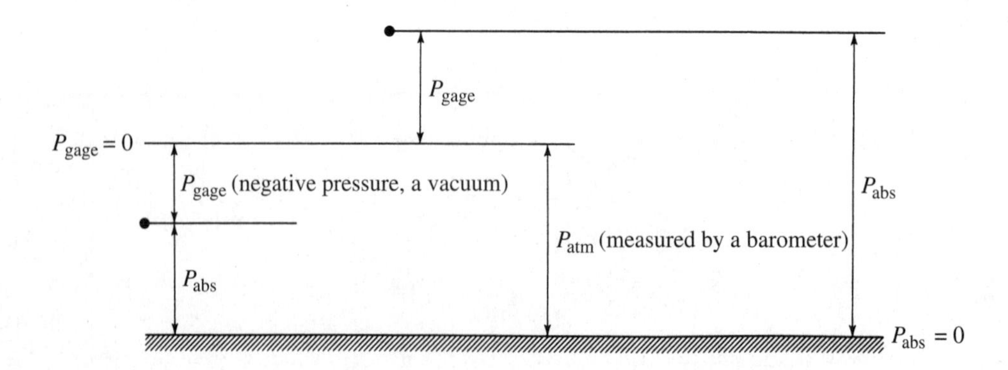

Figure 1.8 Absolute and gage pressure.

on elevation; in Denver, Colorado, it is about 84 kPa; in a mountain city with elevation 3000 m, it is only 70 kPa. In Table B.1, the variation of atmospheric pressure with elevation is listed.

EXAMPLE 1.4

Express a pressure gage reading of 20 mm Hg in absolute pascals at an elevation of 2000 m. Use $\gamma_{Hg} = 13.6\gamma_{water}$.

Solution

First we convert the pressure reading into pascals. We have

$$P = \gamma_{Hg} h = (13.6 \times 9810)\ \frac{N}{m^3} \times 0.020\ m = 2668\ \frac{N}{m^2} \quad \text{or} \quad 2668\ Pa$$

To find the absolute pressure we simply add the atmospheric pressure to the above value. Referring to Table B.1, $P_{atm} = 0.7846 \times 101.3 = 79.48$ kPa. The absolute pressure is then

$$P = P_{gage} + P_{atm} = 2.668 + 79.48 = 82.15\ \text{kPa}$$

Note: We express an answer to either 3 or 4 significant digits, seldom, if ever, more than 4. Information in a problem is assumed known to at most 4 significant digits. For example, if the diameter of a pipe is stated as 2 cm, it is assumed that it is 2.000 cm. Material properties, such as density or a gas constant, are seldom known to even 4 significant digits. So, it is not appropriate to state an answer to more than 4 significant digits.

EXAMPLE 1.5

A 10-cm-diameter cylinder contains a gas pressurized to 600 kPa. A frictionless piston is held in position by a stationary spring with a spring constant of 4.8 kN/m. How much is the spring compressed?

Solution

The pressure given is assumed to be absolute (this is the case unless otherwise stated). A force balance on the piston provides

$$PA - P_{atm} A = K_{spring} \Delta x$$

$$(600\,000 - 100\,000)\pi \times 0.05^2 = 4800 \Delta x \qquad \therefore \Delta x = 0.818\ m$$

Make sure you check the units. Pressure typically has units of pascals when input into equations and area has units of square meters.

1.8 Temperature

Temperature is actually a measure of molecular activity. However, in classical thermodynamics the quantities of interest are defined in terms of macroscopic observations only, and a definition of temperature using molecular measurements is not useful. Thus we must proceed without actually defining temperature. What we shall do instead is discuss *equality of temperatures.*

EQUALITY OF TEMPERATURES

Let two bodies be isolated from the surroundings but placed in contact with each other. If one is hotter than the other, the hotter body will become cooler and the cooler body will become hotter; both bodies will undergo change until all properties (e.g., pressure) of the bodies cease to change. When this occurs, *thermal equilibrium* is said to have been established between the two bodies. Hence, we state that two systems have equal temperatures if no change occurs in any of their properties when the systems are brought into contact with each other. In other words, if two systems are in thermal equilibrium, their temperatures are postulated to be equal.

A rather obvious observation is referred to as the *zeroth law of thermodynamics:* if two systems are equal in temperature to a third, they are equal in temperature to each other.

RELATIVE TEMPERATURE SCALE

To establish a temperature scale, we choose the number of subdivisions, called degrees, between the ice point and the steam point. The *ice point* exists when ice and water are in equilibrium at a pressure of 101 kPa; the *steam point* exists when liquid water and its vapor are in a state of equilibrium at a pressure of 101 kPa. On the Fahrenheit scale there are 180° between these two points; on the Celsius scale, 100°. On the Fahrenheit scale the ice point is assigned the value of 32 and on the Celsius scale it is assigned the value 0. These selections allow us to write

$$T_C = \frac{5}{9}(T_F - 32) \tag{1.15}$$

ABSOLUTE TEMPERATURE SCALE

The second law of thermodynamics will allow us to define an absolute temperature scale; however, since we have not introduced the second law at this point and we have

immediate use for absolute temperature, an empirical absolute temperature scale will be presented. The relations between absolute and relative temperatures are

$$T_K = T_C + 273.15 \tag{1.16}$$

The value 273 is used where precise accuracy is not required, which is the case for most engineering situations. The absolute temperature on the Celsius scale is given in kelvins (K). Note: We do not use the degree symbol when writing kelvins, e.g., $T_1 = 400$ K.

1.9 Energy

A system may possess several different forms of energy. Assuming uniform properties throughout the system, its *kinetic energy* is given by

$$KE = \frac{1}{2}mV^2 \tag{1.17}$$

where[1] V is the velocity of each particle of substance, assumed constant over the entire system. If the velocity is not constant for each particle, then the kinetic energy is found by integrating over the system.

The energy that a system possesses due to its elevation h above some arbitrarily selected datum is its *potential energy;* it is determined from the equation

$$PE = mgh \tag{1.18}$$

Other forms of energy include the energy stored in a battery, energy stored in an electrical condenser, electrostatic potential energy, and surface energy. In addition, there is the energy associated with the translation, rotation, and vibration of the molecules, electrons, protons, and neutrons, and the chemical energy due to bonding between atoms and between subatomic particles. All of these forms of energy will be referred to as *internal energy* and designated by the letter U. In combustion, energy is released when the chemical bonds between atoms are rearranged. In this book, our attention will be primarily focused on the internal energy associated with the motion of molecules, i.e., temperature. In Chap. 9, the combustion process is presented.

Internal energy, like pressure and temperature, is a property of fundamental importance. A substance always has internal energy; if there is molecular activity,

[1]The context will make it obvious if V refers to volume or velocity. A textbook may use a rather clever symbol for one or the other, but that is really not necessary.

there is internal energy. We need not know, however, the absolute value of internal energy, since we will be interested only in its increase or decrease.

We now come to an important law, which is often of use when considering isolated systems. The law of *conservation of energy* states that the energy of an isolated system remains constant. Energy cannot be created or destroyed in an isolated system; it can only be transformed from one form to another. This is expressed as

$$KE + PE + U = \text{const} \qquad \text{or} \qquad \frac{1}{2}mV^2 + mgh + U = \text{const} \tag{1.19}$$

Consider a system composed of two automobiles that hit head on and are at rest after the collision. Because the energy of the system is the same before and after the collision, the initial total kinetic energy KE must simply have been transformed into another kind of energy, in this case, internal energy U, stored primarily in the deformed metal.

EXAMPLE 1.6

A 2200-kg automobile traveling at 90 km/h (25 m/s) hits the rear of a stationary, 1000-kg automobile. After the collision the large automobile slows to 50 km/h (13.89 m/s), and the smaller vehicle has a speed of 88 km/h (24.44 m/s). What has been the increase in internal energy, taking both vehicles as the system?

Solution

The kinetic energy before the collision is ($V = 25$ m/s)

$$KE_1 = \frac{1}{2}m_a V_a^2 = \frac{1}{2} \times 2200 \times 25^2 = 687\,500 \text{ J}$$

where the subscript a refers to the first automobile; the subscript b refers to the second one. After the collision the kinetic energy is

$$KE_2 = \frac{1}{2}m_a V_a^2 + \frac{1}{2}m_b V_b^2 = \frac{1}{2} \times 2200 \times 13.89^2 + \frac{1}{2} \times 1000 \times 24.44^2 = 510\,900 \text{ J}$$

The conservation of energy requires that

$$E_1 = E_2 \qquad \text{or} \qquad KE_1 + U_1 = KE_2 + U_2$$

Thus,

$$U_2 - U_1 = KE_1 - KE_2 = 687\,500 - 510\,900 = 176\,600 \text{ J} \quad \text{or} \quad 176.6 \text{ kJ}$$

Quiz No. 1

1. Engineering thermodynamics does not include energy
   (A) transfer
   (B) utilization
   (C) storage
   (D) transformation
2. Which of the following would be identified as a control volume?
   (A) Compression of the air-fuel mixture in a cylinder
   (B) Filling a tire with air at a service station
   (C) Compression of the gases in a cylinder
   (D) The flight of a dirigible
3. Which of the following is a quasiequilibrium process?
   (A) Mixing a fluid
   (B) Combustion
   (C) Compression of the air-fuel mixture in a cylinder
   (D) A balloon bursting
4. The standard atmosphere in meters of gasoline ($\gamma = 6660\ \text{N/m}^3$) is nearest
   (A) 24.9 m
   (B) 21.2 m
   (C) 18.3 m
   (D) 15.2 m
5. A gage pressure of 400 kPa acting on a 4-cm-diameter piston is resisted by a spring with a spring constant of 800 N/m. How much is the spring compressed? Neglect the piston weight and friction.
   (A) 63 cm
   (B) 95 cm
   (C) 1.32 m
   (D) 1.98 m
6. Which of the following processes can be approximated by a quasiequilibrium process?
   (A) The expansion of combustion gases in the cylinder of an automobile engine
   (B) The rupturing of a balloon

(C) The heating of the air in a room with a radiant heater

(D) The cooling of a hot copper block brought into contact with ice cubes

7. Determine the weight of a mass at a location where $g = 9.77\ \text{m/s}^2$ (on the top of Mt. Everest) if it weighed 40 N at sea level.

   (A) 39.62 N

   (B) 39.64 N

   (C) 39.78 N

   (D) 39.84 N

8. Determine γ if $g = 9.81\ \text{m/s}^2$, $V = 10\ \text{m}^3$, and $v = 20\ \text{m}^3/\text{kg}$.

   (A) $2.04\ \text{N/m}^3$

   (B) $1.02\ \text{N/m}^3$

   (C) $0.49\ \text{N/m}^3$

   (D) $0.05\ \text{N/m}^3$

9. If $P_{atm} = 100$ kPa, the pressure at a point where the gage pressure is 300 mmHg is nearest ($\gamma_{Hg} = 13.6\ \gamma_{water}$)

   (A) 40 kPa

   (B) 140 kPa

   (C) 160 kPa

   (D) 190 kPa

10. A large chamber is separated into compartments 1 and 2, as shown, that are kept at different pressures. Pressure gage *A* reads 400 kPa and pressure gage *B* reads 180 kPa. If the barometer reads 720 mmHg, determine the absolute pressure of *C*.

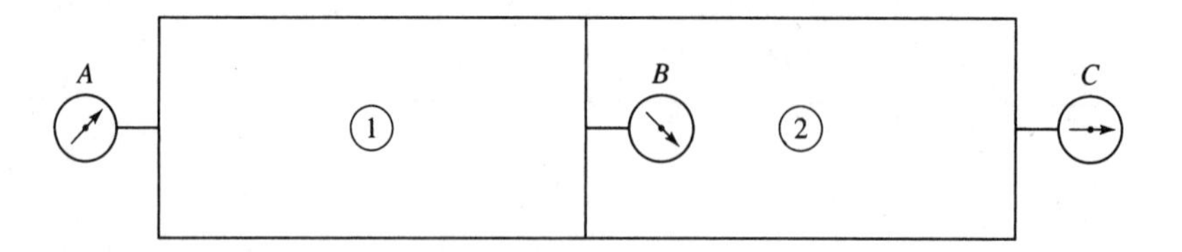

   (A) 320 kPa

   (B) 300 kPa

   (C) 280 kPa

   (D) 260 kPa

11. A 10-kg body falls from rest, with negligible interaction with its surroundings (no friction). Determine its velocity after it falls 5 m.
    (A) 19.8 m/s
    (B) 15.2 m/s
    (C) 12.8 m/s
    (D) 9.9 m/s
12. The potential energy stored in a spring is given by $Kx^2/2$, where K is the spring constant and x is the distance the spring is compressed. Two springs are designed to absorb the kinetic energy of an 1800-kg vehicle. Determine the spring constant necessary if the maximum compression is to be 100 mm for a vehicle speed of 16 m/s.
    (A) 23 MN/m
    (B) 25 MN/m
    (C) 27 MN/m
    (D) 29 MN/m

Quiz No. 2

1. In a quasiequilibrium process, the pressure
    (A) remains constant
    (B) varies with location
    (C) is everywhere constant at an instant
    (D) depends only on temperature
2. Which of the following is not an extensive property?
    (A) Momentum
    (B) Internal energy
    (C) Temperature
    (D) Volume
3. The joule unit can be converted to which of the following?
    (A) $kg \cdot m^2/s$
    (B) $kg \cdot m/s^2$
    (C) $Pa \cdot m^3$
    (D) Pa/m^2

4. Convert 178 kPa gage of pressure to absolute millimeters of mercury ($\rho_{hg} = 13.6\rho_{water}$).
   (A) 2080 mm
   (B) 1820 mm
   (C) 1640 mm
   (D) 1490 mm
5. Calculate the pressure in the 240-mm-diameter cylinder shown. The spring is compressed 60 cm. Neglect friction.

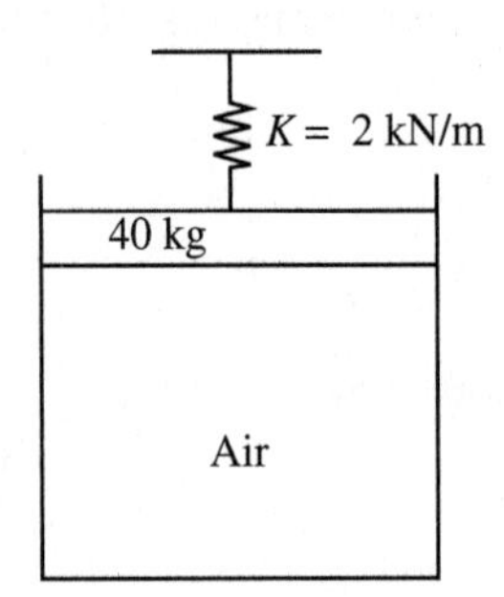

   (A) 198 kPa
   (B) 135 kPa
   (C) 110 kPa
   (D) 35 kPa
6. A cubic meter of a liquid has a weight of 9800 N at a location where $g = 9.79$ m/s^2. What is its weight at a location where $g = 9.83$ m/s^2?
   (A) 9780 N
   (B) 9800 N
   (C) 9820 N
   (D) 9840 N
7. Calculate the force necessary to accelerate a 900-kg rocket vertically upward at the rate of 30 m/s^2.
   (A) 18.2 kN
   (B) 22.6 kN
   (C) 27.6 kN
   (D) 35.8 kN

8. Calculate the weight of a body that occupies 200 m^3 if its specific volume is 10 m^3/kg.
   (A) 20 N
   (B) 92.1 N
   (C) 132 N
   (D) 196 N
9. The pressure at a point where the gage pressure is 70 cm of water is nearest
   (A) 169 kPa
   (B) 107 kPa
   (C) 69 kPa
   (D) 6.9 kPa
10. A bell jar 200 mm in diameter sits on a flat plate and is evacuated until a vacuum of 720 mmHg exists. The local barometer reads 760 mmHg. Estimate the force required to lift the jar off the plate. Neglect the weight of the jar.
    (A) 3500 N
    (B) 3000 N
    (C) 2500 N
    (D) 2000 N
11. An object that weighs 4 N traveling at 60 m/s enters a viscous liquid and is essentially brought to rest before it strikes the bottom. What is the increase in internal energy, taking the object and the liquid as the system? Neglect the potential energy change.
    (A) 734 J
    (B) 782 J
    (C) 823 J
    (D) 876 J
12. A 1700-kg vehicle traveling at 82 km/h collides head-on with a 1400-kg vehicle traveling at 90 km/h. If they come to rest immediately after impact, determine the increase in internal energy, taking both vehicles as the system.
    (A) 655 kJ
    (B) 753 kJ
    (C) 879 kJ
    (D) 932 kJ

CHAPTER 2

Properties of Pure Substances

In this chapter the relationships between pressure, specific volume, and temperature will be presented for a pure substance. A pure substance is homogeneous, but may exist in more than one phase, with each phase having the same chemical composition. Water is a pure substance; the various combinations of its three phases (vapor, liquid, ice) have the same chemical composition. Air in the gas phase is a pure substance, but liquid air has a different chemical composition. Air is not a pure substance if it exists in more than one phase. In addition, only a *simple compressible substance*, one that is essentially free of magnetic, electrical, or surface tension effects, will be considered.

2.1 The P-v-T Surface

It is well known that a substance can exist in three different phases: solid, liquid, and gas. Assume that a solid is contained in a piston-cylinder arrangement such that

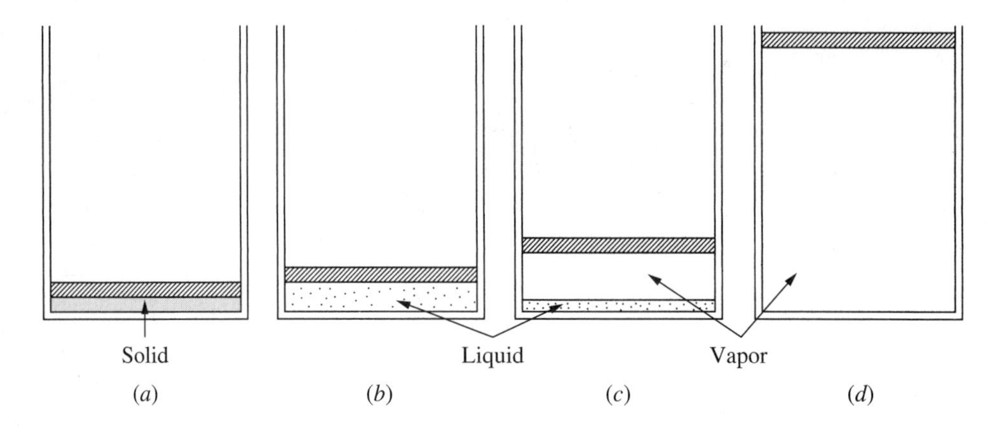

Figure 2.1 The solid, liquid, and vapor phases of a substance.

the pressure is maintained at a constant value; heat is added to the cylinder, causing the substance to experience all three phases, as in Fig. 2.1. We will record the temperature T and specific volume v during the experiment. Start with the solid at some low temperature, as in Fig. 2.2*a*; then add heat until it is all liquid (v does not increase very much). After all the solid is melted, the temperature of the liquid again rises until vapor just begins to form; this state is called the *saturated liquid* state. During the phase change from liquid to vapor,[1] called *vaporization*, the temperature remains constant as heat is added. Finally, all the liquid is vaporized and the state of *saturated vapor* exists, after which the temperature again rises with heat addition. Note, the specific volumes of the solid and liquid are much less than the specific volume of vapor at relatively low pressures.

If the experiment is repeated a number of times using different pressures, a T-v diagram results, shown in Fig. 2.2*b*. At pressures that exceed the pressure of the *critical point*, the liquid simply changes to a vapor without a constant-temperature vaporization process. Property values of the critical point for various substances are included in Table B.3.

The experiment could also be run by holding the temperature fixed and decreasing the pressure, as in Fig. 2.3*a* (the solid is not displayed). The solid would change to a liquid, and the liquid to a vapor, as in the experiment that led to Fig. 2.2. The T-v diagram, with only the liquid and vapor phases shown, is displayed in Fig. 2.3*b*.

The process of melting, vaporization, and *sublimation* (the transformation of a solid directly to a vapor) are shown in Fig. 2.3*c*. Distortions are made in all three diagrams so that the various regions are displayed. The *triple point* is the point where all three phases exist in equilibrium together. A constant pressure line is

[1]A phase change from vapor to a liquid is *condensation*.

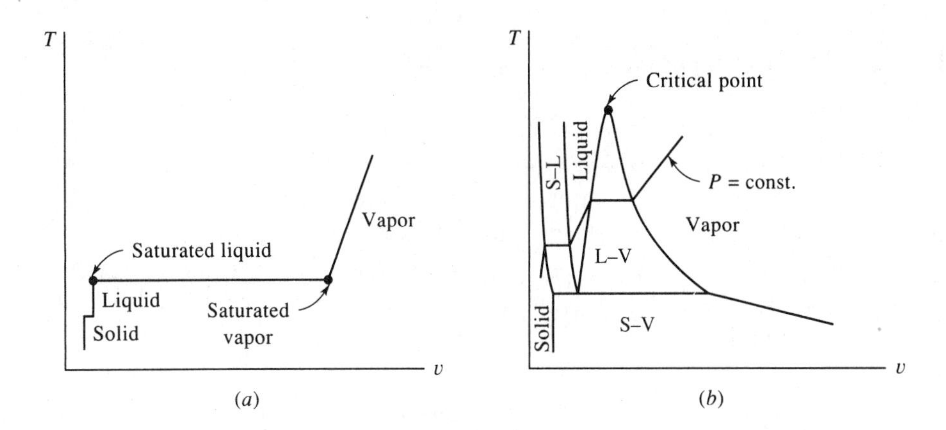

Figure 2.2 The T-v diagram.

shown on the T-v diagram and a constant temperature line on the P-v diagram; one of these two diagrams is often sketched in problems involving a phase change from a liquid to a vapor.

Primary practical interest is in situations involving the liquid, liquid-vapor, and vapor regions. A saturated vapor lies on the *saturated vapor line* and a saturated liquid on the *saturated liquid line*. The region to the right of the saturated vapor line is the *superheated region*; the region to the left of the saturated liquid line is the *compressed liquid region* (also called the *subcooled liquid region*). A *supercritical state* is encountered when the pressure and temperature are greater than the critical values.

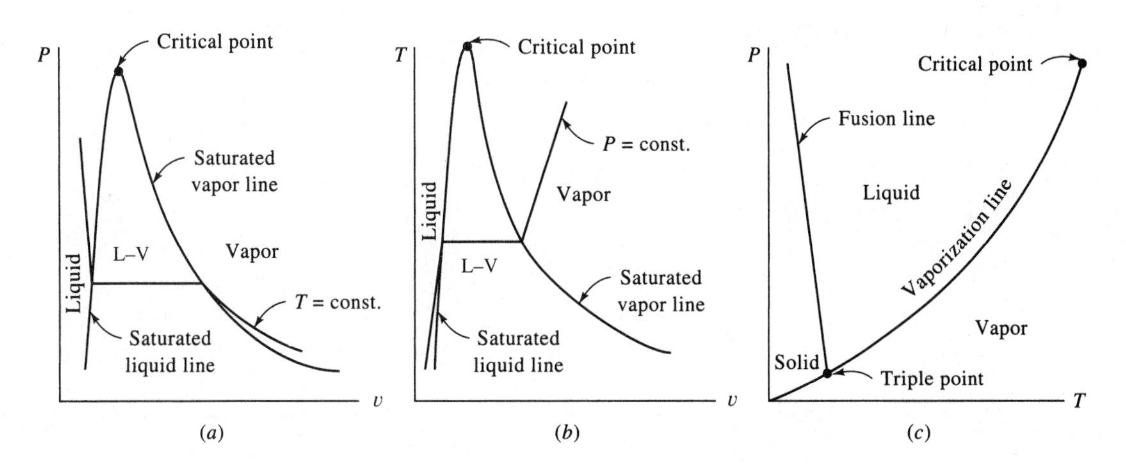

Figure 2.3 The (a) P-v, (b) T-v, and (c) P-T diagrams.

2.2 The Liquid-Vapor Region

At any state (T, v) between saturated points f (state 1) and g (state 2), shown in Fig. 2.4, liquid and vapor exist in equilibrium. Let v_f and v_g represent, respectively, the specific volumes of the saturated liquid and the saturated vapor. Let m be the total mass of a system, m_f the amount of mass in the liquid phase, and m_g the amount of mass in the vapor phase. Then, for a state of the system represented by any (T, v), such as state 3, the total volume of the mixture is the sum of the volume occupied by the liquid and that occupied by the vapor, or

$$V = V_f + V_g \qquad \text{or} \qquad mv = m_f v_f + m_g v_g \tag{2.1}$$

The ratio of the mass of saturated vapor to the total mass is called the *quality* of the mixture, designated by the symbol x; it is

$$x = \frac{m_g}{m} \tag{2.2}$$

We often refer to the region under the saturation lines as the *quality region*, or the *mixture region*, or the *wet region*; it is the only region where quality x has a meaning.

Recognizing that $m = m_f + m_g$ we may write Eq. (2.1), using our definition of quality, as

$$v = v_f + x(v_g - v_f) \tag{2.3}$$

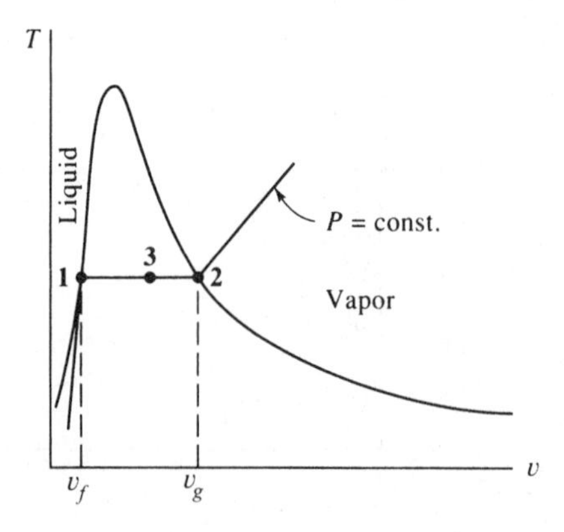

Figure 2.4 A T-v diagram showing the saturated liquid and saturated vapor points.

Because the difference in saturated vapor and saturated liquid values frequently appears in calculations, we often let the subscript "*fg*" denote this difference; that is,

$$v_{fg} = v_g - v_f \tag{2.4}$$

Thus, Eq. (2.3) can be written as

$$v = v_f + x v_{fg} \tag{2.5}$$

The percentage liquid by mass in a mixture is $100(1 - x)$, and the percentage vapor is $100x$.

2.3 The Steam Tables

Tabulations have been made for many substances of the thermodynamic properties P, v, and T and additional properties to be identified in subsequent chapters. Values are presented in the appendix in both tabular and graphical form. Table C.1 gives the saturation properties of water as a function of saturation temperature; Table C.2 gives these properties as a function of saturation pressure. The information contained in the two tables is essentially the same, the choice of which to use being a matter of convenience. We should note, however, that in the mixture region pressure and temperature are dependent. Thus, to establish the state of a mixture, if we specify the pressure, we need to specify a property other than temperature. Conversely, if we specify temperature, we must specify a property other than pressure.

Table C.3 lists the properties of superheated steam. To establish the state in the superheated region, it is necessary to specify two properties. While any two may be used, the most common procedure is to use P and T since these are easily measured. Thus, properties such as v are given in terms of the set of independent properties P and T.

Table C.4 lists data pertaining to compressed liquid. At a given T the specific volume v of a liquid is essentially independent of P. For example, for a temperature of 100°C in Table C.1, the specific volume v_f of liquid is 0.001044 m^3/kg at a pressure of 100 kPa. At a pressure of 10 MPa, the specific volume is 0.001038 m^3/kg, less than 1 percent decrease in specific volume. Thus, it is common in calculations to assume that v (and other properties, as well) of a compressed liquid is equal to v_f at the same temperature. Note, however, that v_f increases significantly with temperature, especially at higher temperatures.

Table C.5 gives the properties of a saturated solid and a saturated vapor for an equilibrium condition. Note that the value of the specific volume of ice is relatively insensitive to temperature and pressure for the saturated-solid line. Also, it has a greater value (almost 10 percent greater) than the minimum value on the saturated-liquid line.

Appendix D provides the properties of refrigerant R134a, a refrigerant used often in air conditioners and commercial coolers.

EXAMPLE 2.1

Determine the volume change when 10 kg of saturated water is completely vaporized at a pressure of (a) 1 kPa, (b) 260 kPa, and (c) 10 000 kPa.

Solution

Table C.2 provides the necessary values. The quantity being sought is $\Delta V = mv_{fg}$ where $v_{fg} = v_g - v_f$. Note that P is given in MPa.

(a) 1 kPa = 0.001 MPa. Thus, $v_{fg} = 129.2 - 0.001 = 129.2\ \text{m}^3/\text{kg}$.

$$\therefore \Delta V = 1292\ \text{m}^3$$

(b) At 0.26 MPa we must interpolate[2] if we use the tables. The tabulated values are used at 0.2 MPa and 0.3 MPa:

$$v_g = \frac{0.26 - 0.2}{0.3 - 0.2}(0.6058 - 0.8857) + 0.8857 = 0.718\ \text{m}^3/\text{kg}$$

The value for v_f is, to four decimal places, 0.0011 m³/kg at 0.2 MPa and at 0.3 MPa; hence, no need to interpolate for v_f. We then have

$$v_{fg} = 0.718 - 0.0011 = 0.717\ \text{m}^3/\text{kg}. \qquad \therefore \Delta V = 7.17\ \text{m}^3$$

(c) At 10 MPa, $v_{fg} = 0.01803 - 0.00145 = 0.01658\ \text{m}^3/\text{kg}$ so that

$$\Delta V = 0.1658\ \text{m}^3$$

Notice the large value of v_{fg} at low pressure compared with the small value of v_{fg} as the critical point is approached. This underscores the distortion of the T-v diagram in Fig. 2.4.

[2]Make sure you can quickly interpolate. It will be required of you often when using the tables.

EXAMPLE 2.2

Four kilograms of water are placed in an enclosed volume of 1 m³. Heat is added until the temperature is 150°C. Find the (a) pressure, (b) mass of the vapor, and (c) volume of the vapor.

Solution

Table C.1 is used. The volume of 4 kg of saturated vapor at 150°C is $0.3928 \times 4 = 1.5712$ m³. Since the given volume is less than this, we assume the state to be in the quality region.

(a) In the quality region, the pressure is given as $P = 475.8$ kPa, the value next to the temperature.

(b) To find the mass of the vapor we must determine the quality. It is found from Eq. (2.3), using $v = 1/4 = 0.25$ m³/kg:

$$v = v_f + x(v_g - v_f)$$
$$0.25 = 0.00109 + x(0.3928 - 0.00109) \qquad \therefore x = 0.6354$$

Using Eq. (2.2), the vapor mass is

$$m_g = mx = 4 \times 0.6354 = 2.542 \text{ kg}$$

(c) Finally, the volume of the vapor is found from

$$V_g = v_g m_g = 0.3928 \times 2.542 = 0.998 \text{ m}^3$$

Note: In mixtures where the quality is not close to zero, the vapor phase occupies most of the volume. In this example, with a quality of 63.5 percent, it occupies 99.8 percent of the volume.

EXAMPLE 2.3

Two kilograms of water are heated at a pressure of 220 kPa to produce a mixture with quality $x = 0.8$. Determine the final volume occupied by the mixture.

Solution

Use Table C.2. To determine the appropriate values at 220 kPa, linearly interpolate between 0.2 and 0.25 MPa. This provides, at 220 kPa,

$$v_g = \frac{220 - 200}{300 - 200}(0.6058 - 0.8857) + 0.8857 = 0.830 \text{ m}^3/\text{kg}$$
$$\therefore v_f = 0.0011 \text{ m}^3/\text{kg}$$

Note, no interpolation is necessary for v_f since, for both pressures, v_f is the same to four decimal places. Using Eq. (2.3), we now find

$$v = v_f + x(v_g - v_f) = 0.0011 + 0.8 \times (0.830 - 0.0011) = 0.662 \text{ m}^3/\text{kg}$$

The total volume occupied by 2 kg is

$$V = mv = 2 \text{ kg} \times 0.662 \text{ m}^3/\text{kg} = 1.32 \text{ m}^3$$

EXAMPLE 2.4

Two kilograms of water are contained in a constant-pressure cylinder held at 2.2 MPa. Heat is added until the temperature reaches 800°C. Determine the final volume of the container.

Solution

Use Table C.3. Since 2.2 MPa lies between 2 MPa and 2.5 MPa, the specific volume is interpolated to be

$$v = 0.2467 + 0.4(0.1972 - 0.2467) = 0.227 \text{ m}^3/\text{kg}$$

The final volume is then

$$V = mv = 2 \times 0.227 = 0.454 \text{ m}^3$$

The linear interpolation above results in a less accurate number than the numbers in the table. So, the final number has fewer significant digits.

2.4 Equations of State

When the vapor of a substance has relatively low density, the pressure, specific volume, and temperature are related by an *equation of state*,

$$Pv = RT \tag{2.6}$$

where, for a particular gas, the *gas constant* is R. A gas for which this equation is valid is called an *ideal gas* or sometimes a *perfect gas*. Note that when using the above equation of state the pressure and temperature must be expressed as absolute quantities.

The gas constant R is related to a *universal gas constant* $\bar{R}$, which has the same value for all gases, by the relationship

$$R = \frac{\bar{R}}{M} \tag{2.7}$$

where M is the *molar mass*, values of which are tabulated in Tables B.2 and B.3. The *mole* is that quantity of a substance (i.e., that number of atoms or molecules) having a mass which, measured in grams, is numerically equal to the atomic or molecular weight of the substance. In the SI system it is convenient to use instead the kilomole (kmol), which amounts to x kilograms of a substance of molecular weight x. For instance, 1 kmol of carbon (with a molecular weight of 12) is a mass of 12 kg; 1 kmol of oxygen is 32 kg. Stated otherwise, $M = 32$ kg/kmol for O_2.

The value $\bar{R} = 8.314$ kJ/kmol·K. For air M is 28.97 kg/kmol, so that for air R is 0.287 kJ/ kg·K, or 287 J/kg·K, a value used extensively in calculations involving air.

Other forms of the ideal-gas equation are

$$PV = mRT \qquad P = \rho RT \qquad PV = n\bar{R}T \tag{2.8}$$

where n is the number of moles.

Care must be taken in using this simple convenient equation of state. A low-density can result from either a low pressure or a high temperature. For air, the ideal-gas equation is surprisingly accurate for a wide range of temperatures and pressures; less than 1 percent error is encountered for pressures as high as 3000 kPa at room temperature, or for temperatures as low as −130°C at atmospheric pressure.

The *compressibility factor* Z helps us in determining whether or not the ideal-gas equation can be used. It is defined as

$$Z = \frac{Pv}{RT} \tag{2.9}$$

and is displayed in Fig. 2.5 for nitrogen. This figure is acceptable for air also, since air is composed mainly of nitrogen. If $Z = 1$, or very close to 1, the ideal-gas equation can be used. If Z is not close to 1, then Eq. (2.9) may be used.

The compressibility factor can be determined for any gas by using a generalized compressibility chart presented in App. G. In the generalized chart the *reduced pressure* P_R and *reduced temperature* T_R must be used. They are calculated from

$$P_R = \frac{P}{P_{cr}} \qquad T_R = \frac{T}{T_{cr}} \qquad v_R = v\frac{P_{cr}}{RT_{cr}} \tag{2.10}$$

where P_{cr} and T_{cr} are the critical-point pressure and temperature, respectively, of Table B.3, and v_R is the pseudoreduced volume in App. G.

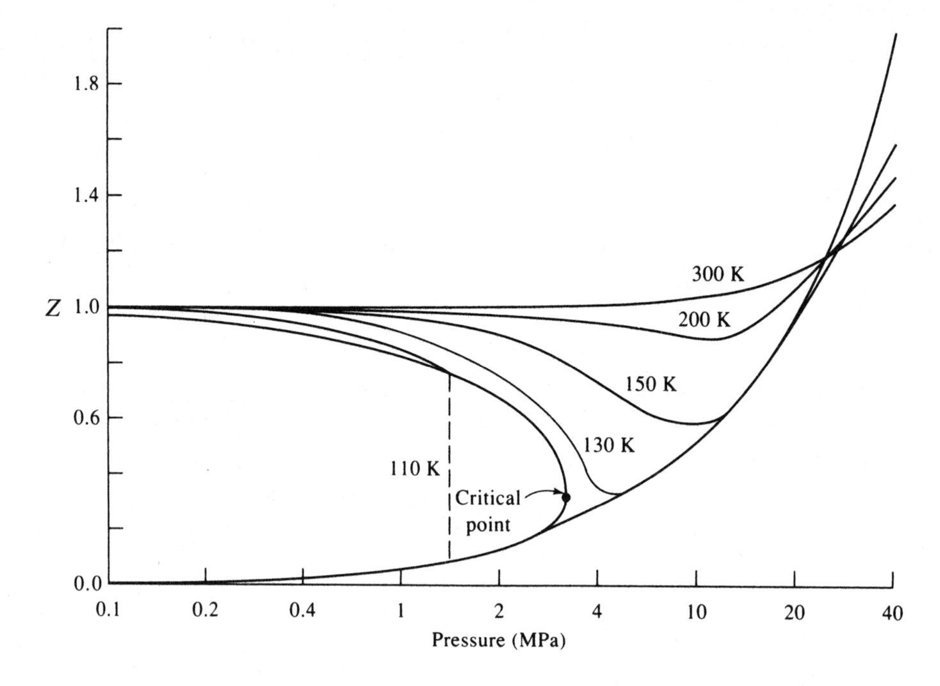

Figure 2.5 The compressibility factor.

EXAMPLE 2.5

An automobile tire with a volume of 0.6 m³ is inflated to a gage pressure of 200 kPa. Calculate the mass of air in the tire if the temperature is 20°C using the ideal-gas equation of state.

Solution

Air is assumed to be an ideal gas at the conditions of this example. In the ideal-gas equation, $PV = mRT$, we use absolute pressure and absolute temperature. Thus, using $P_{atm} = 100$ kPa (to use a pressure of 101 kPa is unnecessary; the difference of 1 percent is not significant in most engineering problems):

$$P_{abs} = P_{gage} + P_{atm} = 200 + 100 = 300 \text{ kPa} \quad \text{and} \quad T = T_C + 273 = 20 + 273 = 293 \text{ K}$$

The mass is then calculated to be

$$m = \frac{PV}{RT} = \frac{(300\,000 \text{ N/m}^2)(0.6 \text{ m}^3)}{(287 \text{ N}\cdot\text{m/kg}\cdot\text{K})(293 \text{ K})} = 2.14 \text{ kg}$$

Be careful to be consistent with the units in the above equation.

2.5 Equations of State for a Nonideal Gas

Nonideal-gas behavior occurs when the pressure is relatively high (> 2 MPa for many gases) or when the temperature is near the saturation temperature. There are no acceptable criteria that can be used to determine if the ideal-gas equation can be used or if nonideal-gas equations must be used. Usually a problem is stated in such a way that it is obvious that nonideal-gas effects must be included; otherwise a problem is solved assuming an ideal gas.

The *van der Waals equation of state* accounts for the volume occupied by the gas molecules and for the attractive forces between molecules. It is

$$P = \frac{RT}{v - b} - \frac{a}{v^2} \tag{2.11}$$

where the constants a and b are related to the critical-point data of Table B.3 by

$$a = \frac{27R^2T_{\text{cr}}^2}{64P_{\text{cr}}} \qquad b = \frac{RT_{\text{cr}}}{8P_{\text{cr}}} \tag{2.12}$$

These constants are also presented in Table B.7 to simplify calculations.

An improved equation is the Redlich-Kwong equation of state:

$$P = \frac{RT}{v - b} - \frac{a}{v(v + b)\sqrt{T}} \tag{2.13}$$

where the constants are also included in Table B.7 and are given by

$$a = 0.4275\frac{R^2T_{\text{cr}}^{2.5}}{P_{\text{cr}}} \qquad b = 0.0867\frac{RT_{\text{cr}}}{P_{\text{cr}}} \tag{2.14}$$

EXAMPLE 2.6

Calculate the pressure of steam at a temperature of 500°C and a density of 24 kg/m^3 using (a) the ideal-gas equation, (b) the van der Waals equation, (c) the Redlich-Kwong equation, (d) the compressibility factor, and (e) the steam table.

Solution

(a) Using the ideal-gas equation,

$$P = \rho RT = 24 \times 0.462 \times 773 = 8570 \text{ kPa}$$

where the gas constant for steam is found in Table B.2.

(b) Using values for a and b from Table B.7, the van der Waals equation provides

$$P = \frac{RT}{v-b} - \frac{a}{v^2} = \frac{0.462 \times 773}{\frac{1}{24} - 0.00169} - \frac{1.703}{\frac{1}{24^2}} = 7950 \text{ kPa}$$

(c) Using values for a and b from Table B.7, the Redlich-Kwong equation gives

$$P = \frac{RT}{v-b} - \frac{a}{v(v+b)\sqrt{T}} = \frac{0.462 \times 773}{\frac{1}{24} - 0.00117} - \frac{43.9}{\frac{1}{24}\left(\frac{1}{24} + 0.00117\right)\sqrt{773}} = 7930 \text{ kPa}$$

(d) The compressibility factor is found from the generalized compressibility chart of App. G. To use the chart we must know the reduced temperature and pressure (using Table B.3):

$$T_R = \frac{T}{T_{cr}} = \frac{773}{647.4} = 1.19 \qquad P_R = \frac{P}{P_{cr}} = \frac{8570}{22\,100} = 0.39$$

where we have used the ideal-gas pressure from part (a). Using the compressibility chart (it is fairly insensitive to the precise values of T_R and P_R so estimates of these values are quite acceptable) and Eq. (2.9), we find

$$P = \frac{ZRT}{v} = \frac{0.93 \times 0.462 \times 773}{1/24} = 7970 \text{ kPa}$$

(e) The steam table provides the best value for the pressure. Using $T = 500°\text{C}$ and $v = 1/24 = 0.0417 \text{ m}^3/\text{kg}$, we find $P = 8000$ kPa.

Note that the ideal-gas law has an error of 7.1 percent, and the error of each of the other three equations is less than 1 percent.

Quiz No. 1

1. The phase change from a vapor to a liquid is referred to as
   (A) vaporization
   (B) condensation
   (C) sublimation
   (D) melting
2. The volume occupied by 4 kg of 200°C steam at a quality of 80 percent is nearest
   (A) 0.004 m^3
   (B) 0.104 m^3
   (C) 0.4 m^3
   (D) 4.1 m^3
3. For a specific volume of 0.2 m^3/kg, the quality of steam, if the absolute pressure is 630 kPa, is nearest
   (A) 0.44
   (B) 0.50
   (C) 0.59
   (D) 0.66
4. Saturated liquid water occupies a volume of 1.2 m^3. Heat is added until it is completely vaporized. If the pressure is held constant at 600 kPa, the final volume is nearest
   (A) 344 m^3
   (B) 290 m^3
   (C) 203 m^3
   (D) 198 m^3
5. Saturated steam is heated in a rigid tank from 70 to 800°C. P_2 is nearest
   (A) 100 kPa
   (B) 200 kPa
   (C) 300 kPa
   (D) 400 kPa

6. Determine the final volume if 3 kg of water is heated at a constant pressure until the quality is 60 percent. The pressure is 270 kPa.
   (A) 1.07 m^3
   (B) 1.24 m^3
   (C) 1.39 m^3
   (D) 2.93 m^3
7. A vertical circular cylinder holds a height of 1 cm of liquid water and 100 cm of vapor. If $P = 200$ kPa, the quality is nearest
   (A) 0.01
   (B) 0.1
   (C) 0.4
   (D) 0.8
8. A rigid vessel with a volume of 10 m^3 contains a water-vapor mixture at 400 kPa at 60 percent quality. The pressure is lowered to 300 kPa by cooling the vessel. Find m_g at state 2.
   (A) 16.5 kg
   (B) 19.5 kg
   (C) 23.8 kg
   (D) 29.2 kg
9. The mass of air in a tire with a volume of 0.2 m^3 with a gage pressure of 280 kPa at 25°C is nearest
   (A) 7.8 kg
   (B) 0.889 kg
   (C) 0.732 kg
   (D) 0.655 kg
10. Estimate the difference in density between the inside and outside of a house in the winter when $P = 100$ kPa, $T_{\text{inside}} = 20°\text{C}$, and $T_{\text{outside}} = -20°\text{C}$. (Would you expect any tendency for air exchange due to this difference?)
    (A) 0.19 kg/m^3
    (B) 0.17 kg/m^3
    (C) 0.15 kg/m^3
    (D) 0.09 kg/m^3

11. Estimate the pressure of nitrogen at a temperature of 220 K and a specific volume of 0.04 m^3/kg using the van der Waals equation.
    (A) 1630 kPa
    (B) 1600 kPa
    (C) 1580 kPa
    (D) 1540 kPa
12. Ten kilograms of 400°C steam are contained in a 734-L tank. Calculate the error in the pressure if it is calculated using the ideal-gas equation (rather than the more accurate steam tables).
    (A) 1%
    (B) 4%
    (C) 6%
    (D) 18%

Quiz No. 2

1. The point that connects the saturated-liquid line to the saturated-vapor line is called the
    (A) triple point
    (B) critical point
    (C) superheated point
    (D) compressed liquid point
2. Estimate the volume occupied by 10 kg of water at 200°C and 2 MPa.
    (A) 0.099 m^3
    (B) 0.016 m^3
    (C) 11.6 L
    (D) 11.8 L
3. The specific volume of water at 200 °C and 80 percent quality is nearest
    (A) 0.06 m^3/kg
    (B) 0.08 m^3/kg
    (C) 0.1 m^3/kg
    (D) 0.12 m^3/kg

4. Calculate the specific volume of water at 221°C if the quality is 85 percent.
   (A) 0.072 m^3/kg
   (B) 0.074 m^3/kg
   (C) 0.76 m^3/kg
   (D) 0.78 m^3/kg
5. Five kilograms of steam occupy a volume of 10 m^3. If the temperature is 86°C, the quality and the pressure are nearest
   (A) 0.76, 60 kPa
   (B) 0.69, 68 kPa
   (C) 0.71, 63 kPa
   (D) 0.72, 61 kPa
6. Water at an initial quality of 50 percent is heated at constant volume from 100 kPa to 200°C. What is P_2?
   (A) 162 kPa
   (B) 260 kPa
   (C) 370 kPa
   (D) 480 kPa
7. Find the quality of steam at 120°C if the vapor occupies 1200 L and the liquid occupies 2 L.
   (A) 42%
   (B) 28%
   (C) 18%
   (D) 4%
8. Two kilograms of steam are contained in a piston-cylinder arrangement. The 20-mm-diameter 48-kg piston is allowed to rise with no friction until the temperature reaches 260°C. The final volume is nearest
   (A) 0.13 m^3
   (B) 0.29 m^3
   (C) 0.34 m^3
   (D) 0.39 m^3

9. Estimate the temperature of 2 kg of air contained in a 40 L volume at 2 MPa.
   (A) 125 K
   (B) 140 K
   (C) 155 K
   (D) 175 K
10. The gage pressure reading on an automobile tire is 240 kPa when the tire temperature is –30°C. The automobile is driven to a warmer climate and the tire temperature increases to 65°C. Estimate the gage pressure in the tire making reasonable assumptions.
    (A) 480 kPa
    (B) 370 kPa
    (C) 320 kPa
    (D) 280 kPa
11. Estimate the pressure of nitrogen at a temperature of 220 K and a specific volume of 0.04 m^3/kg using the compressibility factor.
    (A) 1630 kPa
    (B) 1620 kPa
    (C) 1580 kPa
    (D) 1600 kPa
12. Ten kilograms of 100°C steam are contained in a 17 000-L tank. Calculate the error in the pressure using the ideal-gas equation.
    (A) 1%
    (B) 4%
    (C) 6%
    (D) 18%

CHAPTER 3

Work and Heat

In this chapter we will discuss the two quantities that result from energy transfer across the boundary of a system: work and heat. This will lead into the first law of thermodynamics. Work will be calculated for several common situations. Heat transfer, often simply called "heat," however, is a quantity that requires substantial analysis for its calculation. Heat transferred by conduction, convection, or radiation to systems or control volumes will either be given or information will be provided that it can be determined in our study of thermodynamics; it will not be calculated from temperature information, as is done in a heat transfer course.

3.1 Work

Our definition of work includes the work done by expanding exhaust gases after combustion occurs in the cylinder of an automobile engine, as shown in Fig. 3.1. The energy released during the combustion process is transferred to the crankshaft by means of the connecting rod, in the form of work. Thus, in thermodynamics, *work* can be thought of as energy being transferred across the boundary of a system, where, in this example, the system is the gas in the cylinder.

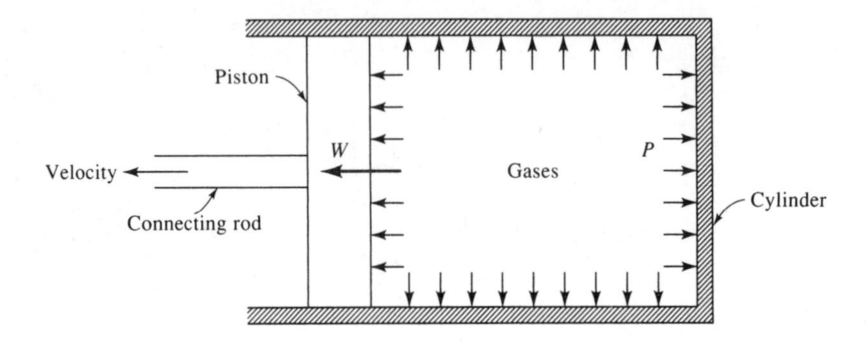

Figure 3.1 Work being done by expanding gases in a cylinder.

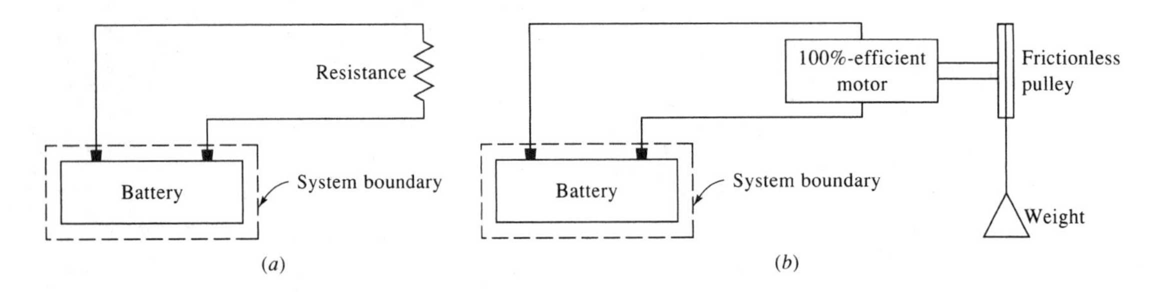

Figure 3.2 Work being done by electrical means.

Work, designated W, is often defined as the product of a force and the distance moved in the direction of the force: the mechanical definition of work. A more general definition of work is the thermodynamic definition: *Work* is done by a system if the sole external effect on the surroundings could be the raising of a weight. The magnitude of the work is the product of that weight and the distance it could be lifted. The schematic of Fig. 3.2*a* qualifies as work in the thermodynamic sense since it could lift the weight of Fig. 3.2*b*.

The convention chosen for positive work is that if the system performs work on the surroundings it is positive.[1] A piston compressing a fluid is doing negative work on the system, whereas a fluid expanding against a piston is doing positive work. The units of work are quickly observed from the units of a force multiplied by a distance: in the SI system, newton-meters (N · m) or joules (J).

The rate of doing work, designated $\dot{W}$, is called *power.* In the SI system, power has units of joules per second (J/s), or watts (W). We will use the unit of horsepower because of its widespread use in rating engines. To convert we simply use 1 hp = 746 W.

---

[1]In some textbooks, it is negative.

The work associated with a unit mass will be designated w:

$$w = \frac{W}{m} \tag{3.1}$$

A final general comment concerning work relates to the choice of the system. Note that if the system in Fig. 3.2 included the entire battery-resistor setup in part (*a*), or the entire battery-motor-pulley-weight setup in part (*b*), no energy would cross the system boundary, with the result that no work would be done. The identification of the system is very important in determining work.

3.2 Work Due to a Moving Boundary

There are a number of work modes that occur in various engineering situations. These include the work needed to stretch a wire, to rotate a shaft, to move against friction, or to cause a current to flow through a resistor. We are primarily concerned with the work required to move a boundary against a pressure force.

Consider the piston-cylinder arrangement shown in Fig. 3.3. There is a seal to contain the gas in the cylinder, the pressure is uniform throughout the cylinder, and there are no gravity, magnetic, or electrical effects. This assures us of a quasiequilibrium process, one in which the gas is assumed to pass through a series of equilibrium states. Now, allow an expansion of the gas to occur by moving the piston upward a small distance dl. The total force acting on the piston is the pressure times the area of the piston. This pressure is expressed as *absolute* pressure since pressure is a result of molecular activity; any molecular activity will yield a pressure that will result in work being done when the boundary moves. The infinitesimal work

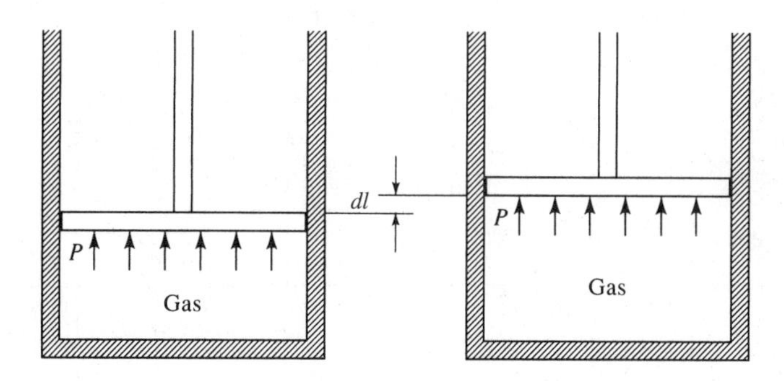

Figure 3.3 Work due to a moving boundary.

that the system (the gas) does on the surroundings (the piston) is then the force multiplied by the distance:

$$\delta W = PAdl \tag{3.2}$$

The symbol δW will be discussed shortly. The quantity Adl is simply dV, the differential volume, allowing Eq. (3.2) to be written in the form

$$\delta W = PdV \tag{3.3}$$

As the piston moves from some position 1 to another position 2, the above expression can be integrated to give

$$W_{1\text{-}2} = \int_{V_1}^{V_2} PdV \tag{3.4}$$

where we assume the pressure is known for each position as the piston moves from volume V_1 to volume V_2. Typical pressure-volume diagrams are shown in Fig. 3.4. The work $W_{1\text{-}2}$ is the area under the P-V curve.

The integration process highlights two important features in Eq. (3.4). First, as we proceed from state 1 to state 2, the area representing the work is very dependent on the path that we follow. That is, states 1 and 2 in Fig. 3.4*a* and *b* are identical, yet the areas under the P-V curves are very different; work depends on the actual path that connects the two states. Thus, work is a path function, as contrasted to a point function (that is dependent only on the end points). The differential of a path function is called an *inexact differential*, whereas the differential of a point function is an *exact differential*. An inexact differential will be denoted with the symbol δ.

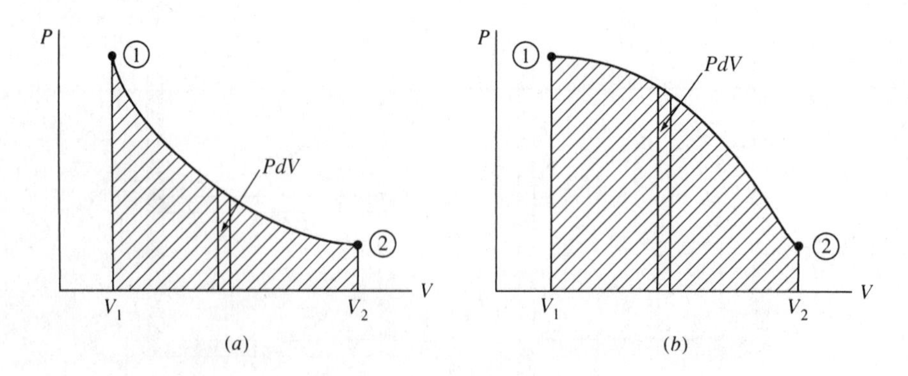

Figure 3.4 Work depends on the path between two states.

The integral of δW is $W_{1\text{-}2}$, where the subscript emphasizes that the work is associated with the path as the process passes from state 1 to state 2; the subscript may be omitted, however, and work done written simply as W. We would never write W_1 or W_2, since work is not associated with a state but with a process. Work is not a property. The integral of an exact differential, for example dT, would be

$$\int_{T_1}^{T_2} dT = T_2 - T_1 \tag{3.5}$$

The second observation to be made from Eq. (3.4) is that the pressure is assumed to be constant throughout the volume at each intermediate position. The system passes through each equilibrium state shown in the *P-V* diagrams of Fig. 3.4. An equilibrium state can usually be assumed even though the variables may appear to be changing quite rapidly. Combustion is a very rapid process that cannot be modeled as a quasiequilibrium process. The other processes in the internal combustion engine—expansion, exhaust, intake, and compression—can be assumed to be quasiequilibrium processes; they occur at a relatively slow rate, thermodynamically.

As a final comment regarding work we may now discuss what is meant by a simple system, as defined in Chap. 1. For a system free of surface, magnetic, and electrical effects, the only work mode is that due to pressure acting on a moving boundary. For such simple systems only two independent variables are necessary to establish an equilibrium state of the system composed of a homogeneous substance. If other work modes are present, such as a work mode due to an electric field, then additional independent variables would be necessary, such as the electric field intensity.

EXAMPLE 3.1

One kilogram of steam with a quality of 20 percent is heated at a constant pressure of 200 kPa until the temperature reaches 400°C. Calculate the work done by the steam.

Solution

The work is given by

$$W = \int_{V_1}^{V_2} P dV = P(V_2 - V_1) = mP(v_2 - v_1)$$

To evaluate the work we must determine v_1 and v_2. Using Table C.2 we find

$$\begin{aligned} v_1 &= v_f + x(v_g - v_f) \\ &= 0.001061 + 0.2(0.8857 - 0.001061) = 0.1780 \text{ m}^3/\text{kg} \end{aligned}$$

From the superheat table C.3 we locate state 2 at $T_2 = 400°C$ and $P_2 = 0.2$ MPa:

$$v_2 = 1.549 \text{ m}^3/\text{kg}$$

The work is then

$$W = 1 \times 200 \times (1.549 - 0.1780) = 274.2 \text{ kJ}$$

Since the pressure has units of kPa, the result is in kJ.

EXAMPLE 3.2
A 110-mm-diameter cylinder contains 100 cm^3 of water at 60°C. A 50-kg piston sits on top of the water. If heat is added until the temperature is 200°C, find the work done.

Solution
The pressure in the cylinder is due to the weight of the piston and remains constant. Assuming a frictionless seal (this is always done unless information is given to the contrary), a force balance provides

$$mg = PA - P_{\text{atm}}A \qquad 50 \times 9.81 = (P - 100\,000)\frac{\pi \times 0.110^2}{4} \qquad \therefore P = 151\,600 \text{ Pa}$$

The atmospheric pressure is included so that absolute pressure results. The volume at the initial state 1 is given as

$$V_1 = 100 \times 10^{-6} = 10^{-4} \text{ m}^3$$

Using v_1 at 60°C, the mass is calculated to be

$$m = \frac{V_1}{v_1} = \frac{10^{-4}}{0.001017} = 0.09833 \text{ kg}$$

At state 2 the temperature is 200°C and the pressure is 0.15 MPa (this pressure is within 1 percent of the pressure of 0.1516 MPa, so it is acceptable). The volume is then

$$V_2 = mv_2 = 0.09833 \times 1.444 = 0.1420 \text{ m}^3$$

Finally, the work is calculated to be

$$W = P(V_2 - V_1) = 151.6(0.1420 - 0.0001) = 21.5 \text{ kJ}$$

EXAMPLE 3.3

Energy is added to a piston-cylinder arrangement, and the piston is withdrawn in such a way that the temperature (i.e., the quantity *PV*) remains constant. The initial pressure and volume are 200 kPa and 2 m^3, respectively. If the final pressure is 100 kPa, calculate the work done by the ideal gas on the piston.

Solution

The work, using Eq. (3.4), can be expressed as

$$W_{1\text{-}2} = \int_{V_1}^{V_2} P dV = \int_{V_1}^{V_2} \frac{C}{V} dV$$

where we have used $PV = C$. To calculate the work we must find C and V_2. The constant C is found from

$$C = P_1 V_1 = 200 \times 2 = 400 \text{ kJ}$$

To find V_2 we use $P_2 V_2 = P_1 V_1$, which is, of course, the equation that would result from an isothermal process (constant temperature) involving an ideal gas. This can be written as

$$V_2 = \frac{P_1 V_1}{p_2} = \frac{200 \times 2}{100} = 4 \text{ m}^3$$

Finally,

$$W_{1\text{-}2} = \int_2^4 \frac{400}{V} dV = 400 \ln \frac{4}{2} = 277 \text{ kJ}$$

This is positive, since work is done during the expansion process by the gas.

3.3 Nonequilibrium Work

It must be emphasized that the area on a *PV* diagram represents the work for a quasiequilibrium process only. For nonequilibrium processes the work cannot be calculated using the integral of *PdV*. Either it must be given for the particular process or it must be determined by some other means. Two examples will be given. Consider a system to be formed by the gas in Fig. 3.5. In part (*a*) work is obviously crossing the boundary of the system by means of the rotating shaft, yet the volume does not change. We could calculate the work input by multiplying the weight by the distance it dropped, neglecting friction in the pulley system. This would not, however, be equal to the integral of *PdV*, which is zero. The paddle wheel provides us with a nonequilibrium work mode.

Suppose the membrane in Fig. 3.5*b* ruptures, allowing the gas to expand and fill the evacuated volume. There is no resistance to the expansion of the gas at the moving boundary as the gas fills the volume; hence, there is no work done, yet there is a change in volume. The sudden expansion is a nonequilibrium process, and again we cannot use the integral of *PdV* to calculate the work for this nonequilibrium work mode.

EXAMPLE 3.4

A 100-kg mass drops 3 m, resulting in an increased volume in the cylinder shown of 0.002 m^3. The weight and the piston maintain a constant gage pressure of 100 kPa. Determine the net work done by the gas on the surroundings. Neglect all friction.

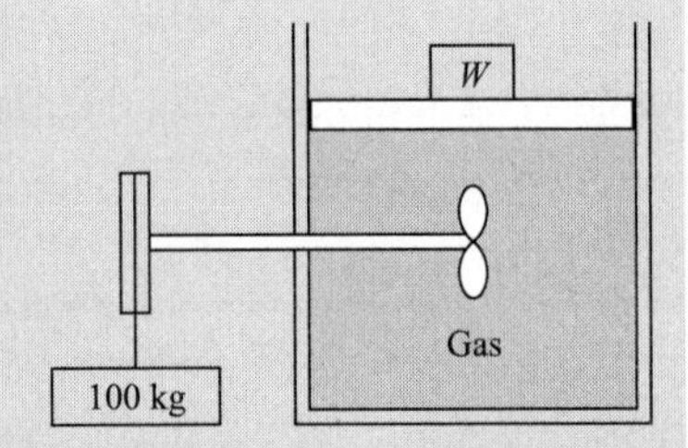

Solution

The paddle wheel does work on the system, the gas, due to the 100-kg mass dropping 3 m. That work is negative and is

$$W = -F \times d = -(100 \times 9.81) \times 3 = -2940 \text{ N} \cdot \text{m}$$

The work done by the system on this frictionless piston is positive since the system is doing the work. It is

$$W = PA \times h = PV = 200\ 000 \times 0.002 = 400 \text{ N} \cdot \text{m}$$

where absolute pressure has been used. The net work done is thus

$$W_{\text{net}} = W_{\text{paddle wheel}} + W_{\text{boundary}} = -2940 + 400 = -2540 \text{ J}$$

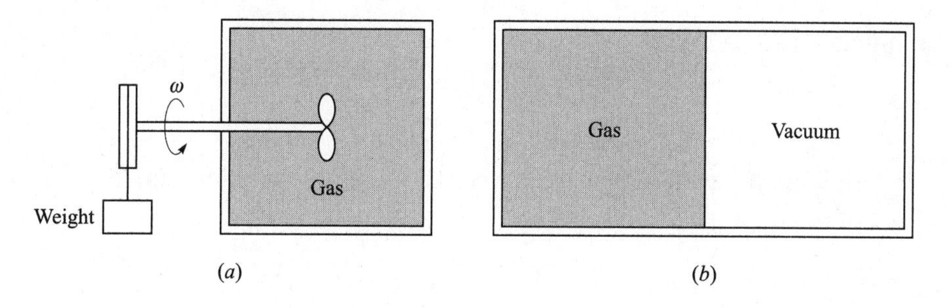

Figure 3.5 Nonequilibrium work.

3.4 Other Work Modes

Work transferred by a rotating shaft (Fig. 3.6) is a common occurrence in mechanical systems. The work results from the shearing forces due to the shearing stress τ, which varies linearly with the radius over the cross-sectional area. The differential shearing force due to the shearing stress is

$$dF = \tau dA = \tau(2\pi r dr) \tag{3.6}$$

The linear velocity with which this force moves is $r\omega$. Hence, the rate of doing work (force times velocity) is

$$W = \int_A r\omega\, dF = \int_0^R r\omega \times \tau \times 2\pi r dr = 2\pi\omega \int_0^R \tau \times r^2 dr \tag{3.7}$$

where R is the radius of the shaft. The torque T is found from the shearing stresses by integrating over the area:

$$T = \int_A r\, dF = 2\pi \int_0^R \tau \times r^2 dr \tag{3.8}$$

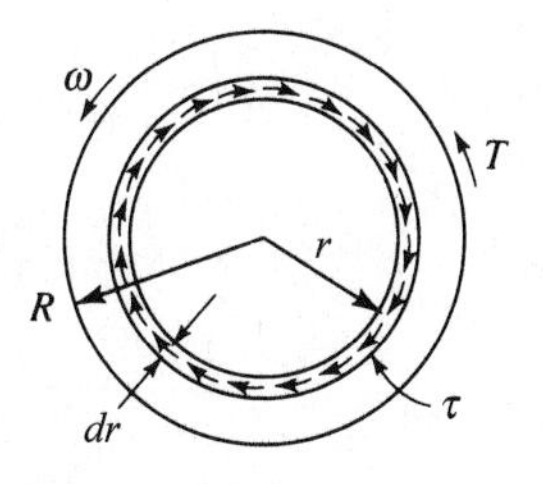

Figure 3.6 Work due to a rotating shaft transmitting a torque.

Combining this with Eq. (3.7) above, we have

$$\dot{W} = T\omega \tag{3.9}$$

To find the work transferred in a given time Δt, we simply multiply Eq. (3.9) by the number of seconds:

$$W = T\omega\Delta t \tag{3.10}$$

The angular velocity must be expressed in rad/s. If it is expressed in rev/min (rpm), we must multiply by $2\pi/60$.

The work necessary to stretch a linear spring (Fig. 3.7) with spring constant K from a length x_1 to x_2 can be found by using the relation for the force:

$$F = Kx \tag{3.11}$$

where x is the distance the spring is stretched from the unstretched position. Note that the force is dependent on the variable x. Hence, we must integrate the force over the distance the spring is stretched; this results in

$$W = \int_{x_1}^{x_2} F\,dx = \int_{x_1}^{x_2} Kx\,dx = \frac{1}{2}K(x_2^2 - x_1^2) \tag{3.12}$$

As a final type let us discuss an electrical work mode, illustrated in Fig. 3.8. The potential difference V across the battery terminals is the "force" that drives the charge through the resistor during the time increment Δt. The current i is related to the charge. The work is given by the expression

$$W = Vi\,\Delta t \tag{3.13}$$

The power would be the rate of doing work, or

$$\dot{W} = Vi \tag{3.14}$$

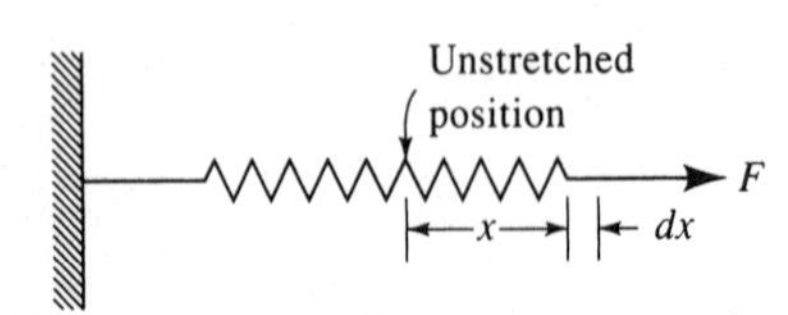

Figure 3.7 Work needed to stretch a spring.

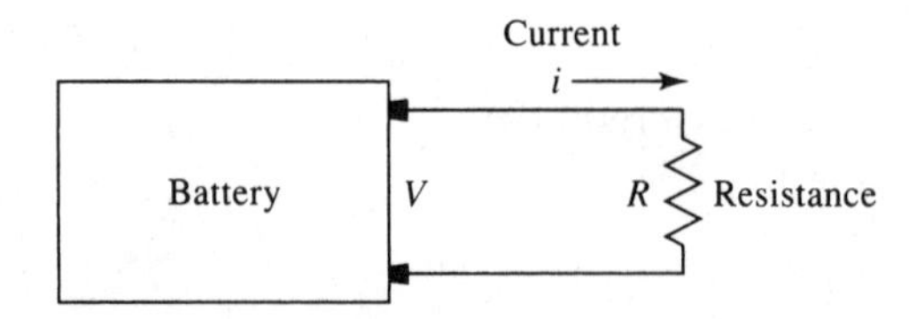

Figure 3.8 Work due to a current flowing through a resistor.

EXAMPLE 3.5
The drive shaft in an automobile delivers 100 N · m of torque as it rotates at 3000 rpm. Calculate the horsepower transmitted.

Solution
The power is found by using $\dot{W} = T\omega$. This requires ω to be expressed in rad/s:

$$\omega = 3000 \times \frac{2\pi}{60} = 314.2 \text{ rad/s}$$

Hence

$$\dot{W} = T\omega = 100 \times 314.2 = 31\,420 \text{ W} \quad \text{or} \quad \frac{31\,420}{746} = 42.1 \text{ hp}$$

EXAMPLE 3.6
The air in a 10-cm-diameter cylinder shown is heated until the spring is compressed 50 mm. Find the work done by the air on the frictionless piston. The spring is initially unstretched, as shown.

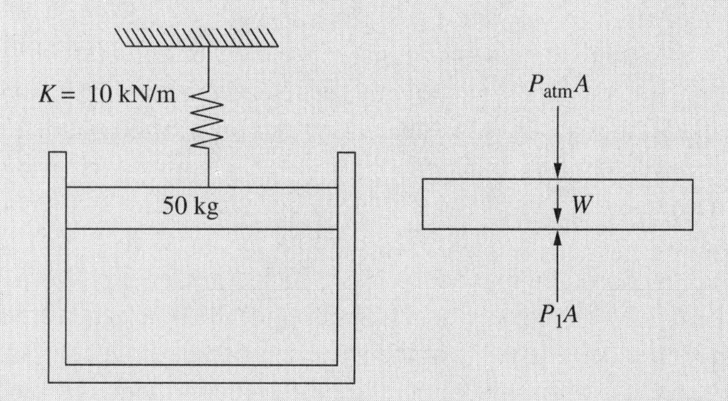

$P_{atm}A$

W

P_1A

Solution
The pressure in the cylinder is initially found from a force balance as shown on the free-body diagram:

$$P_1A = P_{atm}A + W$$

$$P_1\pi \times 0.05^2 = 100\,000 \times \pi \times 0.05^2 + 50 \times 9.81 \qquad \therefore P_1 = 162\,500 \text{ Pa}$$

To raise the piston a distance of 50 mm, without the spring, the pressure would be constant and the work required would be force times distance:

$$W = PA \times d = 162\ 500 \times (\pi \times 0.05^2) \times 0.05 = 63.81 \text{ J}$$

Using Eq. (3.12), the work required to compress the spring is calculated to be

$$W = \frac{1}{2}K(x_2^2 - x_1^2) = \frac{1}{2} \times 10\ 000 \times 0.05^2 = 12.5 \text{ J}$$

The total work is then found by summing the above two values:

$$W_{\text{total}} = 63.81 + 12.5 = 76.31 \text{ J}$$

3.5 Heat Transfer

In the preceding sections we considered several work modes. Transfer of energy that we cannot account for by any of the work modes is called *heat transfer*. We can define heat transfer as energy transferred across the boundary of a system due to a difference in temperature between the system and the surroundings of the system. A process in which there is zero heat transfer is called an *adiabatic process*. Such a process is approximated experimentally by sufficiently insulating the system so that the heat transfer is negligible. An insulated system in thermodynamics will always be assumed to have zero heat transfer.

Consider a system composed of a hot block and a cold block of equal mass. The hot block contains more energy than the cold block due to its higher temperature. When the blocks are brought into contact with each other, energy flows from the hot block to the cold one by means of heat transfer until the blocks attain thermal equilibrium, with both blocks arriving at the same temperature. Once in thermal equilibrium, there are no longer any temperature differences, therefore, there is no longer any heat transfer. Energy in the hot block has simply been transferred to the cold block, so that if the system is insulated, the cold block gains the amount of energy lost by the hot block.

It should be noted that the energy contained in a system might be transferred to the surroundings either by work done by the system or by heat transferred from the system. In either case, heat and work are qualitatively equivalent and are expressed in units of energy. An equivalent reduction in energy is accomplished if 100 J of heat is transferred from a system or if 100 J of work is performed by a system. In Fig. 3.9 the burner illustrates heat being added to the system and the rotating shaft

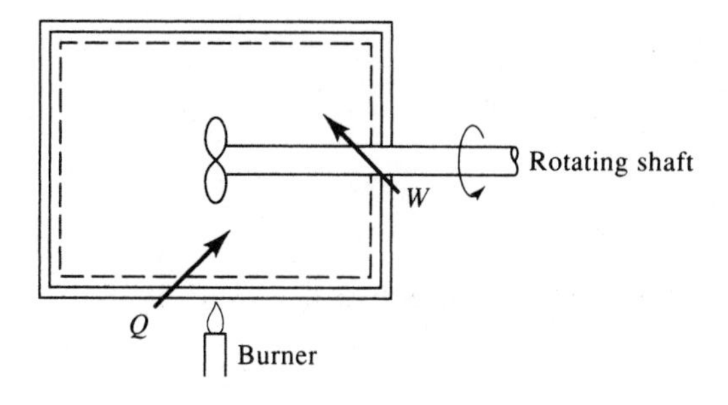

Figure 3.9 Energy added to a system.

illustrates work being done on the system. In either case, energy is being added to the system.

Although heat and work are similar in that they both represent energy crossing a boundary, they differ in that, unlike work, convention prescribes that heat transferred to a system is considered positive. Thus, positive heat transfer adds energy to a system, whereas positive work subtracts energy from a system.

Because a system does not contain heat, heat is not a property. Thus, its differential is inexact and is written as δQ, where Q is the heat transfer. For a particular process between state 1 and state 2 the heat transfer could be written as $Q_{1\text{-}2}$ but it will generally be denoted simply as Q. It is sometimes convenient to refer to the heat transfer per unit mass, or q, defined as

$$q = \frac{Q}{m} \tag{3.15}$$

Often we are interested in the *rate of heat transfer*, or in the *heat flux*, denoted by $\dot{Q}$ with units of J/s = W.

There are three heat transfer modes: conduction, convection, and radiation. In general, each of these three heat transfer modes refers to a separate rate of energy crossing a specific plane, or surface. Often, engineering applications involving heat transfer must consider all three modes in an analysis. In each case, we can write relationships between the rate of heat transfer and temperature. A course in heat transfer is dedicated to calculating the heat transfer in various situations. In thermodynamics, it is given in a problem or it is found from applying the energy equation.

EXAMPLE 3.7

A paddle wheel adds work to a rigid container by rotations caused by dropping a 50-kg weight a distance of 2 m from a pulley. How much heat must be transferred to result in an equivalent effect?

Solution

For this non-quasiequilibrium process the work is given by

$$W = mg \times d = 50 \times 9.8 \times 2 = 980 \text{ J}$$

The heat Q that must be transferred equals the work, 980 J.

Quiz No. 1

1. Which work mode is a nonequilibrium work mode?
   (A) Compressing a spring
   (B) Transmitting torque with a rotating shaft
   (C) Energizing an electrical resistor
   (D) Compressing gas in a cylinder
2. Ten kilograms of saturated steam at 800 kPa are heated at constant pressure to 400°C. The work required is nearest
   (A) 1150 kJ
   (B) 960 kJ
   (C) 660 kJ
   (D) 115 kJ
3. A stop is located 20 mm above the piston at the position shown. If the mass of the frictionless piston is 64 kg, what work must the air do on the piston so that the pressure in the cylinder increases to 500 kPa?

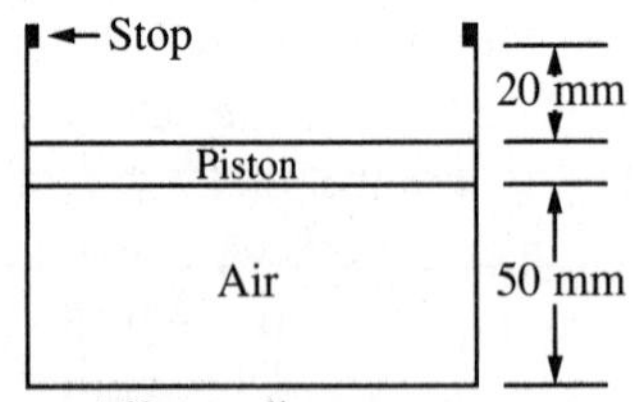

   (A) 22 J
   (B) 28 J
   (C) 41 J
   (D) 53 J

4. Which of the following statements about work for a quasiequilibrium process is incorrect?
   (A) The differential of work is inexact
   (B) Work is the area under a *P-T* diagram
   (C) Work is a path function
   (D) Work is always zero for a constant volume process

Questions 5–8

The frictionless piston shown has a mass of 16 kg. Heat is added until the temperature reaches 400°C. The initial quality is 20 percent. Assume P_{atm} = 100 kPa.

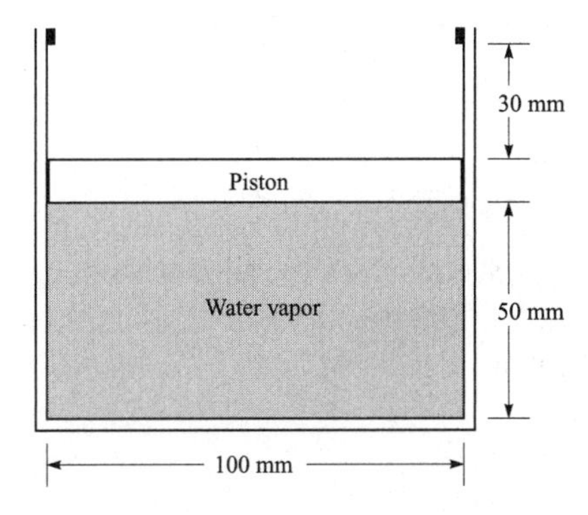

5. The total mass of the water is nearest
   (A) 0.018 kg
   (B) 0.012 kg
   (C) 0.0014 kg
   (D) 0.0010 kg
6. The quality when the piston hits the stops is nearest
   (A) 32%
   (B) 38%
   (C) 44%
   (D) 49%

7. The final pressure is nearest
   (A) 450 kPa
   (B) 560 kPa
   (C) 610 kPa
   (D) 690 kPa
8. The work done on the piston is nearest
   (A) 21 kJ
   (B) 28 kJ
   (C) 36 kJ
   (D) 42 kJ
9. Air is compressed in a cylinder such that the volume changes from 0.2 to 0.02 m^3. The initial pressure is 200 kPa. If the pressure is constant, the work is nearest
   (A) −36 kJ
   (B) −40 kJ
   (C) −46 kJ
   (D) −52 kJ
10. Estimate the work necessary to compress the air in a cylinder from a pressure of 100 kPa to that of 2000 kPa. The initial volume is 1000 cm^3. An isothermal process is to be assumed.
   (A) 0.51 kJ
   (B) 0.42 kJ
   (C) 0.30 kJ
   (D) 0.26 kJ
11. Estimate the work done by a gas during an unknown equilibrium process. The pressure and volume are measured as follows:

P	200	250	300	350	400	450	500	kPa
V	800	650	550	475	415	365	360	cm^3

   (A) 350 J
   (B) 260 J
   (C) 220 J
   (D) 130 J

12. The force needed to compress a nonlinear spring is given by $F = 10x^2$ N, where x is the distance the spring is compressed, measured in meters. Calculate the work needed to compress the spring from 0.2 to 0.8 m.

    (A) 0.54 J

    (B) 0.72 J

    (C) 0.84 J

    (D) 0.96 J

13. A paddle wheel and an electric heater supply energy to a system. If the torque is 20 N·m, the rotational speed is 400 rpm, the voltage is 20 V, and the amperage is 10 A, the work rate is nearest

    (A) −820 W

    (B) −920 W

    (C) −1040 W

    (D) −2340 W

14. A gasoline engine drives a small generator that is to supply sufficient electrical energy for a motor home. What is the minimum horsepower engine that would be necessary if a maximum of 200 A is anticipated from the 12-V system?

    (A) 2.4 hp

    (B) 2.6 hp

    (C) 3.0 hp

    (D) 3.2 hp

Quiz No. 2

1. Which of the following does not transfer work to or from a system?

    (A) A moving piston

    (B) The expanding membrane of a balloon

    (C) An electrical resistance heater

    (D) A membrane that bursts

2. Ten kilograms of air at 800 kPa are heated at constant pressure from 170 to 400°C. The work required is nearest
   (A) 1150 kJ
   (B) 960 kJ
   (C) 660 kJ
   (D) 115 kJ
3. Ten kilograms of saturated liquid water expands until $T_2 = 200°C$ while the pressure remains constant at 400 kPa. Find W_{1-2}.
   (A) 2130 kJ
   (B) 1960 kJ
   (C) 1660 kJ
   (D) 1115 kJ
4. A mass of 0.025 kg of steam at a quality of 10 percent and a pressure of 200 kPa is heated in a rigid container until the temperature reaches 300°C. The pressure at state 2 is nearest
   (A) 8.6 MPa
   (B) 2.8 MPa
   (C) 1.8 MPa
   (D) 0.4 MPa

Questions 5–8

The frictionless piston shown in equilibrium has a mass of 64 kg. Energy is added until the temperature reaches 220°C. Assume $P_{atm} = 100$ kPa.

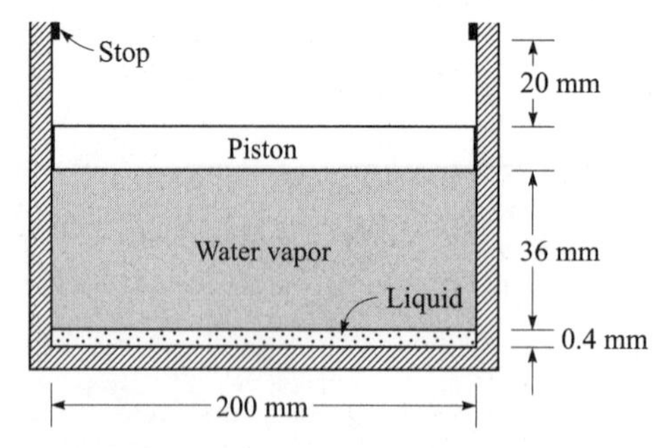

5. The initial quality is nearest
   (A) 12.2%
   (B) 8.3%

(C) 7.8%

(D) 6.2%

6. The quality when the piston just hits the stops is nearest

   (A) 73%

   (B) 92%

   (C) 96%

   (D) 99%

7. The final quality (or pressure if superheat) is nearest

   (A) 1.48 MPa

   (B) 1.52 MPa

   (C) 1.58 MPa

   (D) 1.62 MPa

8. The work done on the piston is nearest

   (A) 75 J

   (B) 96 J

   (C) 66 J

   (D) 11 J

9. Air is compressed in a cylinder such that the volume changes from 0.2 to 0.02 m^3. The pressure at the beginning of the process is 200 kPa. If the temperature is constant at 50°C, the work is nearest

   (A) −133 kJ

   (B) −126 kJ

   (C) −114 kJ

   (D) −92 kJ

10. Air is expanded in a piston-cylinder arrangement at a constant pressure of 200 kPa from a volume of 0.1 m^3 to a volume of 0.3 m^3. Then the temperature is held constant during an expansion of 0.5 m^3. Determine the total work done by the air.

    (A) 98.6 kJ

    (B) 88.2 kJ

    (C) 70.6 kJ

    (D) 64.2 kJ

11. Air undergoes a three-process cycle. Find the net work done for 2 kg of air if the processes are

    1 → 2: constant-pressure expansion

    2 → 3: constant volume

    3 → 1: constant-temperature compression

    The necessary information is $T_1 = 100°C$, $T_2 = 600°C$, and $P_1 = 200$ kPa.

    (A) 105 kJ

    (B) 96 kJ

    (C) 66 kJ

    (D) 11.5 kJ

12. A 200-mm-diameter piston is lowered by increasing the pressure from 100 to 800 kPa such that the *P-V* relationship is PV^2 = const. If $V_1 = 0.1$ m^3, the work done on the system is nearest

    (A) −18.3 kJ

    (B) −24.2 kJ

    (C) −31.6 kJ

    (D) −42.9 kJ

13. A 120-V electric resistance heater draws 10 A. It operates for 10 min in a rigid volume. Calculate the work done on the air in the volume.

    (A) 720 000 kJ

    (B) 720 kJ

    (C) 12 000 J

    (D) 12 kJ

14. An electrical voltage of 110 V is applied across a resistor providing a current of 12 A through the resistor. The work done during a period of 10 min is nearest

    (A) 792 000 kJ

    (B) 792 kJ

    (C) 792 MJ

    (D) 792 mJ

CHAPTER 4

The First Law of Thermodynamics

The first law of thermodynamics is commonly called the law of conservation of energy. In previous courses, the study of conservation of energy may have emphasized changes in kinetic and potential energy and their relationship to work. A more general form of conservation of energy includes the effects of heat transfer and internal energy changes. This more general form is usually called the *first law of thermodynamics.* Other forms of energy could also be included, such as electrostatic, magnetic, strain, and surface energy, but, in a first course in thermodynamics, problems involving these forms of energy are typically not included. We will present the first law for a system and then for a control volume.

4.1 The First Law Applied to a Cycle

We are now ready to present the first law of thermodynamics. A law is not derived or proved from basic principles but is simply a statement that we write based on our observations. Historically, the *first law of thermodynamics*, simply called the *first law*, was stated for a cycle: the net heat transfer is equal to the net work done for a system undergoing a cycle. This is expressed in equation form by

$$\sum W = \sum Q \tag{4.1}$$

or, using the symbol $\oint$ to represent integration around a complete cycle,

$$\oint \delta W = \oint \delta Q \tag{4.2}$$

The first law can be illustrated by considering the following experiment. Let a weight be attached to a pulley-paddle-wheel setup, such as that shown in Fig. 4.1*a*. Let the weight fall a certain distance thereby doing work on the system, contained in the tank shown, equal to the weight multiplied by the distance dropped. The temperature of the system (the substance in the tank) will immediately rise an amount ΔT. Now, the system is returned to its initial state (the completion of the cycle) by transferring heat to the surroundings, as implied by the Q in Fig. 4.1*b*. This reduces the temperature of the system to its initial temperature. The first law states that this heat transfer will be exactly equal to the work that was done by the falling weight.

EXAMPLE 4.1

A linear spring, with spring constant $K = 100$ kN/m, is stretched a distance of 0.8 m and attached to a paddle wheel. The paddle wheel then rotates until the spring is unstretched. Calculate the heat transfer necessary to return the system to its initial state.

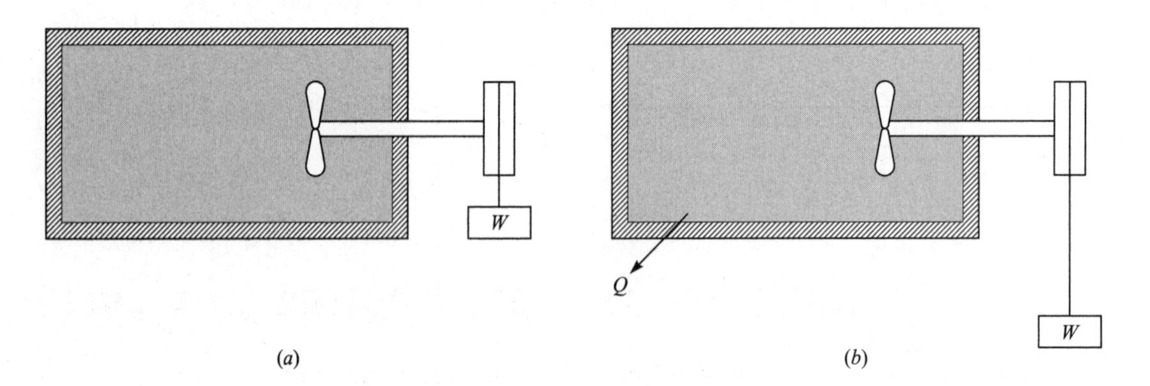

Figure 4.1 The first law applied to a cycle.

Solution

The work done by the linear spring on the system is given by

$$W_{1\text{-}2} = \int_0^{0.8} F\,dx = \int_0^{0.8} 100x\,dx = 100 \times \frac{0.8^2}{2} = 32\ \text{N}\cdot\text{m}$$

Since the heat transfer returns the system to its initial state, a cycle results. The first law then states that

$$Q_{1\text{-}2} = W_{1\text{-}2} = 32\ \text{J}$$

4.2 The First Law Applied to a Process

The first law of thermodynamics is often applied to a process as the system changes from one state to another. Realizing that a cycle results when a system undergoes two or more processes and returns to the initial state, we could consider a cycle composed of the two processes represented by A and B in Fig. 4.2. Applying the first law to this cycle, Eq. (4.2) takes the form

$$\int_1^2 \delta Q_A + \int_2^1 \delta Q_B = \int_1^2 \delta W_A + \int_2^1 \delta W_B \tag{4.3}$$

We interchange the limits on the process from 1 to 2 along B and write this as

$$\int_1^2 \delta(Q - W)_A = \int_1^2 \delta(Q - W)_B \tag{4.4}$$

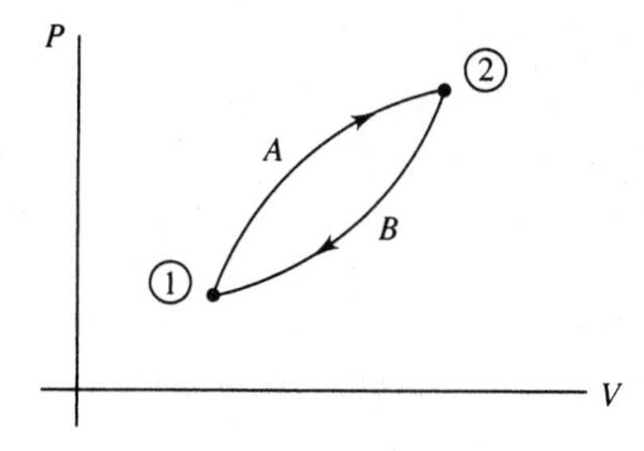

Figure 4.2 A cycle composed of two processes.

That is, the change in the quantity $Q - W$ from state 1 to state 2 is the same along path *A* as along path *B*; since this change is independent of the path between states 1 and 2 we let

$$dQ - dW = dE \tag{4.5}$$

where dE is an exact differential.

The quantity E is an extensive property of the system and represents the *energy* of the system at a particular state. Equation (4.5) can be integrated to yield

$$Q_{1\text{-}2} - W_{1\text{-}2} = E_2 - E_1 \tag{4.6}$$

where $Q_{1\text{-}2}$ is the heat transferred to the system during the process from state 1 to state 2, $W_{1\text{-}2}$ is the work done by the system on the surroundings during the process, and E_2 and E_1 are the values of the property E. More often than not the subscripts will be dropped on Q and W when working problems.

The property E represents all of the energy: kinetic energy KE, potential energy PE, and internal energy U, which includes chemical energy and the energy associated with the molecules and atoms. Any other form of energy is also included in the total energy E. Its associated intensive property is designated e.

The first law of thermodynamics then takes the form

$$\begin{aligned} Q_{1\text{-}2} - W_{1\text{-}2} &= KE_2 - KE_1 + PE_2 - PE_1 + U_2 - U_1 \\ &= \frac{1}{2}m(V_2^2 - V_1^2) + mg(z_2 - z_1) + U_2 - U_1 \end{aligned} \tag{4.7}$$

If we apply the first law to an isolated system, one which is not in contact with its surroundings so that $Q_{1\text{-}2} = W_{1\text{-}2} = 0$, the first law becomes the conservation of energy; that is,

$$E_2 = E_1 \tag{4.8}$$

The internal energy U is an extensive property. Its associated intensive property is the specific internal energy u; that is, $u = U/m$. For simple systems in equilibrium, only two properties are necessary to establish the state of a pure substance, such as air or steam. Since internal energy is a property, it depends only on, say, pressure and temperature; or, for steam in the quality region, it depends on quality and temperature (or pressure). For steam, its value for a particular quality would be

$$u = u_f + x(u_g - u_f) \tag{4.9}$$

We can now apply the first law to systems involving working fluids with tabulated property values. Before we apply the first law to systems involving substances such as ideal gases or solids, we will introduce several additional properties that will simplify that task in the next three articles, but first some examples.

EXAMPLE 4.2

A 5-hp fan is used in a large room to provide for air circulation. Assuming a well insulated, sealed room, determine the internal energy increase after 1 hour of operation.

Solution

By assumption, $Q = 0$. With $\Delta PE = \Delta KE = 0$ the first law becomes $-W = \Delta U$. The work input is

$$W = (-5 \text{ hp})(1 \text{ hr})(746 \text{ W/hp})(3600 \text{ s/hr}) = -1.343 \times 10^7 \text{ J}$$

where $W = \text{J/s}$.

The negative sign results because the work is input to the system. Finally, the internal energy increase is

$$\Delta U = -(-1.343 \times 10^7) = 1.343 \times 10^7 \text{ J}$$

EXAMPLE 4.3

A frictionless piston is used to provide a constant pressure of 400 kPa in a cylinder containing steam originally at 200°C with a volume of 2m³. Calculate the final temperature if 3500 kJ of heat is added.

Solution

The first law of thermodynamics, using $\Delta PE = \Delta KE = 0$, is $Q - W = \Delta U$. The work done during the motion of the piston is

$$W = \int P dV = P(V_2 - V_1) = 400(V_2 - V_1)$$

The mass before and after remains unchanged. Using steam table C.3, this is expressed as

$$m = \frac{V_1}{v_1} = \frac{2}{0.5342} = 3.744 \text{ kg}$$

The volume V_2 is written as $V_2 = mv_2 = 3.744\, v_2$. The first law is then, finding u_1 from the steam tables in App. C,

$$3500 - 400(3.744v_2 - 2) = (u_2 - 2647) \times 3.744$$

This requires a trial-and-error process. One plan for obtaining a solution is to guess a value for v_2 and calculate u_2 from the equation above. If this value checks with the u_2 from the steam tables at the same temperature, then the guess is the correct one. For example, guess v_2 = 1.0 m³/kg. Then the equation gives u_2 = 3395 kJ/kg. From the steam tables, with P = 0.4 MPa, the u_2 value allows us to interpolate T_2 = 653°C and the v_2 gives T_2 = 600°C. Therefore, the guess must be revised. Try v_2 = 1.06 m³/kg. The equation gives u_2 = 3372 kJ/kg. The tables are interpolated to give T_2 = 640°C; for v_2, T_2 = 647°C. The actual v_2 is a little less than 1.06 m³/kg, with the final temperature being approximately

$$T_2 = 644°\text{C}$$

4.3 Enthalpy

In the solution of problems involving systems, certain products or sums of properties occur with regularity. One such combination of properties we define to be *enthalpy H*:

$$H = U + PV \tag{4.10}$$

This property will come in handy, especially in a constant-pressure process, but also in other situations, as we shall see in examples and applications in future chapters.

The specific enthalpy h is found by dividing by the mass: $h = H/m$. From Eq. (4.10), it is

$$h = u + Pv \tag{4.11}$$

Enthalpy is a property of a system and is also found in the steam tables. The energy equation can now be written, for a constant-pressure equilibrium process, as

$$Q_{1\text{-}2} = H_2 - H_1 \tag{4.12}$$

In a nonequilibrium constant-pressure process ΔH would not equal the heat transfer.

It is only the change in enthalpy or internal energy that is important; hence, we can arbitrarily choose the datum from which to measure h and u. We choose saturated liquid at 0°C to be the datum point for water; there $h = 0$ and $u = 0$.

EXAMPLE 4.4
Using the concept of enthalpy, solve the problem presented in Example 4.3.

Solution
The energy equation for a constant-pressure process is (with the subscript on the heat transfer omitted),

$$Q = H_2 - H_1 \qquad \text{or} \qquad 3500 = m(h_2 - 2860)$$

Using steam table C.3, as in Example 4.3, the mass is

$$m = \frac{V}{v} = \frac{2}{0.5342} = 3.744 \text{ kg}$$

Thus,

$$h_2 = \frac{3500}{3.744} + 2860 = 3795 \text{ kJ/kg}$$

From the steam tables this interpolates to

$$T_2 = 600 + \frac{92.6}{224} \times 100 = 641°\text{C}$$

Obviously, enthalpy was very useful in solving this constant-pressure problem. Trial and error was unnecessary, and the solution was rather straightforward. We illustrated that the quantity we invented, enthalpy, is not necessary, but it is quite handy. We will use it often in calculations.

4.4 Latent Heat

The amount of energy that must be transferred in the form of heat to a substance held at constant pressure in order that a phase change occurs is called *latent heat.* It is the change in enthalpy of the substance at the saturated conditions of the two phases. The heat that is necessary to melt a unit mass of a substance at constant pressure is the *heat of fusion* and is equal to $h_{if} = h_f - h_i$, where h_i is the enthalpy of saturated solid (see Table C.5 for ice) and h_f is the enthalpy of saturated liquid. The *heat of vaporization* is the heat required to completely vaporize a unit mass of saturated liquid; it is equal to $h_{fg} = h_g - h_f$. When a solid changes phase directly to a gas, sublimation occurs; the *heat of sublimation* is equal to $h_{ig} = h_g - h_i$.

The heat of fusion and the heat of sublimation are relatively insensitive to pressure or temperature changes. For ice, the heat of fusion is approximately 330 kJ/kg and the heat of sublimation is about 2040 kJ/kg. The heat of vaporization of water is included as h_{fg} in Tables C.1 and C.2 and is very sensitive to pressure and temperature.

4.5 Specific Heats

Since only two independent variables are necessary to establish the state of a simple system, the specific internal energy can be expressed as a function of temperature and specific volume; that is,

$$u = u(T, v) \tag{4.13}$$

Using the chain rule from calculus, the differential can be stated as

$$du = \left.\frac{\partial u}{\partial T}\right|_v dT + \left.\frac{\partial u}{\partial v}\right|_T dv \tag{4.14}$$

Joule's classical experiment showed that $u = u(T)$ for an ideal gas, so that $\partial u/\partial v|_T = 0$. We define the *constant-volume specific heat* C_v as

$$C_v = \left.\frac{\partial u}{\partial T}\right|_v \tag{4.15}$$

so that Eq. (4.14) becomes

$$du = C_v dT \tag{4.16}$$

This can be integrated to give

$$u_2 - u_1 = \int_{T_1}^{T_2} C_v \, dT \tag{4.17}$$

For a known $C_v(T)$, this can be integrated to find the change in internal energy over any temperature interval for an ideal gas.

Likewise, considering specific enthalpy to be dependent on the two variables T and P, we have

$$dh = \left.\frac{\partial h}{\partial T}\right|_P dT + \left.\frac{\partial h}{\partial P}\right|_T dP \tag{4.18}$$

The *constant-pressure specific heat* C_p is defined as

$$C_p = \left.\frac{\partial h}{\partial T}\right|_P \tag{4.19}$$

For an ideal gas we have, returning to Eq. (4.11),

$$h = u + Pv = u + RT \tag{4.20}$$

where we have used the ideal-gas equation of state. Since u is only a function of T, we see that h is also only a function of T for an ideal gas. Hence, $\partial h/\partial P|_T = 0$ and we have, from Eq. (4.18), for an ideal gas,

$$dh = C_p dT \qquad \text{or} \qquad h_2 - h_1 = \int_{T_1}^{T_2} C_p dT \tag{4.21}$$

It is often convenient to specify specific heats on a per-mole, rather than a per-mass, basis; these *molar specific heats* are $\bar{C}_v$ and $\bar{C}_p$. Clearly, we have the relations

$$\bar{C}_v = MC_v \qquad \text{and} \qquad \bar{C}_p = MC_p \tag{4.22}$$

where M is the molar mass. Thus, $\bar{C}_v$ and $\bar{C}_p$ may be simply derived from the values of M and C_v and C_p listed in Table B.2. (The common "overbar notation" for a molar quantity is used throughout thermodynamics.)

The equation for enthalpy can be used to relate, for an ideal gas, the specific heats and the gas constant. In differential form Eq. (4.20) takes the form

$$dh = du + d(Pv) \tag{4.23}$$

Introducing the specific heat relations and the ideal-gas equation, we have

$$C_p = C_v + R \tag{4.24}$$

This relationship, or its molar equivalent, allows C_v to be determined from tabulated values or expressions for C_p. Note that the difference between C_p and C_v for an ideal gas is always a constant, even though both are functions of temperature.

The *specific heat ratio* k is also a property of particular interest; it is defined as

$$k = \frac{C_p}{C_v} \tag{4.25}$$

This can be substituted into Eq. (4.24) to give

$$C_p = R\frac{k}{k-1} \qquad \text{and} \qquad C_v = R\frac{1}{k-1} \tag{4.26}$$

Obviously, since R is a constant for an ideal gas, the specific heat ratio k will depend only on the temperature.

For gases, the specific heats slowly increase with increasing temperature. Since they do not vary significantly over fairly large temperature differences, it is often acceptable to treat C_v and C_p as constants. For such situations there results

$$u_2 - u_1 = C_v(T_2 - T_1) \qquad \text{and} \qquad h_2 - h_1 = C_p(T_2 - T_1) \tag{4.27}$$

For air, we will use $C_v = 0.717$ kJ/kg·K and $C_p = 1.00$ kJ/kg·K, unless otherwise stated.[1] For more accurate calculations with air, or other gases, one should consult ideal-gas tables, such as those in App. E, which tabulate $h(T)$ and $u(T)$, or integrate, using expressions for $C_p(T)$ found in text books. Once C_p is found, C_v can be calculated using $C_v = R - C_p$.

For liquids and solids the specific heat C_p is tabulated in Table B.4. Since it is quite difficult to maintain constant volume while the temperature is changing, C_v values are usually not tabulated for liquids and solids; the difference $C_p - C_v$ is quite small. For most liquids the specific heat is relatively insensitive to temperature change. For water we will use the nominal value of 4.18 kJ/kg·K. For ice, the specific heat in kJ/kg·K is approximately $C_p = 2.1 + 0.0069T$, where T is measured in °C. The variation of specific heat with pressure is usually quite slight except for special situations.

[1]The units on C_p and C_v are often written as kJ/kg·°C since ΔT is the same using either unit.

EXAMPLE 4.5

The specific heat of superheated steam at approximately 150 kPa can be determined by the equation

$$C_p = 2.07 + \frac{T - 400}{1480} \text{ kJ/kg}\cdot{}^\circ\text{C}$$

(a) What is the enthalpy change between 300 and 700°C for 3 kg of steam? Compare with the steam tables.

(b) What is the average value of C_p between 300 and 700°C based on the equation and based on the tabulated data?

Solution

(a) The enthalpy change is found to be

$$\Delta H = m\int_{T_1}^{T_2} C_p\, dT = 3\int_{300}^{700}\left(2.07 + \frac{T-400}{1480}\right)dT = 2565 \text{ kJ}$$

From the tables we find, using $P = 150$ kPa,

$$\Delta H = 3 \times (3928 - 3073) = 2565 \text{ kJ}$$

(b) The average value $C_{p,\text{avg}}$ is found by using the relation

$$mC_{p,\text{avg}}\Delta T = m\int_{T_1}^{T_2} C_p\, dT \qquad \text{or} \qquad 3C_{p,\text{avg}} \times 400 = 3\int_{300}^{700}\left(2.07 + \frac{T-400}{1480}\right)dT$$

The integral was evaluated in part (a); hence, we have

$$C_{p,\text{avg}} = \frac{2565}{3 \times 400} = 2.14 \text{ kJ/kg}\cdot{}^\circ\text{C}$$

Using the values from the steam table, we have

$$C_{p,\text{avg}} = \frac{\Delta h}{\Delta T} = \frac{3928 - 3073}{400} = 2.14 \text{ kJ/kg}\cdot{}^\circ\text{C}$$

Because the steam tables give the same values as the linear equation of this example, we can safely assume that the $C_p(T)$ relationship for steam over this temperature range is closely approximated by a linear relation. This linear relation would change, however, for each pressure chosen; hence, the steam tables are essential.

EXAMPLE 4.6

Determine the enthalpy change for 1 kg of nitrogen which is heated from 300 to 1200 K by (a) using the gas tables, and (b) assuming constant specific heat. Use $M = 28$ kg/kmol.

Solution

(a) Using the gas table in App. E, find the enthalpy change to be

$$\Delta h = 36\,777 - 8723 = 28\,054 \text{ kJ/kmol} \qquad \text{or} \qquad 28\,054/28 = 1002 \text{ kJ/kg}$$

(b) Assuming constant specific heat (found in Table B.2) the enthalpy change is found to be

$$\Delta h = C_p \Delta T = 1.042 \times (1200 - 300) = 938 \text{ kJ/kg}$$

Note: the enthalpy change found by assuming constant specific heat is in error by over 6 percent. If T_2 were closer to 300 K, say 600 K, the error would be much smaller. So, for large temperature differences, the tables should be used.

4.6 The First Law Applied to Various Processes

THE CONSTANT-TEMPERATURE PROCESS

For the isothermal process, tables may be consulted for substances for which tabulated values are available. Internal energy and enthalpy vary slightly with pressure for the isothermal process, and this variation must be accounted for in processes involving many substances. The energy equation is

$$Q - W = \Delta U \tag{4.28}$$

For a gas that approximates an ideal gas, the internal energy depends only on the temperature and thus $\Delta U = 0$ for an isothermal process; for such a process

$$Q = W \tag{4.29}$$

Using the ideal-gas equation $PV = mRT$, the work for a quasiequilibrium isothermal process can be found to be

$$\begin{aligned} W &= \int_{V_1}^{V_2} P\,dV \\ &= mRT \int_{V_1}^{V_2} \frac{dV}{V} = mRT \ln \frac{V_2}{V_1} = mRT \ln \frac{P_1}{P_2} \end{aligned} \tag{4.30}$$

where we used $P_1V_1 = P_2V_2$.

THE CONSTANT-VOLUME PROCESS

The work for a constant-volume quasiequilibrium process is zero, since dV is zero. For such a process the first law becomes

$$Q = \Delta U \tag{4.31}$$

If tabulated values are available for a substance, we may directly determine ΔU.

For a gas, approximated by an ideal gas with constant C_v, we would have

$$Q = mC_v(T_2 - T_1) \qquad \text{or} \qquad q = C_v(T_2 - T_1) \tag{4.32}$$

If nonequilibrium work, such as paddle-wheel work, is present, that work must be accounted for in the first law.

THE CONSTANT-PRESSURE PROCESS

The first law, for a constant-pressure quasiequilibrium process, was shown in Sec. 4.3 to be

$$Q = \Delta H \tag{4.33}$$

Hence, the heat transfer for such a process can easily be found using tabulated values, if available.

For a gas that behaves as an ideal gas, we have

$$Q = m\int_{T_1}^{T_2} C_p\, dT \tag{4.34}$$

For a process involving an ideal gas for which C_p is constant there results

$$Q = mC_p(T_2 - T_1) \tag{4.35}$$

For a nonequilibrium process the work must be accounted for directly in the first law and it cannot be expressed as $P(V_2 - V_1)$. For such a process the above three equations would not be valid.

THE ADIABATIC PROCESS

There are numerous examples of processes for which there is no, or negligibly small, heat transfer. The study of such processes is, however, often postponed until after the second law of thermodynamics is presented. This postponement is not necessary for an ideal gas, and because the adiabatic quasiequilibrium process is quite common, it is presented here.

The differential form of the first law for the adiabatic process is

$$-\delta W = du \tag{4.36}$$

or, for a quasiequilibrium process,

$$du + P dv = 0 \tag{4.37}$$

To utilize this equation for substances such as water or refrigerants, we must use tabulated properties, an impossible task at this point because we do not know a property that remains constant for this process. However, if an ideal gas can be assumed, Eq. (4.37) takes the form

$$C_v dT + \frac{RT}{v} dv = 0 \tag{4.38}$$

Rearrange and integrate between states 1 and 2, assuming constant C_v:

$$\frac{C_v}{R} \ln \frac{T_2}{T_1} = -\ln \frac{v_2}{v_1} \tag{4.39}$$

which can be put in the forms

$$\frac{T_2}{T_1} = \left(\frac{v_1}{v_2}\right)^{R/C_v} = \left(\frac{v_1}{v_2}\right)^{k-1} = \left(\frac{P_2}{P_1}\right)^{\frac{k-1}{k}} \quad \text{and} \quad \frac{P_2}{P_1} = \left(\frac{v_1}{v_2}\right)^k \tag{4.40}$$

The adiabatic process for a real substance such as water, steam, or a solid will be treated in detail in Chap. 5.

THE POLYTROPIC PROCESS

A careful inspection of the special quasiequilibrium processes for an ideal gas presented in this chapter suggests that each process can be expressed as

$$PV^n = \text{const.} \tag{4.41}$$

The work is calculated as follows:

$$\begin{aligned} W &= \int_{V_1}^{V_2} P\,dV = P_1V_1^n \int_{V_1}^{V_2} V^{-n}dV \\ &= \frac{P_1V_1^n}{1-n}(V_2^{1-n} - V_1^{1-n}) = \frac{P_2V_2 - P_1V_1}{1-n} \end{aligned} \tag{4.42}$$

except Eq. (4.30) is used if $n = 1$, since the above is undefined for $n = 1$. The heat transfer follows from the first law.

Each quasiequilibrium process of an ideal gas is associated with a particular value for n as follows:

Isothermal:	$n = 1$
Constant-volume:	$n = \infty$
Constant-pressure:	$n = 0$
Adiabatic (C_p = const):	$n = k$

If the constant n [in Eq. (4.41)] is none of the values ∞, k, 1, or 0, then the process can be referred to as a *polytropic process*. For such a process involving an ideal gas, any of the equations in (4.40) can be used with k simply replaced by n; this is convenient in processes in which there is some heat transfer but which do not maintain constant temperature, pressure, or volume.

EXAMPLE 4.7

Determine the heat transfer necessary to increase the pressure of 70 percent quality steam from 200 to 800 kPa, maintaining the volume constant at 2 m^3. Assume a quasiequilibrium process.

Solution

For the constant-volume quasiequilibrium process the work is zero. The first law reduces to $Q = m(u_2 - u_1)$. The mass must be found. It is

$$m = \frac{V}{v_1} = \frac{2}{0.0011 + 0.7(0.8857 - 0.0011)} = 3.224 \text{ kg}$$

The internal energy at state 1 is

$$u_1 = 504.5 + 0.7(2529.5 - 504.5) = 1922 \text{ kJ/kg}$$

The constant-volume process demands that $v_2 = v_1 = 0.6203$ m^3/kg. From the steam tables at 800 kPa we find, by interpolation, that

$$u_2 = \frac{0.6761 - 0.6203}{0.6761 - 0.6181}(3661 - 3853) = 3668 \text{ kJ/kg}$$

The heat transfer is then

$$Q = m(u_2 - u_1) = 3.224 \times (3668 - 1922) = 5629 \text{ kJ}$$

EXAMPLE 4.8

A piston-cylinder arrangement contains 0.02 m^3 of air at 50°C and 400 kPa. Heat is added in the amount of 50 kJ and work is done by a paddle wheel until the temperature reaches 700°C. If the pressure is held constant, how much paddle-wheel work must be added to the air? Assume constant specific heats.

Solution

The process cannot be approximated by a quasiequilibrium process because of the paddle-wheel work. Thus, the heat transfer is not equal to the enthalpy change. The first law may be written as

$$Q - W = m(u_2 - u_1) \qquad \text{or} \qquad Q - W_{\text{paddle}} = m(h_2 - h_1)$$
$$= mC_p(T_2 - T_1)$$

To find m we use the ideal-gas equation. It gives us

$$m = \frac{PV}{RT} = \frac{400\,000 \times 0.02}{287 \times (50+273)} = 0.0863 \text{ kg}$$

From the first law the paddle-wheel work is found to be

$$W_{\text{paddle}} = Q - mC_p(T_2 - T_1)$$
$$= 50 - 0.0863 \times 1.0 \times (700 - 50) = -6.096 \text{ kJ}$$

EXAMPLE 4.9

Calculate the work necessary to compress air in an insulated cylinder from a volume of 2 m^3 to a volume of 0.2 m^3. The initial temperature and pressure are 20°C and 200 kPa, respectively.

Solution

We will assume that the compression process is approximated by a quasiequilibrium process, which is acceptable for most compression processes, and that the process is adiabatic due to the presence of the insulation (we usually assume an adiabatic process anyhow since heat transfer is assumed to be negligible). The first law is then written as

$$-W = m(u_2 - u_1) = mC_v(T_2 - T_1)$$

The mass is found from the ideal-gas equation to

$$m = \frac{PV}{RT} = \frac{200 \times 2}{0.287(20+273)} = 4.757 \text{ kg}$$

The final temperature T_2 is found for the adiabatic quasiequilibrium process from Eq. (4.40); it is

$$T_2 = T_1\left(\frac{V_1}{V_2}\right)^{k-1} = 293\left(\frac{2}{0.2}\right)^{1.4-1} = 736 \text{ K}$$

Finally,

$$W = -4.757 \text{ kg}\left(0.717 \frac{\text{kJ}}{\text{kg}\cdot\text{K}}\right)(736 - 293) \text{ K} = -1510 \text{ kJ}$$

4.7 The First Law Applied to Control Volumes

We have thus far restricted ourselves to systems; no mass crosses the boundary of a system. This restriction is acceptable for many problems of interest and may, in fact, be imposed on the power plant schematic shown in Fig. 4.3. However, if the first law is applied to this system, only an incomplete analysis can be accomplished. For a more complete analysis we must relate W_{in}, Q_{in}, W_{out}, and Q_{out} to the pressure and temperature changes for the pump, boiler, turbine, and condenser, respectively. To do this we must consider each device of the power plant as a control volume into which and from which a fluid flows. For example, water flows into the pump at a low pressure and leaves the pump at a high pressure; the work input into the pump is obviously related to this pressure rise. We must formulate equations that allow us to make the necessary calculations. For most applications that we will consider it will be acceptable to assume both a *steady flow* (the flow variables do not change with time) and a *uniform flow* (the velocity, pressure, and density are constant over the cross-sectional area). Fluid mechanics treats the more general unsteady, nonuniform situations in much greater detail.

THE CONSERVATION OF MASS

Many devices have an inlet (usually a pipe) and an outlet (also typically a pipe). Consider a device, a control volume, to be operating in a steady-flow mode with uniform profiles in the inlet and outlet pipes. During some time increment Δt, a small amount of mass Δm_1 leaves the inlet pipe and enters the device, and the same

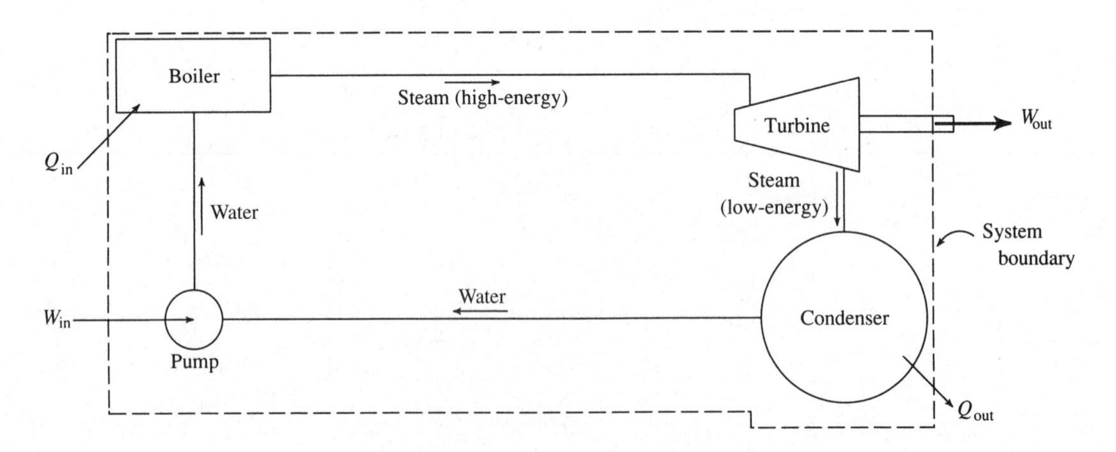

Figure 4.3 A schematic for a power plant.

amount of mass Δm_2 leaves the device and enters the outlet pipe. The amount of mass that enters the device is expressed as

$$\begin{aligned}\Delta m_1 &= \textit{volume} \times \textit{density} = \textit{area} \times \textit{length} \times \textit{density} \\ &= \textit{area} \times (\textit{velocity} \times \textit{time}) \times \textit{density} \\ &= A_1 V_1 \Delta t \rho_1 \end{aligned} \tag{4.43}$$

And, that which leaves the device is

$$\Delta m_2 = A_2 V_2 \Delta t \rho_2 \tag{4.44}$$

Since $\Delta m_1 = \Delta m_2$ for this steady flow, we see that

$$\rho_1 A_1 V_1 = \rho_2 A_2 V_2 \tag{4.45}$$

The units on ρAV are kg/s and is referred to as the *mass flow rate* (or the *mass flux*) $\dot{m}$. Equation (4.45) is the *continuity equation* and is often used in the solution of problems. For an incompressible flow $(\rho_1 = \rho_2)$, we often introduce the *volume flow rate* Q_f defined by AV.

EXAMPLE 4.10

Steam at 2000 kPa and 600°C flows through a 60-mm-diameter pipe into a device and exits through a 120-mm-diameter pipe at 600 kPa and 200°C. If the steam in the 60-mm section has a velocity of 20 m/s, determine the velocity in the 120-mm section. Also calculate the mass flow rate.

Solution

From the superheat Table C.3 we find

$$\rho_1 = \frac{1}{v_1} = \frac{1}{0.1996} = 5.01 \text{ kg/m}^3 \qquad \rho_2 = \frac{1}{v_2} = \frac{1}{0.3520} = 2.84 \text{ kg/m}^3$$

The continuity equation (4.45) is used to write

$$\rho_1 A_1 V_1 = \rho_2 A_2 V_2 \qquad 5.01 \times \pi \times 0.03^2 \times 20 = 2.84 \times \pi \times 0.06^2 \times V_2$$

$$\therefore V_2 = 8.82 \text{ m/s}$$

The mass flow rate is

$$\dot{m} = \rho_1 A_1 V_1 = 5.01 \times \pi \times 0.03^2 \times 20 = 0.283 \text{ kg/s}$$

THE ENERGY EQUATION

Consider again a fixed control volume, a device, with one inlet and one outlet. At some time t the system occupies a small volume 1 (that enters the device from the inlet pipe over a time increment Δt) plus the device; then at $t + \Delta t$ the system occupies the device plus the small volume 2 that leaves the device. The first law for the steady-state device (the energy in the device does not change with time) can be stated as

$$
\begin{aligned}
Q - W &= E_{\text{sys}}(t+\Delta t) - E_{\text{sys}}(t) \\
&= \Delta E_2(t+\Delta t) + E_{\text{device}}(t+\Delta t) - E_{\text{device}}(t) - \Delta E_1(t) \\
&= \rho_2 A_2 V_2 \Delta t\left(u_2 + \frac{1}{2}V_2^2 + gz_2\right) - \rho_1 A_1 V_1 \Delta t\left(u_1 + \frac{1}{2}V_1^2 + gz_1\right)
\end{aligned}
\tag{4.46}
$$

where $\rho AV\Delta t$ is the mass in a small volume and Q and W are transferred to and from the system during the time increment Δt. Also, $E_{\text{device}}(t+\Delta t) = E_{\text{device}}(t)$ due to the steady-state operation. Note that the energy consists of internal energy, kinetic energy, and potential energy.

The work W is composed of two parts: the work, sometimes called *flow work*, due to the pressure needed to move the fluid into and from the device, and the work that results from a shaft that is usually rotating, called *shaft work* W_S, that operates inside the device. This is expressed as

$$
W = P_2 A_2 V_2 \Delta t - P_1 A_1 V_1 \Delta t + W_S \tag{4.47}
$$

where PA is the pressure force and $V\Delta t$ is the distance it moves during the time increment Δt. The negative sign results because the work done on the system is negative when moving the fluid into the control volume. Substitute the expression for work W into Eq. (4.46) and express the flow work term as $\rho AV(P/\rho)\Delta t$. The first law is then arranged as

$$
Q - W_S = \rho_2 A_2 V_2\left(\frac{V_2^2}{2} + gz_2 + u_2 + \frac{P_2}{\rho_2}\right)\Delta t - \rho_1 A_1 V_1\left(\frac{V_1^2}{2} + gz_1 + u_1 + \frac{P_1}{\rho_1}\right)\Delta t \tag{4.48}
$$

Divide through by Δt to obtain the more general *energy equation*:

$$
\dot{Q} - \dot{W}_S = \dot{m}\left(\frac{V_2^2 - V_1^2}{2} + g(z_2 - z_1) + u_2 - u_1 + \frac{P_2}{\rho_2} - \frac{P_1}{\rho_1}\right) \tag{4.49}
$$

where we have used $\dot{m} = \dot{m}_1 = \dot{m}_2$ for this steady flow, and

$$\dot{Q} = \frac{Q}{\Delta t} \qquad \dot{W}_S = \frac{W_S}{\Delta t} \qquad \dot{m} = \rho A V \tag{4.50}$$

For many devices of interest in thermodynamics, the potential energy and gravity effects do not influence its operation, so we write the energy equation as

$$\dot{Q} - \dot{W}_S = \dot{m}(h_2 - h_1) \tag{4.51}$$

since $h = u + Pv = u + P/\rho$. This energy equation is often used in the form

$$q - w_S = h_2 - h_1 \tag{4.52}$$

where $q = \dot{Q}/\dot{m}$ and $w_S = \dot{W}_S/\dot{m}$. This simplified form of the energy equation has a surprisingly large number of applications. A nozzle, or a diffuser, is a device in which the kinetic energy change cannot be neglected so Eq. (4.52) could not be used for such devices.

For a control volume through which a liquid flows, it is most convenient to return to Eq. (4.49). For a steady flow with $\rho_2 = \rho_1 = \rho$, neglecting heat transfer and changes in internal energy, the energy equation takes the form

$$-\dot{W}_S = \dot{m}\left[\frac{P_2 - P_1}{\rho} + \frac{V_2^2 - V_1^2}{2} + g(z_2 - z_1)\right] \tag{4.53}$$

This is the form to use for a pump or a hydroturbine. If $\dot{Q}$ and Δu are not zero, simply include them.

4.8 Applications of the Energy Equation

There are several points that must be considered in the analysis of most problems in which the energy equation is used. As a first step, it is very important to identify the control volume selected in the solution of the problems. If at all possible, the control surface should be chosen so that the flow variables are uniform or known functions over the areas where the fluid enters and exits the control volume. The control surface should be chosen sufficiently far downstream from an abrupt area change (an entrance or a sudden contraction) that the velocity and pressure can be approximated by uniform distributions.

It is also necessary to specify the process by which the flow variables change. Is it incompressible? Isothermal? Constant-pressure? Adiabatic? A sketch of the process on a suitable diagram is often of use in the calculations. If the working substance behaves as an ideal gas, then the appropriate equations may be used; if not, tabulated values must be used, such as those provided for steam. For real gases that do not behave as ideal gases, properties can be found in App. E.

Often heat transfer from a device or the internal energy change across a device, such as a pump, is not desired. For such situations, the heat transfer and internal energy change may be lumped together as *losses*. In a pipeline, losses occur because of friction; in a pump, losses occur because of separated fluid flow around the rotating blades. For many devices the losses are included as an efficiency of the device. Examples will illustrate.

THROTTLING DEVICES

A throttling device involves a steady-flow adiabatic process that provides a pressure drop with no significant potential energy changes, kinetic energy changes, heat transfer, or work. Two such devices are sketched in Fig. 4.4. For this process [see Eq. (4.52)],

$$h_2 = h_1 \tag{4.54}$$

where section 1 is upstream and section 2 is downstream. Most valves are throttling devices, for which the energy equation takes the form of Eq. (4.54). They are also used in many refrigeration units in which the sudden drop in pressure causes a change in phase of the working substance.

EXAMPLE 4.11

Steam enters a throttling valve at 8000 kPa and 300°C and leaves at a pressure of 2000 kPa. Determine the final temperature and specific volume of the steam.

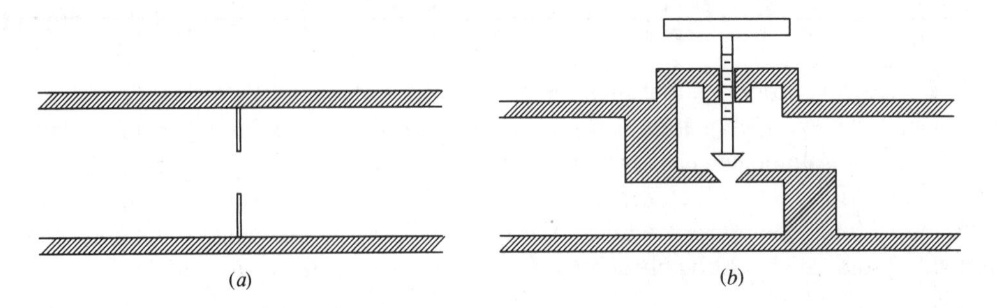

Figure 4.4 Throttling devices. (*a*) Orifice plate. (*b*) Globe value.

Solution

The enthalpy of the steam as it enters is found from the superheat steam table to be $h_1 = 2785$ kJ/kg. This must equal the exiting enthalpy as demanded by Eq. (4.54). The exiting steam is in the quality region, since at 2000 kPa $h_g = 2799.5$ kJ/kg. Thus the final temperature is $T_2 = 212.4$°C.

To find the specific volume we must know the quality. It is found from

$$h_2 = h_f + xh_{fg} \qquad 2785 = 909 + 1891x \qquad \therefore x_2 = 0.992$$

The specific volume is then

$$\begin{aligned} v_2 &= v_f + x(v_g - v_f) \\ &= 0.0012 + 0.992 \times (0.0992 - 0.0012) \\ &= 0.0988 \text{ m}^3/\text{kg} \end{aligned}$$

COMPRESSORS, PUMPS, AND TURBINES

A pump is a device that transfers energy to a liquid with the result that the pressure is increased. Compressors and blowers also fall into this category but have the primary purpose of increasing the pressure in a gas. A turbine, on the other hand, is a device in which work is done by the fluid on a set of rotating blades. As a result there is a pressure drop from the inlet to the outlet of the turbine. In some situations there may be heat transferred from the device to the surroundings, but often the heat transfer is negligible. In addition, the kinetic and potential energy changes are negligible. For such devices operating in a steady-state mode, the energy equation takes the form [see Eq. (4.51)]

$$-\dot{W}_S = \dot{m}(h_2 - h_1) \qquad \text{or} \qquad -w_S = h_2 - h_1 \tag{4.55}$$

where $\dot{W}_S$ is negative for a compressor and positive for a gas or steam turbine.

For liquids, such as water, the energy equation (4.53), neglecting kinetic and potential energy changes, becomes

$$-\dot{W}_S = \dot{m}\frac{P_2 - P_1}{\rho} \tag{4.56}$$

which is used for a pump or a hydroturbine.

EXAMPLE 4.12

Steam enters a turbine at 4000 kPa and 500°C and leaves as shown. For an inlet velocity of 200 m/s calculate the turbine power output. Neglect any heat transfer and kinetic energy change. Show that the kinetic energy change is negligible.

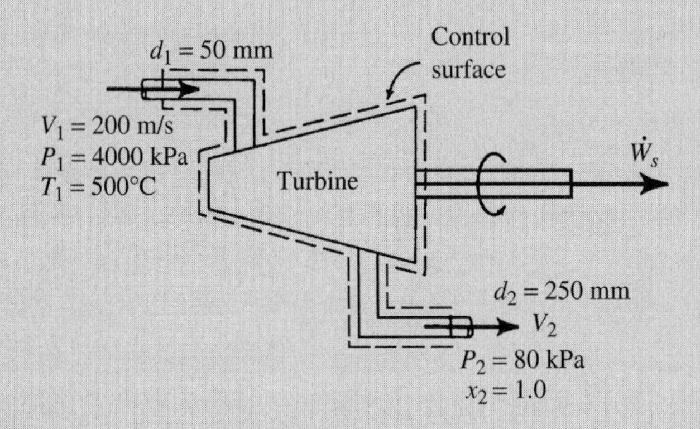

Solution

The energy equation in the form of Eq. (4.55) is $-\dot{W}_T = \dot{m}(h_2 - h_1)$. We find $\dot{m}$ as follows:

$$\dot{m} = \rho_1 A_1 V_1 = \frac{1}{v_1} A_1 V_1 = \frac{1}{0.08643} \times \pi \times 0.025^2 \times 200 = 4.544 \text{ kg/s}$$

The enthalpies are found from Table C.3 to be

$$h_1 = 3445 \text{ kJ/kg} \qquad h_2 = 2666 \text{ kJ/kg}$$

The maximum power output is then

$$\dot{W}_T = -4.544 \times (2666 - 3445) = 3540 \text{ kJ/s} \quad \text{or} \quad 3.54 \text{ MW}$$

To show that the kinetic energy change is negligible we must calculate the exiting velocity:

$$V_2 = \frac{A_1 V_1 \rho_1}{A_2 \rho_2} = \frac{\pi \times 0.025^2 \times 200/0.08643}{\pi \times 0.125^2/2.087} = 193 \text{ m/s}$$

The kinetic energy change is then

$$\Delta KE = \dot{m}\left(\frac{V_2^2 - V_1^2}{2}\right) = 4.544 \times \frac{193^2 - 200^2}{2} = -6250 \text{ J/s} \quad \text{or} \quad -6.25 \text{ kJ/s}$$

This is less than 0.2 percent of the power output and is indeed negligible. Kinetic energy changes are usually omitted in the analysis of devices (but not in a nozzle or a diffuser).

EXAMPLE 4.13

Determine the maximum pressure increase provided by a 10-hp pump with a 6-cm-diameter inlet and a 10-cm-diameter outlet. The inlet velocity of the water is 10 m/s.

Solution

The energy equation (4.53) is used. By neglecting the heat transfer and assuming no increase in internal energy, we establish the maximum pressure rise. Neglecting the potential energy change, the energy equation takes the form

$$-\dot{W}_S = \dot{m}\left(\frac{P_2 - P_1}{\rho} + \frac{V_2^2 - V_1^2}{2}\right)$$

The velocity V_1 is given and V_2 is found from the continuity equation as follows, recognizing that $\rho_1 = \rho_2$,

$$A_1V_1 = A_2V_2 \qquad \pi \times 0.03^2 \times 10 = \pi \times 0.05^2 \times V_2 \qquad \therefore V_2 = 3.6 \text{ m/s}$$

The mass flow rate, needed in the energy equation, is then, using $\rho = 1000$ kg/m^3,

$$\dot{m} = \rho A_1 V_1 = 1000 \times \pi \times 0.03^2 \times 10 = 28.27 \text{ kg/s}$$

Recognizing that the pump work is negative, the energy equation is

$$-(-10) \times 746 = 28.27\left[\frac{\Delta P}{1000} + \frac{3.6^2 - 10^2}{2 \times 9.81}\right]. \qquad \therefore \Delta P = 268\,000 \text{ Pa}$$

or a pressure rise of 268 kPa.

Note that in this example the kinetic energy terms are retained because of the difference in inlet and exit areas; if they were omitted, about a 2 percent error would result. In most applications the inlet and exit areas are not given; but even if they are, as in this example, kinetic energy changes can be ignored in a pump or turbine.

NOZZLES AND DIFFUSERS

A nozzle is a device that is used to increase the velocity of a flowing fluid. It does this by reducing the pressure. A diffuser is a device that increases the pressure in a flowing fluid by reducing the velocity. There is no work input into the devices and usually negligible heat transfer. With the additional assumptions of negligible internal energy and potential energy changes, the energy equation takes the form

$$0 = \frac{V_2^2 - V_1^2}{2} + h_2 - h_1 \tag{4.57}$$

Three equations may be used for nozzle and diffuser flow: energy, continuity, and a process equation, such as for an adiabatic quasiequilibrium flow. Thus, we may have three unknowns at the exit, given the entering conditions. There may also be shock waves in supersonic flows or "choked" subsonic flows. These more complicated flows are included in a compressible flow course. Only the more simple situations will be included here.

EXAMPLE 4.14

Air flows through the supersonic nozzle shown. The inlet conditions are 7 kPa and 420°C. The nozzle exit diameter is adjusted such that the exiting velocity is 700 m/s. Calculate (a) the exit temperature, (b) the mass flow rate, and (c) the exit diameter. Assume an adiabatic quasiequilibrium flow.

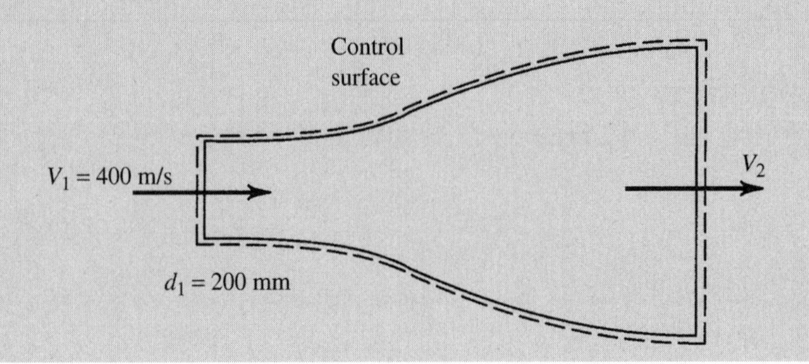

Solution

(a) To find the exit temperature the energy equation (4.57) is used. It is, using $\Delta h = C_p \Delta T$,

$$\frac{V_1^2}{2} + C_p T_1 = \frac{V_2^2}{2} + C_p T_2$$

We then have, using $C_p = 1000$ J/kg·K,

$$T_2 = \frac{V_1^2 - V_2^2}{2C_p} + T_1 = \frac{400^2 - 700^2}{2 \times 1000} + 420 = 255°\text{C}$$

Note: If T_1 is expressed in kelvins, then the answer would be in kelvins.

(b) To find the mass flow rate we must find the density at the entrance. From the inlet conditions we have

$$\rho_1 = \frac{P_1}{RT_1} = \frac{7000}{287 \times 693} = 0.0352 \text{ kg/m}^3$$

The mass flow rate is then

$$\dot{m} = \rho_1 A_1 V_1 = 0.0352 \times \pi \times 0.1^2 \times 400 = 0.442 \text{ kg/s}$$

(c) To find the exit diameter we would use $\rho_1 A_1 V_1 = \rho_2 A_2 V_2$, the continuity equation. This requires the density at the exit. It is found by assuming adiabatic quasiequilibrium flow. Referring to Eq. (4.40), with $\rho = 1/v$, we have

$$\rho_2 = \rho_1 \left(\frac{T_2}{T_1}\right)^{1/(k-1)} = 0.0352 \left(\frac{528}{693}\right)^{1/(1.4-1)} = 0.01784 \text{ kg/m}^3$$

Hence,

$$d_2^2 = \frac{\rho_1 d_1^2}{\rho_2 V_2} = \frac{.0352 \times .2^2 \times 400}{0.01784 \times 700} = .0451. \qquad \therefore d_2 = 0.212 \text{ m} \qquad \text{or} \qquad 212 \text{ mm}$$

HEAT EXCHANGERS

Heat exchangers are used to transfer energy from a hotter body to a colder body or to the surroundings by means of heat transfer. Energy is transferred from the hot gases after combustion in a power plant to the water in the pipes of the boiler, and from the hot water that leaves an automobile engine to the atmosphere by use of a radiator. Many heat exchangers utilize a flow passage into which a fluid enters and from which the fluid exits at a different temperature. The velocity does not normally change, the pressure drop through the passage is usually neglected, and the potential energy change is assumed zero. The energy equation then results in

$$\dot{Q} = \dot{m}(h_2 - h_1) \tag{4.58}$$

Energy may be exchanged between two moving fluids, as shown schematically in Fig. 4.5. For a control volume including the combined unit, which is assumed to be insulated, the energy equation, as applied to the control volume of Fig. 4.5*a*, would be

$$0 = \dot{m}_A(h_{A2} - h_{A1}) + \dot{m}_B(h_{B2} - h_{B1}) \tag{4.59}$$

The energy that leaves fluid A is transferred to fluid B by means of the heat transfer $\dot{Q}$. For the control volumes shown in Fig. 4.5*b* we have

$$\dot{Q} = \dot{m}_B(h_{B2} - h_{B1}) \qquad -\dot{Q} = \dot{m}_A(h_{A2} - h_{A1}) \tag{4.60}$$

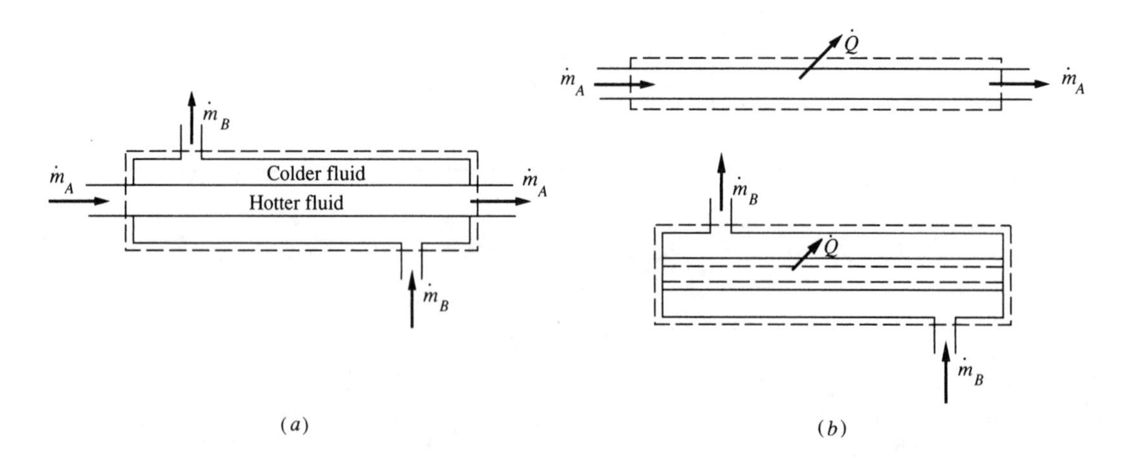

Figure 4.5 A heat exchanger. (*a*) Combined unit. (*b*) Separated control volumes.

EXAMPLE 4.15

Liquid sodium, flowing at 100 kg/s, enters a heat exchanger at 450°C and exits at 350°C. The C_p of sodium is 1.25 kJ/kg·°C. Liquid water enters at 5000 kPa and 20°C. Determine the minimum mass flow rate of the water so that the water just vaporizes. Also, calculate the rate of heat transfer.

Solution

The energy equation (4.60) is used as

$$\dot{m}_s(h_{s1} - h_{s2}) = \dot{m}_w(h_{w2} = h_{w1}) \qquad \text{or} \qquad \dot{m}_s C_p(T_{s1} - T_{s2}) = \dot{m}_w(h_{w2} - h_{w1})$$

Using the given values, we use Table C.4 to find h_{w1} and C.3 to find h_{w2} (or interpolate in Table C.2),

$$100 \times 1.25 \times (450 - 350) = \dot{m}_w(2794 - 88.65). \qquad \therefore \dot{m}_w = 4.62 \text{ kg/s}$$

The heat transfer is found using the energy equation (4.60) applied to one of the separate control volumes:

$$\dot{Q} = \dot{m}_w(h_{w2} - h_{w1}) = 4.62 \times (2794 - 88.65) = 12\,500 \text{ kW} \qquad \text{or} \qquad 12.5 \text{ MW}$$

Quiz No. 1

1. Select a correct statement of the first law if kinetic and potential energy changes are neglected.
   (A) Heat transfer equals the work done for a process.
   (B) Net heat transfer equals the net work for a cycle.
   (C) Net heat transfer minus net work equals internal energy change for a cycle.
   (D) Heat transfer minus work equals internal energy for a process.
2. Saturated water vapor at 400 kPa is heated in a rigid volume until $T_2 = 400°C$. The heat transfer is nearest
   (A) 406 kJ/kg
   (B) 508 kJ/kg
   (C) 604 kJ/kg
   (D) 702 kJ/kg

3. How much heat must be added to a 0.3-m^3 rigid volume containing water at 200°C in order that the final temperature is raised to 800°C? The initial pressure is 1 MPa.
   (A) 1207 kJ
   (B) 1308 kJ
   (C) 1505 kJ
   (D) 1702 kJ
4. A piston-cylinder arrangement provides a constant pressure of 800 kPa on steam which has an initial quality of 0.95 and an initial volume of 1200 cm^3. Determine the heat transfer necessary to raise the temperature to 400°C.
   (A) 97 kJ
   (B) 108 kJ
   (C) 121 kJ
   (D) 127 kJ
5. One kilogram of steam in a cylinder accepts 170 kJ of heat transfer while the pressure remains constant at 1000 kPa. Estimate the temperature T_2 if $T_1 = 320$°C.
   (A) 420°C
   (B) 410°C
   (C) 400°C
   (D) 390°C
6. Estimate the work required for the process of Prob. 5.
   (A) 89 kJ
   (B) 85 kJ
   (C) 45 kJ
   (D) 39 kJ
7. The pressure of steam at 400°C and $u = 2949$ kJ/kg is most nearly
   (A) 2000 kPa
   (B) 1900 kPa
   (C) 1800 kPa
   (D) 1700 kPa

8. Sixteen ice cubes, each with a temperature of −10°C and a volume of 8 mL, are added to 1 L of water at 20°C in an insulated container. What is the equilibrium temperature? Use $(C_p)_{ice} = 2.1$ kJ/kg·°C.

   (A) 9°C

   (B) 10°C

   (C) 11°C

   (D) 12°C

9. Two kilograms of air are compressed from 210 to 2000 kPa while maintaining the temperature constant at 30°C. The required heat transfer is nearest

   (A) 392 kJ

   (B) −392 kJ

   (C) −438 kJ

   (D) 438 kJ

10. Find the work needed to compress 2 kg of air in an insulated cylinder from 100 to 600 kPa if $T_1 = 20$°C.

    (A) −469 kJ

    (B) −394 kJ

    (C) −281 kJ

    (D) −222 kJ

11. Energy is added to 5 kg of air with a paddle wheel until $\Delta T = 100$°C. Find the paddle wheel work if the rigid container is insulated.

    (A) −358 kJ

    (B) −382 kJ

    (C) −412 kJ

    (D) −558 kJ

12. Methane is heated at constant pressure from 0 to 300°C. How much heat is needed if $P_1 = 200$ kPa?

    (A) 731 kJ/kg

    (B) 692 kJ/kg

    (C) 676 kJ/kg

    (D) 623 kJ/kg

13. The air in the cylinder of an air compressor is compressed from 100 kPa to 10 MPa. Estimate the work required if the air is initially at 100°C.
    (A) –250 kJ/kg
    (B) –395 kJ/kg
    (C) –543 kJ/kg
    (D) –729 kJ/kg
14. Heat is added to a fixed 0.15-m³ volume of steam initially at a pressure of 400 kPa and a quality of 0.5. Estimate the final temperature if 800 kJ of heat is added.
    (A) 200°C
    (B) 250°C
    (C) 300°C
    (D) 380°C
15. Select an assumption that is made when deriving the continuity equation $\dot{m}_1 = \dot{m}_2$.
    (A) Constant density
    (B) Steady flow
    (C) Uniform flow
    (D) Constant velocity
16. Steam enters a valve at 10 MPa and 550°C and exits at 0.8 MPa. What exiting temperature is expected?
    (A) 505°C
    (B) 510°C
    (C) 520°C
    (D) 530°C
17. A nozzle accelerates air from 20 to 200 m/s. What temperature change is expected?
    (A) 10°C
    (B) 20°C
    (C) 30°C
    (D) 40°C

18. The minimum power needed by a water pump that increases the pressure of 4 kg/s from 100 kPa to 6 MPa is nearest
    (A) 250 kW
    (B) 95 kW
    (C) 24 kW
    (D) 6 kW
19. Air enters a device at 4 MPa and 300°C with a velocity of 150 m/s. The inlet area is 10 cm^2 and the outlet area is 50 cm^2. Determine the mass flow rate and the outlet velocity if the air exits at 0.4 MPa and 100°C.
    (A) 195 m/s
    (B) 185 m/s
    (C) 175 m/s
    (D) 165 m/s
20. A pump is to increase the pressure of 200 kg/s of water by 4 MPa. The water enters through a 20-cm-diameter pipe and exits through a 12-cm-diameter pipe. Calculate the minimum horsepower required to operate the pump.
    (A) 70 kW
    (B) 85 kW
    (C) 94 kW
    (D) 107 kW
21. A turbine at a hydroelectric plant accepts 20 m^3/s of water at a gage pressure of 300 kPa and discharges it to the atmosphere. Determine the maximum power output.
    (A) 30 MW
    (B) 25 MW
    (C) 14 MW
    (D) 6 MW
22. An air compressor draws air from the atmosphere and discharges it at 500 kPa through a 100-mm-diameter outlet at 100 m/s. Determine the minimum power required to drive the insulated compressor. Assume atmospheric conditions of 25°C and 80 kPa.
    (A) 560 kW
    (B) 450 kW
    (C) 324 kW
    (D) 238 kW

23. Air enters a compressor at atmospheric conditions of 20°C and 80 kPa and exits at 800 kPa and 200°C through a 10-cm-diameter pipe at 20 m/s. Calculate the rate of heat transfer if the power input is 400 kW.
    (A) –127 kJ/s
    (B) –187 kJ/s
    (C) –233 kJ/s
    (D) –343 kJ/s
24. Nitrogen enters a diffuser at 200 m/s with a pressure of 80 kPa and a temperature of –20°C. It leaves with a velocity of 15 m/s at an atmospheric pressure of 95 kPa. If the inlet diameter is 100 mm, the exit temperature is nearest:
    (A) 0°C
    (B) 10°C
    (C) 20°C
    (D) 30°C

Quiz No. 2

1. Select the incorrect statement of the first law (neglect kinetic and potential energy changes).
    (A) The heat transfer equals the internal energy change for an adiabatic process.
    (B) The heat transfer and the work have the same magnitude for a constant volume quasiequilibrium process in which the internal energy remains constant.
    (C) The total energy input must equal the total work output for an engine operating on a cycle.
    (D) The internal energy change plus the work must equal zero for an adiabatic quasiequilibrium process.
2. A system undergoes a cycle consisting of the three processes listed in the table. Compute the missing values (a, b, c, d). All quantities are in kJ.

Process	Q	W	ΔE
$1 \rightarrow 2$	a	100	100
$2 \rightarrow 3$	b	–50	c
$3 \rightarrow 1$	100	d	–200

(A) (200, 50, 100, 300)
(B) (0, 50, 100, 300)
(C) (200, −50, 100, 300)
(D) (0, 50, −100, 300)

3. A 0.2-m^3 rigid volume contains steam at 600 kPa and a quality of 0.8. If 1000 kJ of heat are added, the final temperature is nearest
   (A) 720°C
   (B) 710°C
   (C) 690°C
   (D) 670°C
4. A 2-m^3 rigid volume contains water at 80°C with a quality of 0.5. Calculate the final temperature if 800 kJ of heat are added.
   (A) 120°C
   (B) 100°C
   (C) 90°C
   (D) 80°C
5. Steam is contained in a 4-L volume at a pressure of 1.5 MPa and a temperature of 400°C. If the pressure is held constant by expanding the volume while 20 kJ of heat is added, the final temperature is nearest
   (A) 875°C
   (B) 825°C
   (C) 805°C
   (D) 725°C
6. Saturated water is heated at constant pressure of 400 kPa until $T_2 = 400°C$. How much heat must be added?
   (A) 2070 kJ/kg
   (B) 2370 kJ/kg
   (C) 2670 kJ/kg
   (D) 2870 kJ/kg
7. Estimate C_p for steam at 4 MPa and 350°C.
   (A) 2.48 kJ/kg·°C
   (B) 2.71 kJ/kg·°C
   (C) 2.53 kJ/kg·°C
   (D) 2.31 kJ/kg·°C

8. Estimate the equilibrium temperature if 20 kg of copper at 0°C and 10 L of water at 30°C are placed in an insulated container.
   (A) 27.2°C
   (B) 25.4°C
   (C) 22.4°C
   (D) 20.3°C
9. One kilogram of air is compressed at a constant temperature of 100°C until the volume is halved. How much heat is rejected?
   (A) 42 kJ
   (B) 53 kJ
   (C) 67 kJ
   (D) 74 kJ
10. Energy is added to 5 kg of air with a paddle wheel until $\Delta T = 100°C$. Find the paddle wheel work if the rigid container is insulated.
    (A) 524 kJ
    (B) 482 kJ
    (C) 412 kJ
    (D) 358 kJ
11. Helium is contained in a 2-m^3 rigid volume at 50°C and 200 kPa. Calculate the heat transfer needed to increase the pressure to 800 kPa.
    (A) 1800 kJ
    (B) 1700 kJ
    (C) 1600 kJ
    (D) 1500 kJ
12. Air is compressed adiabatically from 100 kPa and 20°C to 800 kPa. The temperature T_2 is nearest
    (A) 440°C
    (B) 360°C
    (C) 290°C
    (D) 260°C

13. The initial temperature and pressure of 8000 cm^3 of air are 100°C and 800 kPa, respectively. Determine the necessary heat transfer if the volume does not change and the final pressure is 200 kPa.

    (A) −12 kJ

    (B) −32 kJ

    (C) −52 kJ

    (D) −72 kJ

14. Nitrogen at 100°C and 600 kPa expands in such a way that it can be approximated by a polytropic process with $n = 1.2$. Calculate the heat transfer if the final pressure is 100 kPa.

    (A) 76.5 kJ/kg

    (B) 66.5 kJ/kg

    (C) 56.5 kJ/kg

    (D) 46.5 kJ/kg

15. The term $\dot{m}\Delta h$ in the control volume equation $\dot{Q} - \dot{W}_S = \dot{m}\Delta h$

    (A) accounts for the rate of change in energy in the control volume.

    (B) represents the rate of change of energy between the inlet and outlet.

    (C) is often neglected in control volume applications.

    (D) includes the work rate due to the pressure forces.

16. Air enters an insulated compressor at 100 kPa and 20°C and exits at 800 kPa. Estimate the exiting temperature.

    (A) 530°C

    (B) 462°C

    (C) 323°C

    (D) 258°C

17. If $\dot{m}_1 = 2$ kg/s for the compressor of Prob. 16 and $d_1 = 20$ cm, V_1 is nearest

    (A) 62 m/s

    (B) 53 m/s

    (C) 41 m/s

    (D) 33 m/s

18. Ten kilograms of saturated steam at 10 kPa are to be completely condensed using 400 kg/s of cooling water. ΔT of the cooling water is nearest
    (A) 14°C
    (B) 18°C
    (C) 24°C
    (D) 32°C
19. Steam at 9000 kPa and 600°C passes through a throttling process so that the pressure is suddenly reduced to 400 kPa. What is the expected temperature after the throttle?
    (A) 570°C
    (B) 540°C
    (C) 510°C
    (D) 480°C
20. The inlet conditions on an air compressor are 50 kPa and 20°C. To compress the air to 400 kPa, 5 kW of energy is needed. Neglecting heat transfer and kinetic and potential energy changes, estimate the mass flow rate.
    (A) 0.094 kg/s
    (B) 0.053 kg/s
    (C) 0.021 kg/s
    (D) 0.016 kg/s
21. Superheated steam enters an insulated turbine at 4000 kPa and 500°C and leaves at 20 kPa with $x_2 = 0.9$. If the mass flow rate is 6 kg/s, the power output is nearest
    (A) 5.22 MW
    (B) 6.43 MW
    (C) 7.77 MW
    (D) 8.42 MW
22. Air enters a turbine at 600 kPa and 100°C and exits at 140 kPa and –20°C. Calculate the power output, neglecting heat transfer.
    (A) 140 kJ/kg
    (B) 120 kJ/kg
    (C) 100 kJ/kg
    (D) 80 kJ/kg

23. Air enters a nozzle with $P_1 = 585$ kPa, $T_1 = 195°C$, and $V_1 = 100$ m/s. If the air exits to the atmosphere where the pressure is 85 kPa, find exiting velocity, assuming an adiabatic process.

    (A) 523 m/s

    (B) 694 m/s

    (C) 835 m/s

    (D) 932 m/s

24. Water is used in a heat exchanger to cool 5 kg/s of air from 400 to 200°C. Calculate the minimum mass flow rate of the water if $\Delta T_{\text{water}} = 10°C$.

    (A) 24 kg/s

    (B) 32 kg/s

    (C) 41 kg/s

    (D) 53 kg/s

CHAPTER 5

The Second Law of Thermodynamics

Water flows down a hill, heat flows from a hot body to a cold one, rubber bands unwind, fluid flows from a high-pressure region to a low-pressure region, and we all get old! Our experiences in life suggest that processes have a definite direction. The first law of thermodynamics relates the several variables involved in a physical process, but does not give any information as to the direction of the process. It is the second law of thermodynamics that helps us establish the direction of a particular process.

Consider, for example, the work done by a falling weight as it turns a paddle wheel thereby increasing the internal energy of air contained in a fixed volume. It would not be a violation of the first law if we postulated that an internal energy decrease of the air can turn the paddle and raise the weight. This, however, would be a violation of the second law and would thus be impossible.

In the first part of this chapter, we will state the second law as it applies to a cycle. It will then be applied to a process and finally a control volume; we will treat the second law in the same way we treated the first law.

5.1 Heat Engines, Heat Pumps, and Refrigerators

We refer to a device operating on a cycle as a heat engine, a heat pump, or a refrigerator, depending on the objective of the particular device. If the objective of the device is to perform work it is a *heat engine;* if its objective is to transfer heat to a body it is a *heat pump;* if its objective is to transfer heat from a body, it is a *refrigerator.* Generically, a heat pump and a refrigerator are collectively referred to as a refrigerator. A schematic diagram of a simple heat engine is shown in Fig. 5.1.

An engine or a refrigerator operates between two *thermal energy reservoirs,* entities that are capable of providing or accepting heat without changing temperatures. The atmosphere and lakes serve as *heat sinks;* furnaces, solar collectors, and burners serve as *heat sources.* Temperatures T_H and T_L identify the respective temperatures of a source and a sink.

The net work W produced by the engine of Fig. 5.1 in one cycle would be equal to the net heat transfer, a consequence of the first law [see Eq. (4.2)]:

$$W = Q_H - Q_L \tag{5.1}$$

where Q_H is the heat transfer to or from the high-temperature reservoir, and Q_L is the heat transfer to or from the low-temperature reservoir.

If the cycle of Fig. 5.1 were reversed, a net work input would be required, as shown in Fig. 5.2. A heat pump would provide energy as heat Q_H to the warmer body (e.g., a house), and a refrigerator would extract energy as heat Q_L from the cooler body (e.g., a freezer). The work would also be given by Eq. (5.1), where we use magnitudes only.

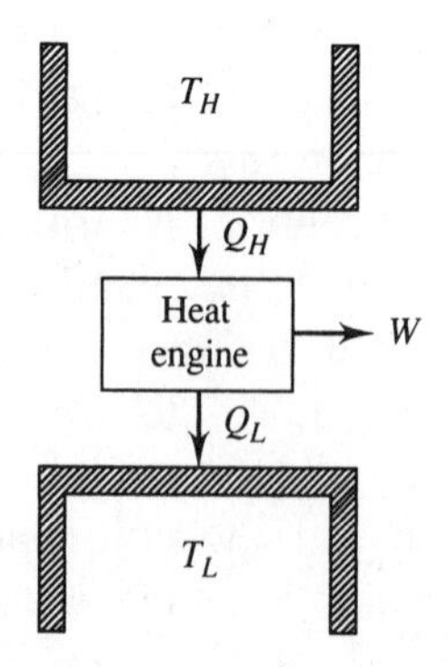

Figure 5.1 A heat engine.

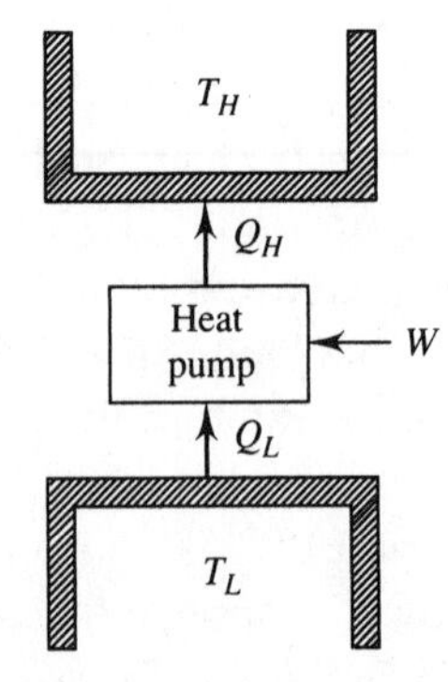

Figure 5.2 A refrigerator.

The *thermal efficiency* of the heat engine and the *coefficients of performance* (abbreviated COP) of the refrigerator and the heat pump are defined as follows:

$$\eta = \frac{W}{Q_H} \qquad \mathrm{COP}_{\mathrm{refrig}} = \frac{Q_L}{W} \qquad \mathrm{COP}_{\mathrm{h.p.}} = \frac{Q_H}{W} \tag{5.2}$$

where W is the net work output of the engine or the work input to the refrigerator. Each of the performance measures represents the desired output divided by the input (energy that is purchased).

The second law of thermodynamics will place limits on the above measures of performance. The first law would allow a maximum of unity for the thermal efficiency and an infinite coefficient of performance. The second law, however, establishes limits that are surprisingly low, limits that cannot be exceeded regardless of the cleverness of proposed designs.

One additional note: There are devices that we will refer to as heat engines that do not strictly meet our definition; they do not operate on a thermodynamic cycle but instead exhaust the working fluid and then intake new fluid. The internal combustion engine is an example. Thermal efficiency, as defined above, remains a quantity of interest for such engines.

5.2 Statements of the Second Law

As with the other basic laws presented, we do not derive a basic law but merely observe that such a law is never violated. The second law of thermodynamics can be stated in a variety of ways. Here we present two: the *Clausius statement* and the *Kelvin-Planck statement.* Neither is presented in mathematical terms. We will, however, provide a property of the system, entropy, which can be used to determine whether the second law is being violated for any particular situation.

Clausius Statement *It is impossible to construct a device that operates in a cycle and whose sole effect is the transfer of heat from a cooler body to a hotter body.*

This statement relates to a refrigerator (or a heat pump). It states that it is impossible to construct a refrigerator that transfers energy from a cooler body to a hotter body without the input of work; this violation is shown in Fig. 5.3*a*.

Kelvin-Planck Statement *It is impossible to construct a device that operates in a cycle and produces no other effect than the production of work and the transfer of heat from a single body.*

In other words, it is impossible to construct a heat engine that extracts energy from a reservoir, does work, and does not transfer heat to a low-temperature reservoir.

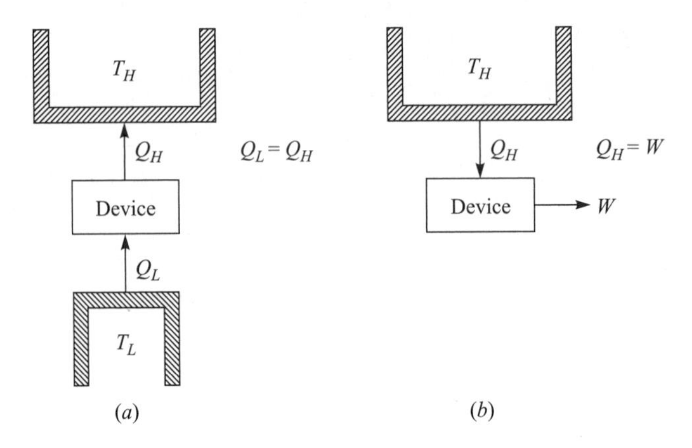

Figure 5.3 Violations of the second law.

This rules out any heat engine that is 100 percent efficient, like the one shown in Fig. 5.3*b*.

Note that the two statements of the second law are negative statements. They are expressions of experimental observations. No experimental evidence has ever been obtained that violates either statement of the second law. An example will demonstrate that the two statements are equivalent.

EXAMPLE 5.1

Show that the Clausius and Kelvin-Planck statements of the second law are equivalent.

Solution

Consider the system shown. The device in (*a*) transfers heat and violates the Clausius statement, since it has no work input. Let the heat engine transfer the same amount of heat Q_L. Then Q'_H is greater than Q_L by the amount W. If we simply

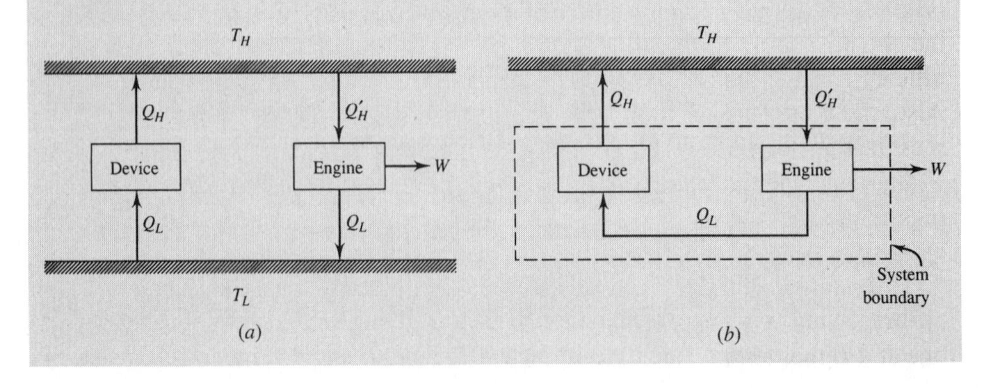

transfer the heat Q_L directly from the engine to the device, as shown in (*b*), there is no need for the low-temperature reservoir and the net result is a conversion of energy $(Q'_H - Q_H)$ from the high-temperature reservoir into an equivalent amount of work, a violation of the Kelvin-Planck statement of the second law. Conversely, a violation of the Kelvin-Planck statement is equivalent to a violation of the Clausius statement.

5.3 Reversibility

In our study of the first law we made use of the concept of equilibrium and we defined equilibrium, or quasiequilibrium, with reference to the system only. We must now introduce the concept of *reversibility* so that we can discuss the most efficient engine that can possibly be constructed; it is an engine that operates with reversible processes only: a *reversible engine*.

A *reversible process* is defined as a process which, having taken place, can be reversed and in so doing leaves no change in either the system or the surroundings. Observe that our definition of a reversible process refers to both the system and the surroundings. The process obviously has to be a quasiequilibrium process; additional requirements are:

1. No friction is involved in the process.
2. Heat transfer occurs due to an infinitesimal temperature difference only.
3. Unrestrained expansion does not occur.

The mixing of different substances and combustion also lead to irreversibilities, but the above three are the ones of concern in the study of the devices of interest.

The fact that friction makes a process irreversible is intuitively obvious. Consider the system of a block on the inclined plane of Fig. 5.4*a*. Weights are added until the block is raised to the position shown in Fig. 5.4*b*. Now, to return the system to its original state some weight must be removed so that the block will slide back down the plane, as shown in Fig. 5.4*c*. Note that the surroundings have experienced a significant change; the removed weights must be raised, which requires a work input. Also, the block and plane are at a higher temperature due to the friction, and heat must be transferred to the surroundings to return the system to its original state. This will also change the surroundings. Because there has been a change in the surroundings as a result of the process and the reversed process, we conclude that the process was irreversible. A reversible process requires that no friction be present.

To demonstrate the less obvious fact that heat transfer across a finite temperature difference makes a process irreversible, consider a system composed of two blocks, one at a higher temperature than the other. Bringing the blocks together results in a

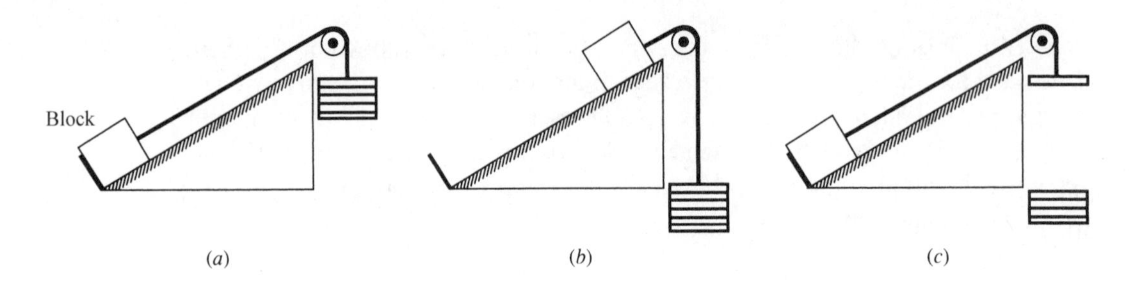

Figure 5.4 Irreversibility due to friction.

heat transfer process; the surroundings are not involved. To return the system to its original state, we must refrigerate the block that had its temperature raised. This will require a work input, demanded by the second law, resulting in a change in the surroundings since the surroundings must supply the work. Hence, heat transfer across a finite temperature difference is an irreversible process. A reversible process requires that all heat transfer occur across an infinitesimal temperature difference so that a reversible process is approached.

For an example of unrestrained expansion, consider the high-pressure gas contained in the cylinder of Fig. 5.5*a*. Pull the pin and let the piston suddenly move to the stops shown. Note that the only work done by the gas on the surroundings is to move the piston against atmospheric pressure. Now, to reverse this process it is necessary to exert a force on the piston. If the force is sufficiently large, we can move the piston to its original position, shown in Fig. 5.5*d*. This will demand a considerable amount of work to be supplied by the surroundings. In addition, the temperature will increase substantially, and this heat must be transferred to the surroundings to return the temperature to its original value. The net result is a significant change in the surroundings, a consequence of irreversibility. Unrestrained expansion cannot occur in a reversible process.

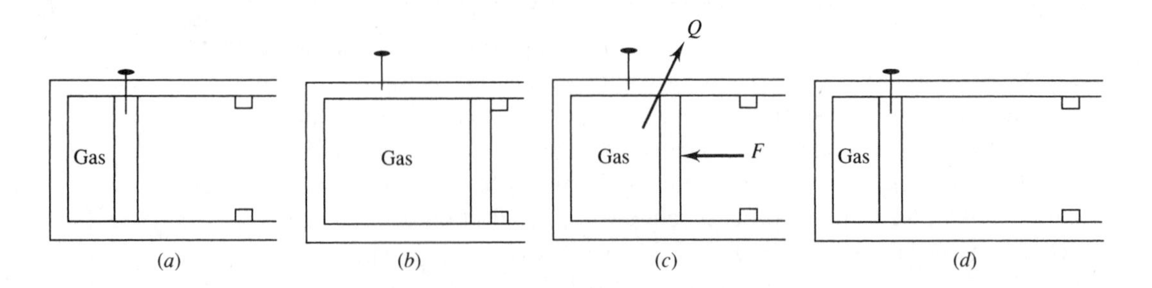

Figure 5.5 Unrestrained expansion.

5.4 The Carnot Engine

The heat engine that operates the most efficiently between a high-temperature reservoir and a low-temperature reservoir is the *Carnot engine.* It is an ideal engine that uses reversible processes to form its cycle of operation; thus it is also called a *reversible engine.* We will determine the efficiency of the Carnot engine and also evaluate its reverse operation. The Carnot engine is very useful, since its efficiency establishes the maximum possible efficiency of any real engine. If the efficiency of a real engine is significantly lower than the efficiency of a Carnot engine operating between the same limits, then additional improvements may be possible.

The cycle associated with the Carnot engine is shown in Fig. 5.6, using an ideal gas as the working substance. It is composed of the following four reversible processes:

1 → 2: *Isothermal expansion.* Heat is transferred reversibly from the high-temperature reservoir at the constant temperature T_H. The piston in the cylinder is withdrawn and the volume increases.

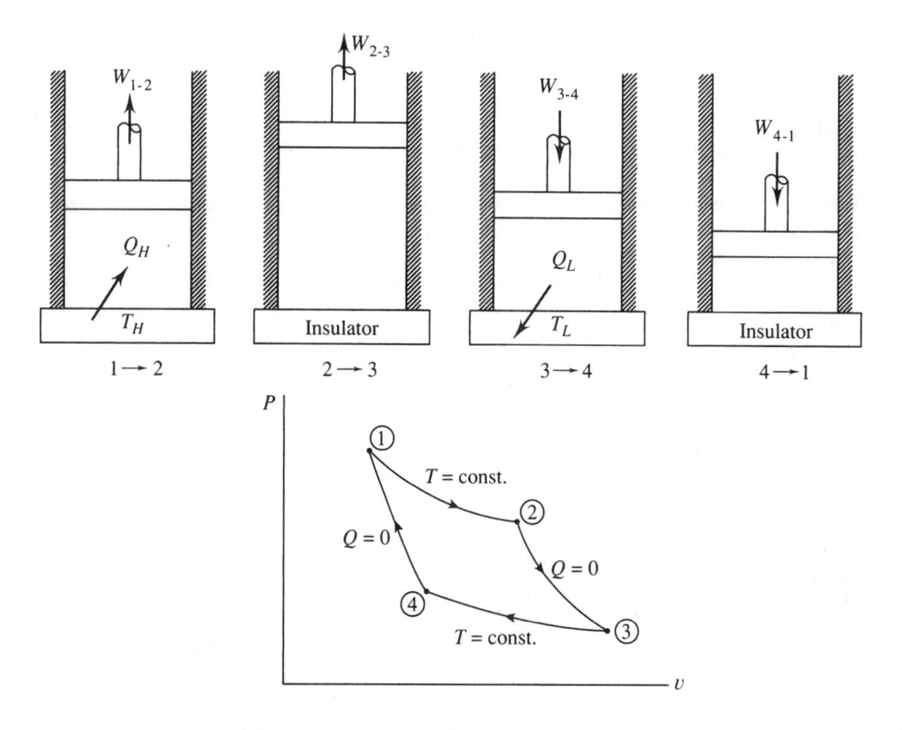

Figure 5.6 The Carnot cycle.

2 → 3: *Adiabatic reversible expansion.* The cylinder is completely insulated so that no heat transfer occurs during this reversible process. The piston continues to be withdrawn, with the volume increasing.

3 → 4: *Isothermal compression.* Heat is transferred reversibly to the low-temperature reservoir at the constant temperature T_L. The piston compresses the working substance, with the volume decreasing.

4 → 1: *Adiabatic reversible compression.* The completely insulated cylinder allows no heat transfer during this reversible process. The piston continues to compress the working substance until the original volume, temperature, and pressure are reached, thereby completing the cycle.

Applying the first law to the cycle, we note that

$$Q_H - Q_L = W_{\text{net}} \tag{5.3}$$

where Q_L is assumed to be a positive value for the heat transfer to the low-temperature reservoir. This allows us to write the thermal efficiency [see Eq. (5.2)] for the Carnot cycle as

$$\eta = \frac{Q_H - Q_L}{Q_H} = 1 - \frac{Q_L}{Q_H} \tag{5.4}$$

The following examples will be used to prove the following three postulates:

Postulate 1 *It is impossible to construct an engine, operating between two given temperature reservoirs, that is more efficient than the Carnot engine.*

Postulate 2 *The efficiency of a Carnot engine is not dependent on the working substance used or any particular design feature of the engine.*

Postulate 3 *All reversible engines, operating between two given temperature reservoirs, have the same efficiency as a Carnot engine operating between the same two temperature reservoirs.*

EXAMPLE 5.2
Show that the efficiency of a Carnot engine is the maximum possible efficiency.

Solution
Assume that an engine exists, operating between two reservoirs, that has an efficiency greater than that of a Carnot engine; also, assume that a Carnot

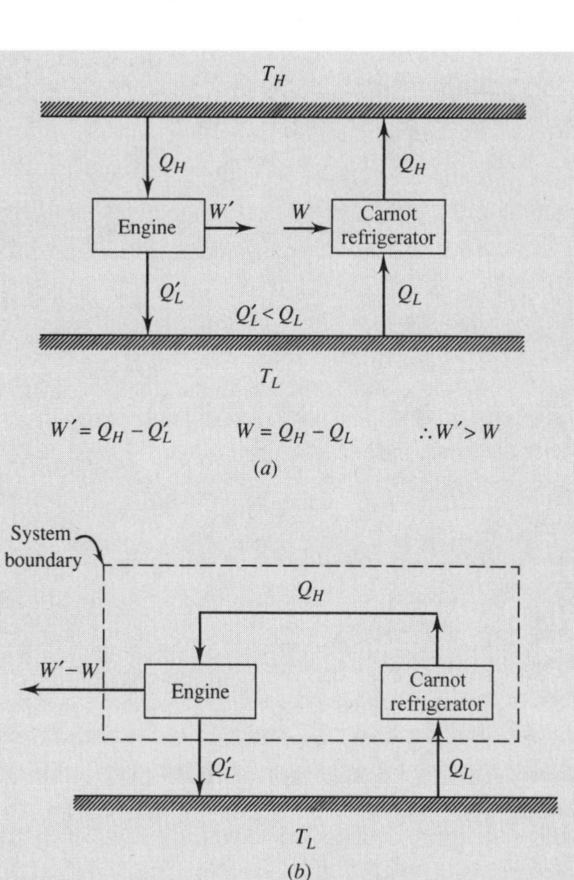

engine operates as a refrigerator between the same two reservoirs, as shown below. Let the heat transferred from the high-temperature reservoir to the engine be equal to the heat rejected by the refrigerator; then the work produced by the engine will be greater than the work required by the refrigerator (i.e., $Q'_L < Q_L$) since the efficiency of the engine is greater than that of a Carnot engine. Now, our system can be organized as shown in (*b*). The engine drives the refrigerator using the rejected heat from the refrigerator. But, there is some net work ($W' - W$) that leaves the system. The net result is the conversion of energy from a single reservoir into work, a violation of the second law. Thus, the Carnot engine is the most efficient engine operating between two particular reservoirs.

EXAMPLE 5.3

Show that the efficiency of a Carnot engine operating between two reservoirs is independent of the working substance used by the engine.

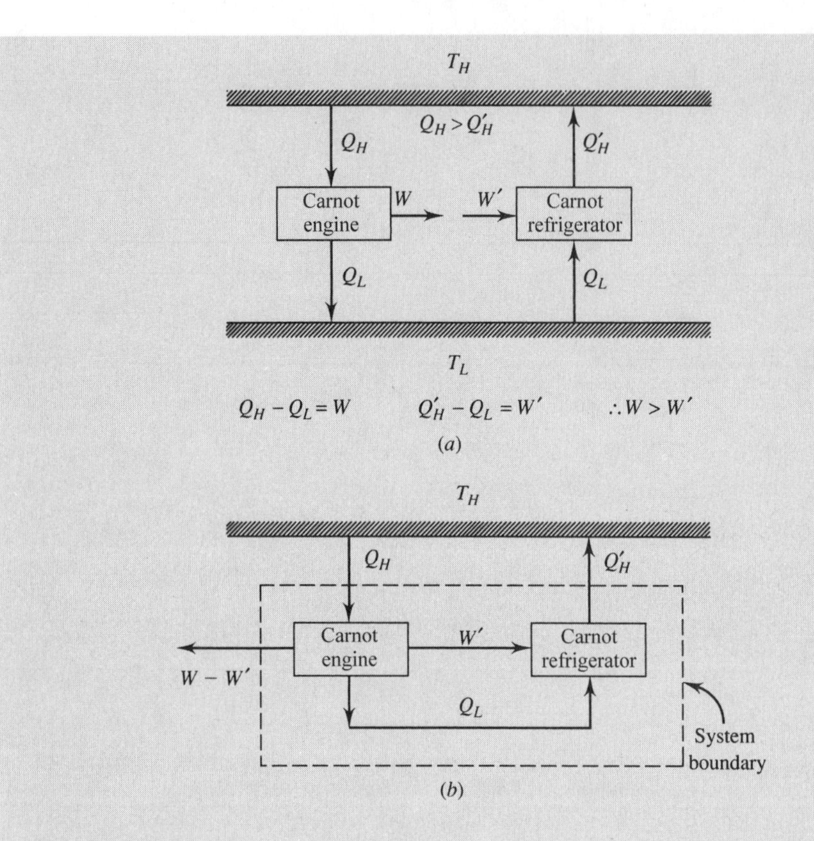

Solution

Suppose that a Carnot engine drives a Carnot refrigerator as shown in (*a*). Let the heat rejected by the engine equal the heat required by the refrigerator. Suppose the working fluid in the engine results in Q_H being greater than Q'_H; then W would be greater than W' (a consequence of the first law) and we would have the equivalent system shown in (*b*). The net result is a transfer of heat ($Q_H - Q'_H$) from a single reservoir and the production of work, a clear violation of the second law. Thus, the efficiency of a Carnot engine is not dependent on the working substance.

5.5 Carnot Efficiency

Since the efficiency of a Carnot engine is dependent only on the two reservoir temperatures, the objective of this article will be to determine that relationship. We will assume the working substance to be an ideal gas (refer to Postulate 2) and simply

perform the required calculations for the four processes of Fig. 5.5. The heat transfer for each of the four processes is as follows:

$$\textbf{1}\rightarrow\textbf{2:}\qquad Q_H = W_{1-2} = \int_{V_1}^{V_2} P dV = mRT_H \ln\frac{V_2}{V_1} \tag{5.5}$$

$$\textbf{2}\rightarrow\textbf{3:}\qquad Q_{2-3} = 0$$

$$\textbf{3}\rightarrow\textbf{4:}\qquad Q_L = -W_{3-4} = -\int_{V_3}^{V_4} P dV = -mRT_L \ln\frac{V_4}{V_3} \tag{5.6}$$

$$\textbf{4}\rightarrow\textbf{1:}\qquad Q_{4-1} = 0$$

Note that we want Q_L to be a positive quantity, as in the thermal efficiency relationship; hence, the negative sign. The thermal efficiency [Eq. (5.4)] is then

$$\eta = 1 - \frac{Q_L}{Q_H} = 1 + \frac{T_L}{T_H}\frac{\ln V_4/V_3}{\ln V_2/V_1} \tag{5.7}$$

During the reversible adiabatic processes $2 \rightarrow 3$ and $4 \rightarrow 1$, we have

$$\frac{T_L}{T_H} = \left(\frac{V_2}{V_3}\right)^{k-1} \qquad\qquad \frac{T_L}{T_H} = \left(\frac{V_1}{V_4}\right)^{k-1} \tag{5.8}$$

Thus we see that

$$\frac{V_2}{V_3} = \frac{V_1}{V_4} \qquad \text{or} \qquad \frac{V_4}{V_3} = \frac{V_1}{V_2} \tag{5.9}$$

Substituting into Eq. (5.7), we obtain the efficiency of the Carnot engine (recognizing that $\ln V_2/V_1 = -\ln V_1/V_2$):

$$\eta = 1 - \frac{T_L}{T_H} \tag{5.10}$$

We have simply replaced Q_L/Q_H with T_L/T_H. We can make this replacement for all reversible engines or refrigerators. We see that the thermal efficiency of a Carnot engine is dependent only on the high and low absolute temperatures of the reservoirs. The fact that we used an ideal gas to perform the calculations is not important since we have shown that Carnot efficiency is independent of the working substance. Consequently, the relationship (5.10) is applicable for all working substances, or for all reversible engines, regardless of the particular design characteristics.

The Carnot engine, when operated in reverse, becomes a heat pump or a refrigerator, depending on the desired heat transfer. The coefficient of performance for a Carnot heat pump becomes

$$\text{COP} = \frac{Q_H}{W_{\text{net}}} = \frac{Q_H}{Q_H - Q_L} = \frac{T_H}{T_H - T_L} \tag{5.11}$$

The coefficient of performance for a Carnot refrigerator takes the form

$$\text{COP} = \frac{Q_L}{W_{\text{net}}} = \frac{Q_L}{Q_H - Q_L} = \frac{T_L}{T_H - T_L} \tag{5.12}$$

The above measures of performance set limits that real devices can only approach. The reversible cycles assumed are obviously unrealistic, but the fact that we have limits that we know cannot be exceeded is often very helpful.

EXAMPLE 5.4

A Carnot engine operates between two temperature reservoirs maintained at 200 and 20°C, respectively. If the desired output of the engine is 15 kW, determine the $\dot{Q}_H$ and $\dot{Q}_L$.

Solution

The efficiency of a Carnot engine is given by

$$\eta = \frac{\dot{W}}{\dot{Q}_L} = 1 - \frac{T_L}{T_H}$$

This gives, converting the temperatures to absolute temperatures,

$$\dot{Q}_H = \frac{\dot{W}}{1 - T_L / T_H} = \frac{15}{1 - 293/473} = 39.42 \text{ kW}$$

Using the first law, we have

$$\dot{Q}_L = \dot{Q}_H - \dot{W} = 39.42 - 15 = 24.42 \text{ kW}$$

EXAMPLE 5.5

A refrigeration unit is cooling a space to −5°C by rejecting energy to the atmosphere at 20°C. It is desired to reduce the temperature in the refrigerated space to −25°C. Calculate the minimum percentage increase in work required, by assuming a Carnot refrigerator, for the same amount of energy removed.

Solution

For a Carnot refrigerator we know that

$$\text{COP} = \frac{Q_L}{W} = \frac{T_L}{T_H - T_L}$$

For the first situation we have

$$W_1 = Q_L\left(\frac{T_H - T_L}{T_L}\right) = Q_L\left(\frac{293 - 268}{268}\right) = 0.0933Q_L$$

For the second situation

$$W_2 = Q_L\left(\frac{293 - 248}{248}\right) = 0.1815Q_L$$

The percentage increase in work is then

$$\frac{W_2 - W_1}{W_1} = \frac{0.1815Q_L - 0.0933Q_L}{0.0933Q_L} \times 100 = 94.5\%$$

Note the large increase in energy required to reduce the temperature in a refrigerated space. And this is a minimum percentage increase, since we have assumed an ideal refrigerator.

5.6 Entropy

To allow us to apply the second law of thermodynamics to a process we will identify a property called entropy. This will parallel our discussion of the first law; first we stated the first law for a cycle and then derived a relationship for a process.

Consider the reversible Carnot engine operating on a cycle consisting of the processes described in Sec. 5.5. The quantity $\oint \delta Q/T$ is the cyclic integral of the heat transfer divided by the absolute temperature at which the heat transfer occurs. Since the temperature T_H is constant during the heat transfer Q_H and T_L is constant during heat transfer Q_L, the integral is given by

$$\oint \frac{\delta Q}{T} = \frac{Q_H}{T_H} - \frac{Q_L}{T_L} \tag{5.13}$$

where the heat Q_L leaving the Carnot engine is considered to be positive. Using Eqs. (5.4) and (5.10) we see that, for the Carnot cycle,

$$\frac{Q_L}{Q_H} = \frac{T_L}{T_H} \quad \text{or} \quad \frac{Q_H}{T_H} = \frac{Q_L}{T_L} \tag{5.14}$$

Substituting this into Eq. (5.13), we find the interesting result

$$\oint \frac{\delta Q}{T} = 0 \tag{5.15}$$

Thus, the quantity $\delta Q/T$ is a perfect differential, since its cyclic integral is zero. We let this differential be denoted by dS, where S depends only on the state of the system. This, in fact, was our definition of a property of a system. We shall call this extensive property *entropy;* its differential is given by

$$dS = \left.\frac{\delta Q}{T}\right|_{\text{rev}} \tag{5.16}$$

where the subscript "rev" emphasizes the reversibility of the process. This can be integrated for a process to give

$$\Delta S = \int_1^2 \left.\frac{\delta Q}{T}\right|_{\text{rev}} \tag{5.17}$$

From the above equation we see that the entropy change for a reversible process can be either positive or negative depending on whether energy is added to or extracted from the system during the heat transfer process. For a reversible adiabatic process ($Q = 0$) the entropy change is zero. If the process is adiabatic but irreversible, it is not generally true that $\Delta S = 0$.

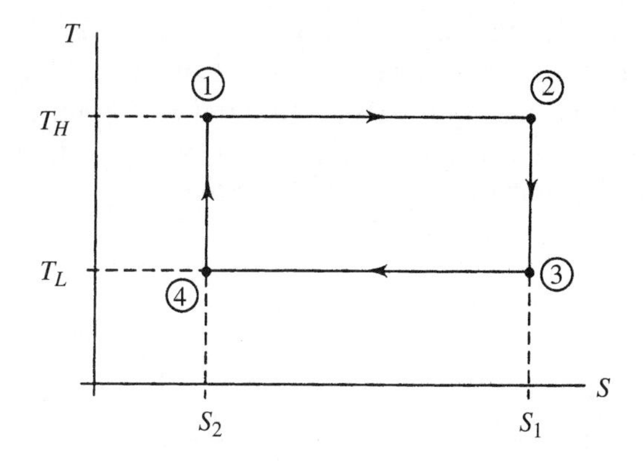

Figure 5.7 The Carnot cycle.

We often sketch a temperature-entropy diagram for cycles or processes of interest. The Carnot cycle provides a simple display when plotting temperature vs. entropy; it is shown in Fig. 5.7. The change in entropy for the first isothermal process from state 1 to state 2 is

$$S_2 - S_1 = \int_1^2 \frac{\delta Q}{T} = \frac{Q_H}{T_H} \tag{5.18}$$

The entropy change for the reversible adiabatic process from state 2 to state 3 is zero. For the isothermal process from state 3 to state 4 the entropy change is negative that of the first process from state 1 to state 2; the process from state 4 to state 1 is also a reversible adiabatic process and is accompanied with a zero entropy change.

The heat transfer during a reversible process can be expressed in differential form [see Eq. (5.16)] as

$$\delta Q = TdS \qquad \text{or} \qquad Q = \int_1^2 TdS \tag{5.19}$$

Hence, the area under the curve in the *T-S* diagram represents the heat transfer during any reversible process. The rectangular area in Fig. 5.6 thus represents the net heat transfer during the Carnot cycle. Since the heat transfer is equal to the work done for a cycle, the area also represents the net work accomplished by the system during the cycle. For this Carnot cycle $Q_{\text{net}} = W_{\text{net}} = \Delta T \Delta S$.

The first law of thermodynamics, for a reversible infinitesimal change, becomes, using Eq. (5.19),

$$TdS - PdV = dU \tag{5.20}$$

This is an important relationship in our study of simple systems. We arrived at it assuming a reversible process. However, since it involves only properties of the system, it holds for any process including any irreversible process. If we have an irreversible process, in general, $\delta W \neq PdV$ and $\delta Q \neq TdS$ but Eq. (5.20) still holds as a relationship between the properties since changes in properties do not depend on the process. Dividing by the mass, we have

$$Tds - Pdv = du \tag{5.21}$$

where the specific entropy is $s = S/m$.

To relate the entropy change to the enthalpy change we differentiate the definition of enthalpy and obtain

$$dh = du + Pdv + vdP \tag{5.22}$$

Substituting into Eq. (5.21) for du, we have

$$Tds = dh - vdP \tag{5.23}$$

Equations (5.21) and (5.23) will be used in subsequent sections of our study of thermodynamics for various reversible and irreversible processes.

ENTROPY FOR AN IDEAL GAS WITH CONSTANT SPECIFIC HEATS

Assuming an ideal gas, Eq. (5.21) becomes

$$ds = \frac{du}{T} + \frac{Pdv}{T} = C_v \frac{dT}{T} + R\frac{dv}{v} \tag{5.24}$$

where we have used $du = C_v dT$ and $Pv = RT$. Equation (5.24) is integrated, assuming constant specific heat, to yield

$$s_2 - s_1 = C_v \ln\frac{T_2}{T_1} + R\ln\frac{v_2}{v_1} \tag{5.25}$$

Similarly, Eq. (5.23) is rearranged and integrated to give

$$s_2 - s_1 = C_p \ln\frac{T_2}{T_1} - R\ln\frac{P_2}{P_1} \tag{5.26}$$

Note again that the above equations were developed assuming a reversible process; however, they relate the change in entropy to other thermodynamic properties at the two end states. Since the change of a property is independent of the process used in going from one state to another, the above relationships hold for any process, reversible or irreversible, providing the working substance can be approximated by an ideal gas with constant specific heats.

If the entropy change is zero, as in a reversible adiabatic process, Eqs. (5.25) and (5.26) can be used to obtain

$$\frac{T_2}{T_1}=\left(\frac{v_1}{v_2}\right)^{k-1} \qquad \frac{T_2}{T_1}=\left(\frac{P_2}{P_1}\right)^{(k-1)/k} \qquad \frac{P_2}{P_1}=\left(\frac{v_1}{v_2}\right)^{k} \tag{5.27}$$

These are, of course, identical to the equations obtained in Chap. 4 for an ideal gas with constant specific heats undergoing a quasiequilibrium adiabatic process. We now refer to such a process as an *isentropic process*.

EXAMPLE 5.6

Air is contained in an insulated, rigid volume at 20°C and 200 kPa. A paddle wheel, inserted in the volume, does 720 kJ of work on the air. If the volume is 2m³, calculate the entropy increase assuming constant specific heats.

Solution

To determine the final state of the process we use the energy equation, assuming zero heat transfer. We have $-W=\Delta U=mC_v\,\Delta T$. The mass m is found from the ideal-gas equation to be

$$m=\frac{PV}{RT}=\frac{200\times 2}{0.287\times 293}=4.76 \text{ kg}$$

The first law, taking the paddle-wheel work as negative, is then

$$-W=mC_v\,\Delta T$$

$$-(-720)=4.76\times 0.717\times(T_2-293) \qquad \therefore T_2=504.0 \text{ K}$$

Using Eq. (5.25) for this constant-volume process there results

$$\Delta S=mC_v\ln\frac{T_2}{T_1}=4.76\times 0.717\times\ln\frac{504}{293}=1.851 \text{ kJ/K}$$

EXAMPLE 5.7

After a combustion process in a cylinder the pressure is 1200 kPa and the temperature is 350°C. The gases are expanded to 140 kPa with a reversible adiabatic process. Calculate the work done by the gases, assuming they can be approximated by air with constant specific heats.

Solution

The first law can be used, with zero heat transfer, to give $-w = \Delta u = C_v(T_2 - T_1)$. The temperature T_2 is found from Eq. (5.27) to be

$$T_2 = T_1\left(\frac{P_2}{P_1}\right)^{(k-1)/k} = 623\left(\frac{140}{1200}\right)^{0.4/1.4} = 337\text{ K}$$

This allows the work to be calculated:

$$w = C_v(T_2 - T_1) = 0.717 \times (623 - 337) = 205\text{ kJ/kg}$$

ENTROPY FOR AN IDEAL GAS WITH VARIABLE SPECIFIC HEATS

If the specific heats for an ideal gas cannot be assumed constant over a particular temperature range we return to Eq. (5.23) and write

$$ds = \frac{dh}{T} - \frac{v\,dP}{T} = \frac{C_p}{T}dT - \frac{R}{P}dP \tag{5.28}$$

The gas constant R can be removed from the integral, but $C_p = C_p(T)$ cannot be removed. Hence, we integrate Eq. (5.28) and obtain

$$s_2 - s_1 = \int_{T_1}^{T_2}\frac{C_p}{T}dT - R\ln\frac{P_2}{P_1} \tag{5.29}$$

The integral in the above equation depends only on temperature, and we can evaluate its magnitude from the gas tables (given in App. E). It is found, using the tabulated function s° from App. E, to be

$$s_2^\circ - s_1^\circ = \int_{T_1}^{T_2}\frac{C_p}{T}dT \tag{5.30}$$

Thus, the entropy change is (in some textbooks ϕ is used rather than s°)

$$s_2 - s_1 = s_2^\circ - s_1^\circ - R\ln\frac{P_2}{P_1} \tag{5.31}$$

This more exact expression for the entropy change is used only when improved accuracy is desired (usually for large temperature differences).

For an isentropic process we cannot use Eqs. (5.25) and (5.26) if the specific heats are not constant. However, we can use Eq. (5.31) and obtain, for an isentropic process,

$$\frac{P_2}{P_1} = \exp\left(\frac{s_2^\circ - s_1^\circ}{R}\right) = \frac{\exp(s_2^\circ/R)}{\exp(s_1^\circ/R)} = \frac{f(T_2)}{f(T_1)} \tag{5.32}$$

Thus, we define a *relative pressure* P_r, which depends only on the temperature, as

$$P_r = e^{s^\circ/R} \tag{5.33}$$

It is included as an entry in air table E.1. The pressure ratio for an isentropic process is then

$$\frac{P_2}{P_1} = \frac{P_{r2}}{P_{r1}} \tag{5.34}$$

The volume ratio can be found using the ideal-gas equation of state. It is

$$\frac{v_2}{v_1} = \frac{P_1}{P_2}\frac{T_2}{T_1} \tag{5.35}$$

where we would assume an isentropic process when using the relative pressure ratio. Consequently, we define a *relative specific volume* v_r, dependent solely on the temperature, as

$$v_r = \frac{T}{P_r} \tag{5.36}$$

Using its value from air table E.1 we find the specific volume ratio for an isentropic process; it is

$$\frac{v_2}{v_1} = \frac{v_{r2}}{v_{r1}} \tag{5.37}$$

With the entries from air table E.1 we can perform the calculations required when working problems involving air with variable specific heats.

EXAMPLE 5.8

Repeat Example 5.6 assuming variable specific heats.

Solution

Using the gas tables, we write the first law as $-W = \Delta U = m(u_2 - u_1)$. The mass is found in Example 5.6 to be 4.76 kg. The first law is written as

$$u_2 = -\frac{W}{m} + u_1 = -\frac{-720}{4.76} + 209.1 = 360.4 \text{ kJ/kg}$$

where u_1 is found at 293 K in air table E.1 by interpolation. Now, using this value for u_2, we can interpolate to find $T_2 = 501.2$ K and $s_2^\circ = 2.222$. The value for s_1° is interpolated to be $s_1^\circ = 1.678$. The pressure at state 2 is found using the ideal-gas equation for our constant-volume process:

$$\frac{P_2}{T_2} = \frac{P_1}{T_1} \qquad P_2 = P_1\frac{T_2}{T_1} = 200 \times \frac{501.2}{293} = 342.1 \text{ kPa}$$

Finally, the entropy change is

$$\Delta S = m\left(s_2^\circ - s_1^\circ - R\ln\frac{P_2}{P_1}\right)$$

$$= 4.76 \times \left(2.222 - 1.678 - 0.287\ln\frac{342.1}{200}\right) = 1.856 \text{ kJ/K}$$

The approximate result of Example 5.6 is seen to be less than 0.3 percent in error. If T_2 were substantially greater than 500 K, the error would be much more significant.

EXAMPLE 5.9

After a combustion process in a cylinder the pressure is 1200 kPa and the temperature is 350°C. The gases are expanded to 140 kPa in a reversible, adiabatic process. Calculate the work done by the gases, assuming they can be approximated by air with variable specific heats.

Solution

First, at 623 K the relative pressure P_{r1} is interpolated to be

$$P_{r1} = \frac{3}{40}(23.13 - 18.36) + 18.36 = 18.72$$

For an isentropic process,

$$P_{r2} = P_{r1}\frac{P_2}{P_1} = 18.72 \times \frac{140}{1200} = 2.184$$

With this value for the relative pressure at state 2,

$$T_2 = \frac{2.184 - 2.149}{3.176 - 2.149} \times 40 + 340 = 341.4 \text{ K}$$

The work is found from the first law to be

$$\begin{aligned} w &= u_1 - u_2 \\ &= \left[\frac{3}{40}(481.0 - 450.1) + 450.1\right] - \left[\frac{2.184 - 2.149}{3.176 - 2.149}(271.7 - 242.8) + 242.8\right] \\ &= 208.6 \text{ kJ/kg} \end{aligned}$$

The approximate result of Example 5.7 is seen to have an error less than 1.5 percent. The temperature difference $(T_2 - T_1)$ is not very large; hence the small error.

ENTROPY FOR STEAM, SOLIDS, AND LIQUIDS

The entropy change has been found for an ideal gas with constant specific heats and for an ideal gas with variable specific heats. For pure substances, such as steam, entropy is included as an entry in the steam tables (given in App. C). In the quality region, it is found using the relation

$$s = s_f + x s_{fg} \tag{5.38}$$

Note that the entropy of saturated liquid water at 0°C is arbitrarily set equal to zero. It is only the change in entropy that is of interest.

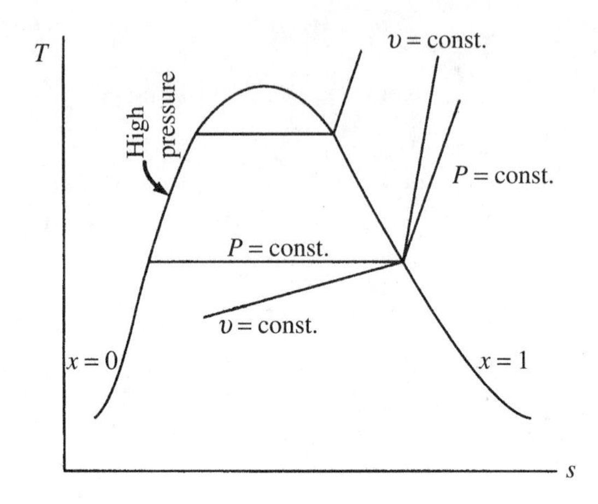

Figure 5.8 The *T-s* diagram for steam.

For a compressed liquid it is included as an entry in Table C.4, the compressed liquid table, or it can be approximated by the saturated liquid values s_f at the given temperature (ignoring the pressure). From the compressed liquid table at 10 MPa and 100°C, $s = 1.30$ kJ/kg·K, and from the saturated steam table C.1 at 100°C, $s_f = 1.31$ kJ/kg·K; this is an insignificant difference.

The temperature-entropy diagram is of particular interest and is often sketched during the problem solution. A *T-s* diagram is shown in Fig. 5.8; for steam it is essentially symmetric about the critical point. Note that the high-pressure lines in the compressed liquid region are indistinguishable from the saturated liquid line. It is often helpful to visualize a process on a *T-s* diagram, since such a diagram illustrates assumptions regarding irreversibilities.

For a solid or a liquid, the entropy change can be found quite easily if we can assume the specific heat to be constant. Returning to Eq. (5.21), we can write, assuming the solid or liquid to be incompressible so that $dv = 0$,

$$Tds = du = CdT \tag{5.39}$$

where we have dropped the subscript on the specific heat since for solids and liquids $C_p \cong C_v$, such as Table B.4, usually list values for C_p; these are assumed to be equal to C. Assuming a constant specific heat, we find that

$$\Delta s = \int_{T_1}^{T_2} C\frac{dT}{T} = C\ln\frac{T_2}{T_1} \tag{5.40}$$

If the specific heat is a known function of temperature, the integration can be performed. Specific heats for solids and liquids are listed in Table B.4.

EXAMPLE 5.10

Steam is contained in a rigid container at an initial pressure of 600 kPa and 300°C. The pressure is reduced to 40 kPa by removing energy via heat transfer. Calculate the entropy change and the heat transfer.

Solution

From the steam tables, $v_1 = v_2 = 0.4344\ \text{m}^3/\text{kg}$. State 2 is in the quality region. Using the above value for v_2 the quality is found as follows:

$$0.4344 = 0.0011 + x(0.4625 - 0.0011) \qquad \therefore x = 0.939$$

The entropy at state 2 is $s_2 = 1.777 + 0.939 \times 5.1197 = 6.584\ \text{kJ/kg} \cdot \text{K}$. The entropy change is then

$$\Delta s = s_2 - s_1 = 6.584 - 7.372 = -0.788\ \text{kJ/kg} \cdot \text{K}$$

The heat transfer is found from the first law using $w = 0$ and $u_1 = 2801$ kJ/kg:

$$q = u_2 - u_1 = (604.3 + 0.939 \times 1949.3) - 2801 = -366\ \text{kJ/kg}$$

5.7 The Inequality of Clausius

The Carnot cycle is a reversible cycle and produces work which we will refer to as W_{rev}. Consider an irreversible cycle operating between the same two reservoirs, shown in Fig. 5.9. Obviously, since the Carnot cycle possesses the maximum possible

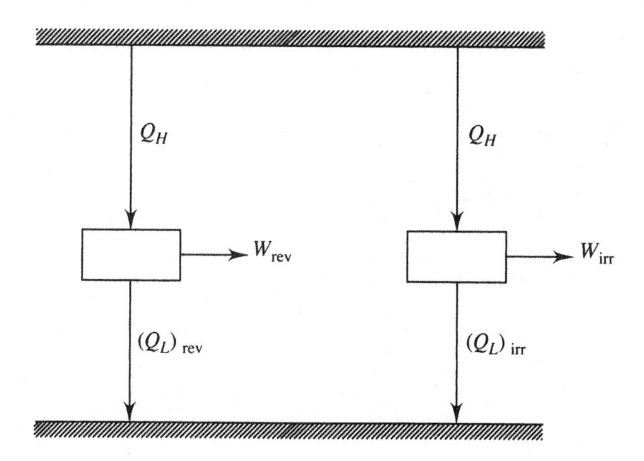

Figure 5.9 A reversible and an irreversible engine operating between two reservoirs.

efficiency, the efficiency of the irreversible cycle must be less than that of the Carnot cycle. In other words, for the same amount of heat addition Q_H, we must have

$$W_{\text{irr}} < W_{\text{rev}} \tag{5.41}$$

From the first law applied to a cycle ($W = Q_H - Q_L$) we see that, assuming that $(Q_H)_{\text{irr}}$ and $(Q_H)_{\text{rev}}$ are the same,

$$(Q_L)_{\text{rev}} < (Q_L)_{\text{irr}} \tag{5.42}$$

This requires, referring to Eqs. (5.13) and (5.15),

$$\oint \left(\frac{\delta Q}{T} \right)_{\text{irr}} < 0 \tag{5.43}$$

since the above integral for a reversible cycle is zero.

If we were considering an irreversible refrigerator rather than an engine, we would require more work for the same amount of refrigeration Q_L. By applying the first law to refrigerators, we would arrive at the same inequality as that of Eq. (5.43). Hence, for all cycles, reversible or irreversible, we can write

$$\oint \left(\frac{\delta Q}{T} \right) \leq 0 \tag{5.44}$$

This is known as *the inequality of Clausius.* It is a consequence of the second law of thermodynamics. It will be used to establish entropy relationships for real processes.

5.8 Entropy Change for an Irreversible Process

Consider a cycle to be composed of two reversible processes, shown in Fig. 5.10. Suppose that we can also return from state 2 to state 1 along the irreversible process marked by path C. For the system experiencing the reversible cycle we have

$$\underset{\text{along } A}{\int_1^2 \frac{\delta Q}{T}} + \underset{\text{along } B}{\int_2^1 \frac{\delta Q}{T}} = 0 \tag{5.45}$$

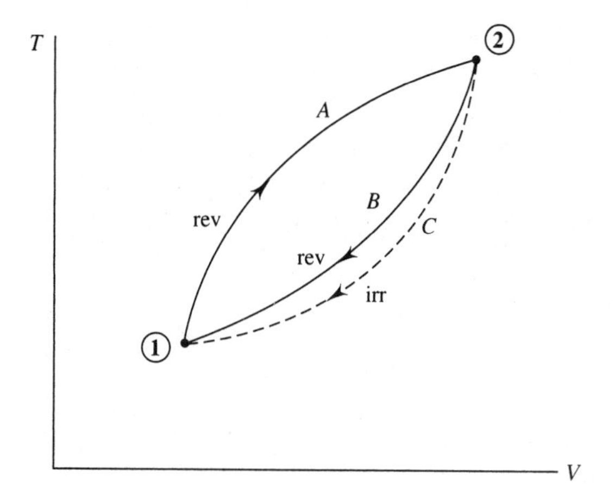

Figure 5.10 A cycle that includes the possibility of an irreversible process.

For the system experiencing the cycle involving the irreversible process, the Clausius inequality demands that

$$\underset{\text{along } A}{\int_1^2 \frac{\delta Q}{T}} + \underset{\text{along } C}{\int_2^1 \frac{\delta Q}{T}} < 0 \tag{5.46}$$

Subtracting Eq. (5.45) from Eq. (5.46) results in

$$\underset{\text{along } B}{\int_2^1 \frac{\delta Q}{T}} > \underset{\text{along } C}{\int_2^1 \frac{\delta Q}{T}} \tag{5.47}$$

But, along the reversible path B, $\delta Q/T = dS$. Thus, for any path representing any process,

$$\Delta S \geq \int \frac{\delta Q}{T} \qquad \text{or} \qquad dS \geq \frac{\delta Q}{T} \tag{5.48}$$

The equality holds for a system experiencing a reversible process and the inequality for an irreversible process.

We can also introduce the *entropy production* σ into Eq. (5.48):

$$\Delta S - \int \frac{\delta Q}{T} = \sigma \qquad \text{or} \qquad dS - \frac{\delta Q}{T} = \delta\sigma \tag{5.49}$$

where we have used $\delta\sigma$ since it is a path dependent function. It is obvious that entropy production is zero for a reversible process and positive for an irreversible process, i.e., any real process.

Relationship (5.48) leads to an important conclusion in thermodynamics. Consider an infinitesimal process of a system at absolute temperature T. If the process is reversible, the differential change in entropy is $\delta Q/T$; if the process is irreversible, the change in entropy is greater than $\delta Q/T$. We thus conclude that the effect of irreversibility (e.g., friction) is to increase the entropy of a system. Alternatively, observe that the entropy production σ in Eq. (5.49), for any real process, must be positive.

Finally, consider an *isolated* system, a system that exchanges no work or heat with its surroundings. For such a system the first law demands that $E_2 = E_1$ for any process. Equation (5.48) takes the form

$$\Delta S \geq 0 \tag{5.50}$$

demanding that the entropy of an isolated system either remains constant or increases, depending on whether the process is reversible or irreversible. Hence, for any real process the entropy of an isolated system must increase. Alternatively, we can state that the entropy production associated with a real process of an isolated system must be positive since for an isolated system $\Delta S = \sigma$.

We can restate Eq. (5.50) by considering an isolated system to be composed of any system under consideration plus the surroundings to that system as a second larger system. The heat and work transferred to the system under consideration are negative the heat and work transferred from the surroundings. This combination of the system and surroundings is often referred to as the universe. For the universe we can write

$$\Delta S_{\text{sys}} + \Delta S_{\text{surr}} = \sigma \geq 0 \tag{5.51}$$

where $\sigma = 0$ applies to a reversible (ideal) process and the inequality to an irreversible (real) process. Relation (5.51), the *principle of entropy increase*, is often used as the mathematical statement of the second law. In Eq. (5.51), σ has also been referred to as ΔS_{gen}, the entropy generated.

EXAMPLE 5.11

Air is contained in one half of an insulated tank. The other half is completely evacuated. The membrane separating the two halves is punctured and the air quickly fills the entire volume. Calculate the specific entropy change of this isolated system.

Solution

The entire tank is chosen as the system boundary. No heat transfer occurs across the boundary and the air does no work. The first law then takes the form $\Delta U = mC_v(T_2 - T_1) = 0$. Hence, the final temperature is equal to the initial temperature. Using Eq. (5.25) for the entropy change, we have, with $T_1 = T_2$,

$$\Delta s = R \ln \frac{v_2}{v_1} = 0.287 \times \ln 2 = 0.199 \text{ kJ/kg} \cdot \text{K}$$

Note that this satisfies Eq. (5.48) since for this example $Q = 0$, so that $\int \delta Q/T = 0 < m\Delta s$.

EXAMPLE 5.12

Two kilograms of steam at 400°C and 600 kPa are cooled at constant pressure by transferring heat from a cylinder until the steam is completely condensed. The surroundings are at 25°C. Determine the net entropy production.

Solution

The entropy of the steam that defines our system decreases since heat is transferred from the system to the surroundings. From the steam tables this change is found to be

$$\Delta S_{\text{sys}} = m(s_2 - s_1) = 2(1.9316 - 7.708) = -11.55 \text{ kJ/K}$$

The heat transfer to the surroundings occurs at constant temperature. Hence, the entropy change of the surroundings [see Eq. (5.18)] is

$$\Delta S_{\text{surr}} = \frac{Q}{T}$$

The heat transfer for the constant-pressure process needed to condense the steam is

$$Q = m\,\Delta h = 2(3270.2 - 670.6) = 5199 \text{ kJ}$$

giving $\Delta S_{\text{surr}} = 5199/298 = 17.45$ kJ/K and

$$\sigma = \Delta S_{\text{surr}} + \Delta S_{\text{sys}} = 17.45 - 11.55 = 5.90 \text{ kJ/K}$$

The entropy production is positive, as it must be.

5.9 The Second Law Applied to a Control Volume

Return once again to a fixed control volume, a device, with one inlet and one outlet. At time t the system occupies a small volume 1 (that enters the device from the inlet pipe over a time increment Δt) plus the device; then at $t + \Delta t$, the system occupies the device plus the small volume 2 that leaves the device. The second law for the steady-state device, can be stated by applying Eq. (5.48):

$$S_{\text{sys}}(t+\Delta t) - S_{\text{sys}}(t) \geq \int \frac{\delta Q}{T}$$

$$S_{\text{device}}(t+\Delta t) + s_2 \Delta m_2 - s_1 \Delta m_1 - S_{\text{device}}(t) \geq \int \frac{\delta Q}{T} \tag{5.52}$$

Recognizing that $S_{\text{device}}(t+\Delta t) = S_{\text{device}}(t)$ due to the steady flow (all quantities are independent of time), we have

$$s_2 \Delta m_2 - s_1 \Delta m_1 - \int_{\text{device}} \frac{\delta Q_{\text{c.s.}}}{T_{\text{c.s.}}} \geq 0 \tag{5.53}$$

where $\delta Q_{\text{c.s.}}$ is the incremental heat transfer into the device at a boundary location where the temperature is $T_{\text{c.s.}}$, the subscript "c.s." representing the control surface surrounding the device. Introducing the entropy production of Eq. (5.49),

$$s_2 \Delta m_2 - s_1 \Delta m_1 - \int_{\text{device}} \frac{\delta Q_{\text{c.s.}}}{T_{\text{c.s.}}} = \Delta \sigma_{\text{device}} \tag{5.54}$$

Let's assume that the entire control surface is at a constant temperature $T_{\text{c.s.}}$, a rather common assumption. If we divide Eq. (5.54) by Δt and use dots to denote rates, we arrive at the steady-state rate equation

$$\dot{m}(s_2 - s_1) - \frac{\dot{Q}}{T_{\text{c.s.}}} = \dot{\sigma}_{\text{device}} \tag{5.55}$$

Zero rate of entropy production is associated with a reversible process whereas positive entropy production is associated with the many irreversibilities that may exist within a control volume; these are due to viscous effects, separation of the

flow from turbine or compressor blades, and large temperature differences over which heat transfer takes place, to name a few.

By transferring energy to the control volume via heat transfer, we can obviously increase the entropy of the fluid flowing from the control volume so that $s_2 > s_1$. If irreversibilities are present, $\dot{\sigma}_{\text{device}}$ will be positive and s_2 will be greater than the exiting entropy for the same rate of heat transfer without irreversibilities.

We also note that for an adiabatic steady-flow process, the entropy also increases from inlet to exit due to irreversibilities since, for that case, Eq. (5.56) shows that

$$s_2 \geq s_1 \tag{5.56}$$

since $\dot{\sigma}_{\text{device}}$ must be positive or zero. For the reversible adiabatic process the inlet entropy and exit entropy are equal since $\dot{\sigma}_{\text{device}} = 0$; this is the isentropic process. We use this fact when studying a reversible adiabatic process involving, for example, steam or a refrigerant flowing through an ideal turbine or compressor.

We have calculated the efficiency of a cycle earlier in this chapter. Now we desire a quantity that can be used as a measure of the irreversibilities that exist in a particular device. The *efficiency* of a device is one such measure; it is defined as the ratio of the actual performance of a device to the ideal performance. The ideal performance is often that associated with an isentropic process. For example, the efficiency of a turbine would be

$$\eta_{\text{turb}} = \frac{w_a}{w_s} \tag{5.57}$$

where w_a is the actual shaft work output and w_s is the maximum shaft work output, i.e., the work associated with an isentropic process. In general, the efficiency is defined using the optimum output; for a nozzle we would use the maximum kinetic energy increase. For a compressor the actual work required is greater than the minimum work requirement of an isentropic process. For a compressor or pump the efficiency is defined to be

$$\eta_{\text{comp}} = \frac{w_s}{w_a} \tag{5.58}$$

Each efficiency above is often called the *adiabatic efficiency* since each efficiency is based on an adiabatic process.

EXAMPLE 5.13

Superheated steam enters a turbine, as shown, and exits at 10 kPa. For a mass flow rate of 2 kg/s, determine the maximum power output.

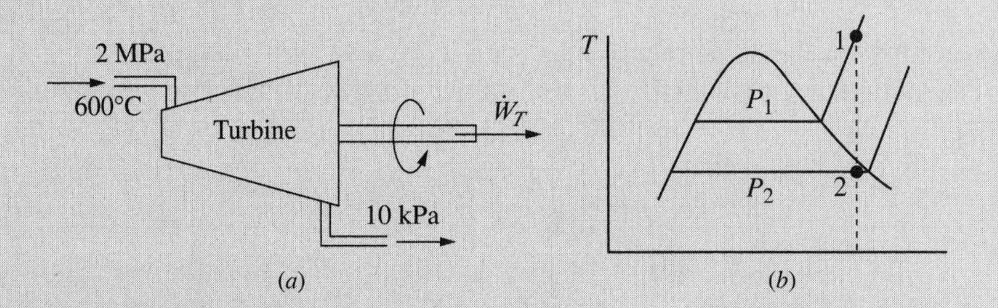

Solution

If we neglect kinetic energy and potential energy changes, the first law, for an adiabatic process, is $-W_T = m(h_2 - h_1)$. Since we desire the maximum power output, the process is assumed to be reversible, the entropy exiting is the same as the entropy entering, as shown (such a sketch is quite useful in visualizing the process). From the steam tables, at 600°C and 2 MPa,

$$h_1 = 3690 \text{ kJ/kg} \qquad s_1 = s_2 = 7.702 \text{ kJ/kg}\cdot\text{K}$$

With the above value for s_2 we see that state 2 is in the quality region. The quality is determined as follows:

$$s_2 = s_f + x_2 s_{fg}$$
$$7.702 = 0.6491 + 7.5019x_2 \qquad \therefore x_2 = 0.9401$$

Then

$$h_2 = h_f + x_2 h_{fg} = 191.8 + 0.9401 \times 2393 = 2442 \text{ kJ/kg}$$

so that

$$\dot{W}_T = \dot{m}(h_2 - h_1)$$
$$= 2 \times (3690 - 2442) = 2496 \text{ kJ/s} \qquad \text{or} \qquad 3346 \text{ hp}$$

EXAMPLE 5.14

The turbine of Example 5.13 is assumed to be 80 percent efficient. Determine the entropy and temperature of the final state. Sketch the real process on a *T-s* diagram.

Solution

Using the definition of efficiency, the actual power output is

$$\dot{W}_a = \eta_T \dot{W}_s = 0.8 \times 2496 = 1997 \text{ kW}$$

From the first law, $-\dot{W}_a = \dot{m}(h_{2'} - h_1)$, we have

$$h_{2'} = h_1 - \frac{\dot{W}_a}{\dot{m}} = 3690 - \frac{1997}{2} = 2692 \text{ kJ/kg}$$

Using this value and $P_{2'} = 10$ kPa, we see that state 2′ lies in the superheated region, since $h_{2'} > h_g$. This is shown schematically below. At $P_2 = 10$ kPa and $h_{2'} = 2770$, we interpolate to find the value of $T_{2'}$:

$$T_{2'} = \frac{2692 - 2688}{2783 - 2688}(150 - 100) + 100 = 102°\text{C}$$

The entropy is interpolated to be $s_{2'} = 8.46$ kJ/kg·K.

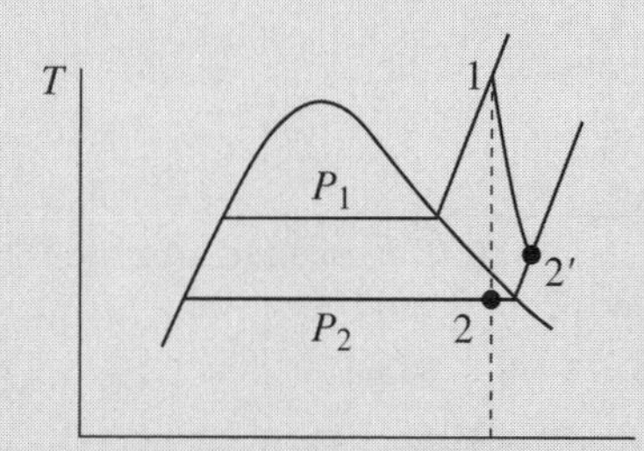

Note that the irreversibility has the desired effect of moving state 2 into the superheated region, thereby eliminating the formation of droplets due to the condensation of moisture. In an actual turbine, moisture formation cannot be tolerated because of damage to the turbine blades, so the irreversibility is somewhat desirable.

EXAMPLE 5.15

A preheater is used to preheat water in a power plant cycle, as shown. The superheated steam is at a temperature of 250°C and the entering water is at 45°C. All pressures are 600 kPa. Calculate the rate of entropy production.

Solution

The conservation of mass requires that

$$\dot{m}_3 = \dot{m}_2 + \dot{m}_1 = 0.5 + 4 = 4.5 \text{ kg/s}$$

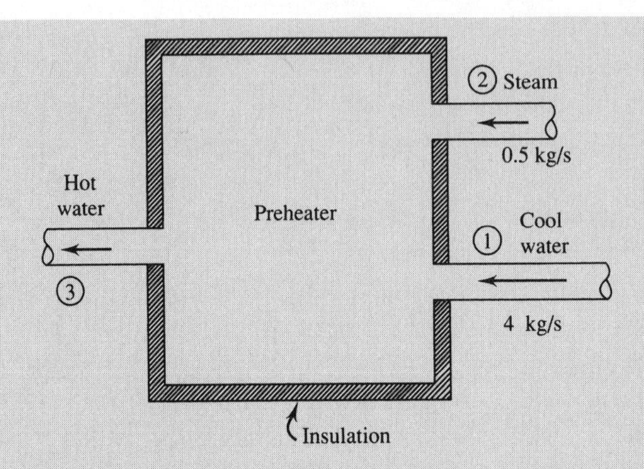

The first law allows us to calculate the temperature of the exiting water. Neglecting kinetic energy and potential energy changes and assuming zero heat transfer, the first law takes the form $\dot{m}_3 h_3 = \dot{m}_2 h_2 + \dot{m}_1 h_1$. Using the steam tables (h_1 is the enthalpy of saturated liquid at 45°C),

$$\dot{m}_3 h_3 = \dot{m}_2 h_2 + \dot{m}_1 h_1$$

$$4.5 h_3 = 0.5 \times 2957 + 4 \times 188.4 \qquad \therefore h_3 = 496 \text{ kJ/kg}$$

This enthalpy is less than that of saturated liquid at 600 kPa. Thus, the exiting water is subcooled. Its temperature is interpolated from the saturated steam tables (find T that gives $h_f = 496$ kJ/kg) to be

$$T_3 = \frac{496 - 461.3}{503.7 - 461.3} \times 10 + 110 = 118°\text{C}$$

The entropy at this temperature is then interpolated (using s_f) to be $s_3 =$ 1.508 kJ/kg·K. The entropy of the entering superheated steam is found to be $s_2 = 7.182$ kJ/kg·K. The entering entropy of the water is s_f at $T_1 = 45°$C, or $s_1 =$ 0.639 kJ/kg·K. Finally, modifying Eq. (5.56), to account for two inlets, we have, with no heat transfer,

$$\dot{\sigma}_{\text{prod}} = \dot{m}_3 s_3 - \dot{m}_2 s_2 - \dot{m}_1 s_1 = 4.5 \times 1.508 - 0.5 \times 7.182 - 4 \times 0.639 = 0.639 \text{ kW/K}$$

This is positive, indicating that entropy is produced, a consequence of the second law. The mixing process between the superheated steam and the subcooled water is indeed an irreversible process.

Quiz No. 1

1. An inventor claims a thermal engine operates between ocean layers at 27 and 10°C. It produces 10 kW and discharges 9900 kJ/min. Such an engine is
   (A) Impossible
   (B) Reversible
   (C) Possible
   (D) Probable
2. A heat pump is to provide 2000 kJ/h to a house maintained at 20°C. If it is −20°C outside, what is the minimum power requirement?
   (A) 385 kJ/h
   (B) 316 kJ/h
   (C) 273 kJ/h
   (D) 184 kJ/h
3. Select an acceptable paraphrase of the Kelvin-Planck statement of the second law.
   (A) No process can produce more work than the heat that it accepts.
   (B) No engine can produce more work than the heat that it intakes.
   (C) An engine cannot produce work without accepting heat.
   (D) An engine has to reject heat.
4. A power plant burns 1000 kg of coal each hour and produces 500 kW of power. Calculate the overall thermal efficiency if each kg of coal produces 6 MJ of energy.
   (A) 0.35
   (B) 0.30
   (C) 0.25
   (D) 0.20

5. Two Carnot engines operate in series between two reservoirs maintained at 327 and 27°C, respectively. The energy rejected by the first engine is input into the second engine. If the first engine's efficiency is 20 percent greater than the second engine's efficiency, calculate the intermediate temperature.

   (A) 106°C

   (B) 136°C

   (C) 243°C

   (D) 408°C

6. A heat pump is to maintain a house at 20°C when the outside air is at –25°C. It is determined that 1800 kJ is required each minute to accomplish this. Calculate the minimum power required.

   (A) 3.87 kW

   (B) 4.23 kW

   (C) 4.61 kW

   (D) 3.99 kW

7. Which of the following entropy relationships is incorrect?

   (A) Air, $V = \text{const}$: $\Delta s = C_v \ln T_2/T_1$

   (B) Water: $\Delta s = C_p \ln T_2/T_1$

   (C) Reservoir: $\Delta s = C_p \ln T_2/T_1$

   (D) Copper: $\Delta s = C_p \ln T_2/T_1$

8. Two kilograms of air are heated at constant pressure of 200 kPa to 500°C. Calculate the entropy change if the initial volume is 0.8 m^3.

   (A) 2.04 kJ/K

   (B) 2.65 kJ/K

   (C) 3.12 kJ/K

   (D) 4.04 kJ/K

9. Two kilograms of air are compressed from 120 kPa and 27°C to 600 kPa in a rigid container. Calculate the entropy change.

   (A) 5.04 kJ/K

   (B) 4.65 kJ/K

   (C) 3.12 kJ/K

   (D) 2.31 kJ/K

10. A paddle wheel provides 200 kJ of work to the air contained in a 0.2-m^3 rigid volume, initially at 400 kPa and 40°C. Determine the entropy change if the volume is insulated.

    (A) 0.504 kJ/K

    (B) 0.443 kJ/K

    (C) 0.312 kJ/K

    (D) 0.231 kJ/K

11. 0.2 kg of air is compressed slowly from 150 kPa and 40°C to 600 kPa, in an adiabatic process. Determine the final volume.

    (A) 0.0445 m^3

    (B) 0.0662 m^3

    (C) 0.0845 m^3

    (D) 0.0943 m^3

12. A rigid, insulated 4-m^3 volume is divided in half by a membrane. One chamber is pressurized with air to 100 kPa and the other is completely evacuated. The membrane is ruptured and after a period of time equilibrium is restored. The entropy change of the air is nearest

    (A) 0.624 kJ/K

    (B) 0.573 kJ/K

    (C) 0.473 kJ/K

    (D) 0.351 kJ/K

13. Ten kilograms of air are expanded isentropically from 500°C and 6 MPa to 400 kPa. The work accomplished is nearest

    (A) 7400 kJ

    (B) 6200 kJ

    (C) 4300 kJ

    (D) 2990 kJ

14. Find the work needed to isentropically compress 2 kg of air in a cylinder at 400 kPa and 400°C to 2 MPa.

    (A) 1020 kJ

    (B) 941 kJ

    (C) 787 kJ

    (D) 563 kJ

15. Calculate the total entropy change if 10 kg of ice at 0°C are mixed in an insulated container with 20 kg of water at 20°C. Heat of melting for ice is 340 kJ/kg.
    (A) 6.19 kJ/K
    (B) 3.95 kJ/K
    (C) 1.26 kJ/K
    (D) 0.214 kJ/K
16. A 5-kg block of copper at 100°C is submerged in 10 kg of water at 10°C, and after a period of time, equilibrium is established. If the container is insulated, calculate the entropy change of the universe.
    (A) 0.082 kJ/K
    (B) 0.095 kJ/K
    (C) 0.108 kJ/K
    (D) 0.116 kJ/K
17. Find w_T of the insulated turbine shown.

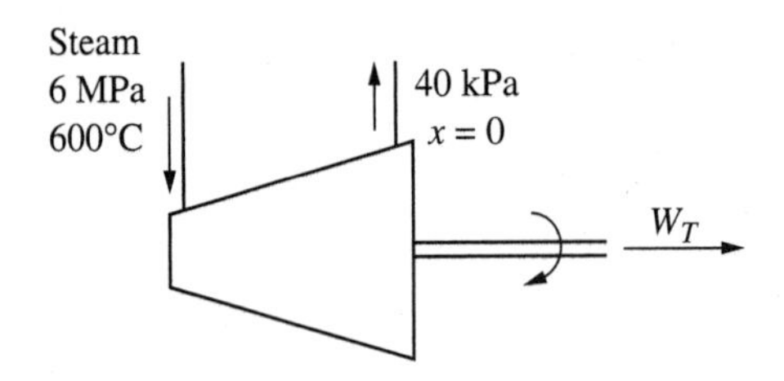

    (A) 910 kJ/kg
    (B) 1020 kJ/kg
    (C) 1200 kJ/kg
    (D) 1430 kJ/kg
18. The efficiency of the turbine of Prob. 17 is nearest
    (A) 92%
    (B) 89%
    (C) 85%
    (D) 81%

19. A nozzle accelerates steam at 120 kPa and 200°C from 20 m/s to the atmosphere. If it is 85 percent efficient, the exiting velocity is nearest
    (A) 290 m/s
    (B) 230 m/s
    (C) 200 m/s
    (D) 185 m/s
20. A turbine produces 4 MW by extracting energy from 4 kg of steam which flows through the turbine every second. The steam enters at 600°C and 1600 kPa and exits at 10 kPa. The turbine efficiency is nearest
    (A) 82%
    (B) 85%
    (C) 87%
    (D) 91%

Quiz No. 2

1. An engine operates on 100°C geothermal water. It exhausts to a 20°C stream. Its maximum efficiency is nearest
   (A) 21%
   (B) 32%
   (C) 58%
   (D) 80%
2. Which of the following can be assumed to be reversible?
   (A) A paddle wheel
   (B) A burst membrane
   (C) A resistance heater
   (D) A piston compressing gas in a race engine
3. A Carnot engine operates between reservoirs at 20 and 200°C. If 10 kW of power is produced, find the rejected heat rate.
   (A) 26.3 kJ/s
   (B) 20.2 kJ/s
   (C) 16.3 kJ/s
   (D) 12.0 kJ/s

4. An automobile that has a gas mileage of 13 km/L is traveling at 100 km/h. At this speed essentially all the power produced by the engine is used to overcome air drag. If the air drag force is given by $\frac{1}{2}\rho V^2 A C_D$ determine the thermal efficiency of the engine at this speed using projected area $A = 3$ m², drag coefficient $C_D = 0.28$, and heating value of gasoline 9000 kJ/kg. Gasoline has a density of 740 kg/m³.
   (A) 0.431
   (B) 0.519
   (C) 0.587
   (D) 0.652
5. A proposed power cycle is designed to operate between temperature reservoirs of 900 and 20°C. It is supposed to produce 43 hp from the 2500 kJ of energy extracted each minute. Is the proposal feasible?
   (A) No
   (B) Yes
   (C) Maybe
   (D) Insufficient information
6. A Carnot refrigeration cycle is used to estimate the energy requirement in an attempt to reduce the temperature of a specimen to absolute zero. Suppose that we wish to remove 0.01 J of energy from the specimen when it is at 2×10^{-6} K. How much work is necessary if the high-temperature reservoir is at 20°C?
   (A) 622 kJ
   (B) 864 kJ
   (C) 1170 kJ
   (D) 1465 kJ
7. One kilogram of air is heated in a rigid container from 20 to 300°C. The entropy change is nearest
   (A) 0.64 kJ/K
   (B) 0.54 kJ/K
   (C) 0.48 kJ/K
   (D) 0.34 kJ/K

8. Which of the following second law statements is incorrect?
   - (A) The entropy of an isolated system must remain constant or increase.
   - (B) The entropy of a hot copper block decreases as it cools.
   - (C) If ice is melted in water in an insulated container, the net entropy decreases.
   - (D) Work must be input if energy is transferred from a cold body to a hot body.
9. A piston allows air to expand from 6 MPa to 200 kPa. The initial volume and temperature are 500 cm^3 and 800°C. If the temperature is held constant, calculate the entropy change.
   - (A) 10.9 kJ/K
   - (B) 9.51 kJ/K
   - (C) 8.57 kJ/K
   - (D) 7.41 kJ/K
10. The entropy change in a certain expansion process is 5.2 kJ/K. The nitrogen, initially at 80 kPa, 27°C, and 4 m^3, achieves a final temperature of 127°C. Calculate the final volume.
    - (A) 255 m^3
    - (B) 223 m^3
    - (C) 158 m^3
    - (D) 126 m^3
11. A piston is inserted into a cylinder causing the pressure in the air to change from 50 to 4000 kPa while the temperature remains constant at 27°C. To accomplish this, heat transfer must occur. Determine the entropy change.
    - (A) −0.92 kJ/kg · K
    - (B) −0.98 kJ/kg · K
    - (C) −1.08 kJ/kg · K
    - (D) −1.26 kJ/kg · K

12. A torque of 40 N·m is needed to rotate a shaft at 40 rad/s. It is attached to a paddle wheel located in a rigid 2-m^3 volume. Initially the temperature is 47°C and the pressure is 200 kPa; if the paddle wheel rotates for 10 min and 500 kJ of heat is transferred to the air in the volume, determine the entropy increase assuming constant specific heats.
    (A) 3.21 kJ/K
    (B) 2.81 kJ/K
    (C) 2.59 kJ/K
    (D) 2.04 kJ/K
13. Four kilograms of air expand in an insulated cylinder from 500 kPa and 227°C to 20 kPa. What is the work output?
    (A) 863 kJ
    (B) 892 kJ
    (C) 964 kJ
    (D) 1250 kJ
14. Steam, at a quality of 85 percent, is expanded in a cylinder at a constant pressure of 800 kPa by adding 2000 kJ/kg of heat. Compute the entropy increase.
    (A) 3.99 kJ/ kg·K
    (B) 3.74 kJ/ kg·K
    (C) 3.22 kJ/ kg·K
    (D) 2.91 kJ/ kg·K
15. Two kilograms of saturated steam at 100°C are contained in a cylinder. If the steam undergoes an isentropic expansion to 20 kPa, determine the work output.
    (A) 376 kJ
    (B) 447 kJ
    (C) 564 kJ
    (D) 666 kJ
16. Ten kilograms of iron at 300°C are chilled in a large volume of ice and water. Find the total entropy change.
    (A) 0.88 kJ/K
    (B) 1.01 kJ/K
    (C) 1.26 kJ/K
    (D) 1.61 kJ/K

17. Five kilograms of ice at –20°C are mixed with 10 kg of water initially at 20°C. If there is no significant heat transfer from the container, determine the net entropy change. It takes 330 kJ to melt a kg of ice.
    (A) 0.064 kJ/K
    (B) 0.084 kJ/K
    (C) 1.04 kJ/K
    (D) 1.24 kJ/K
18. Two kilograms of air are stored in a rigid volume of 2 m^3 with the temperature initially at 300°C. Heat is transferred from the air until the pressure reaches 120 kPa. Calculate the entropy change of the air.
    (A) –0.292 kJ/K
    (B) –0.357 kJ/K
    (C) –0.452 kJ/K
    (D) –0.498 kJ/K
19. Calculate the entropy change of the universe for Prob. 18 if the surroundings are at 27°C.
    (A) 0.252 kJ/K
    (B) 0.289 kJ/K
    (C) 0.328 kJ/K
    (D) 0.371 kJ/K
20. Two hundred kilowatts are to be produced by a steam turbine. The outlet steam is to be saturated at 80 kPa and the steam entering will be at 600°C. For an isentropic process, determine the mass flow rate of steam.
    (A) 0.342 kg/s
    (B) 0.287 kg/s
    (C) 0.198 kg/s
    (D) 0.116 kg/s

CHAPTER 6

Power and Refrigeration Vapor Cycles

The ideal Carnot cycle is used as a model against which all real and all other ideal cycles are compared. The efficiency of a Carnot power cycle is the maximum possible for any power cycle. We observed that the Carnot-cycle efficiency is increased by raising the temperature at which heat is added or by lowering the temperature at which heat is rejected. We will observe that this carries over to real cycles: using the highest maximum temperature and the lowest minimum temperature maximizes cycle efficiency.

We will first discuss vapor cycles that are used to generate power, then vapor cycles that are used to refrigerate or heat a space. In the next chapter we will examine both power and refrigeration cycles that use air as the working fluid.

6.1 The Rankine Cycle

The electric power generating industry uses power cycles that operate in such a way that the working fluid changes phase from a liquid to a vapor. The simplest vapor power cycle is called the *Rankine cycle*, shown schematically in Fig. 6.1*a*. A major feature of such a cycle is that the pump requires very little work to deliver high-pressure water to the boiler. A possible disadvantage is that the expansion process in the turbine usually enters the quality region, resulting in the formation of liquid droplets that may damage the turbine blades.

The Rankine cycle is an idealized cycle in which losses in each of the four components are neglected. The Rankine cycle is composed of the four ideal processes shown on the *T-s* diagram in Fig. 6.1*b*:

1 → 2: Isentropic compression in a pump

2 → 3: Constant-pressure heat addition in a boiler

3 → 4: Isentropic expansion in a turbine

4 → 1: Constant-pressure heat rejection in a condenser

The pump is used to increase the pressure of the saturated liquid. Actually, states 1 and 2 are essentially the same since the high-pressure lines are extremely close to the saturation curve; they are shown separated for illustration only. The *boiler* (also called a *steam generator*) and the *condenser* are heat exchangers that neither require nor produce any work. An energy balance was applied to similar devices of the Rankine cycle in Sec. 4.8.

If we neglect kinetic energy and potential energy changes, the net work output is the area under the *T-s* diagram, represented by area 1-2-3-4-1 of Fig. 6.1*b*; this

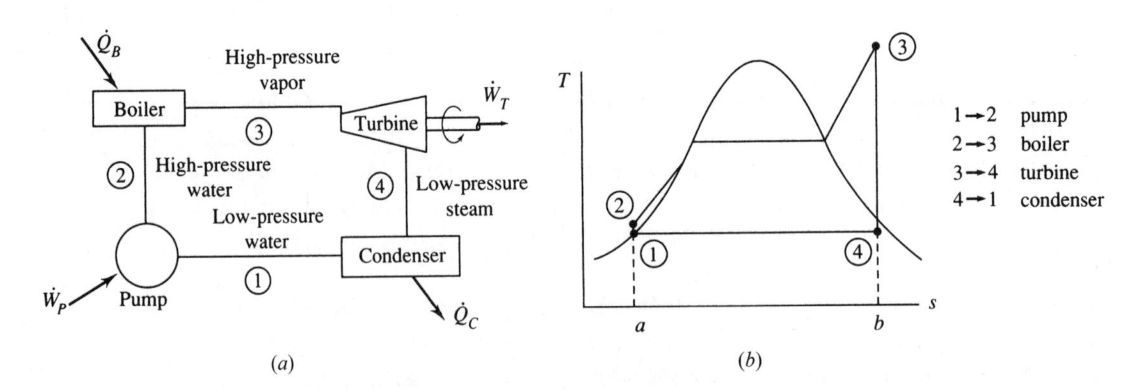

Figure 6.1 The Rankine cycle. (*a*) The major components. (*b*) The *T-s* diagram.

is true since the first law requires that $W_{net} = Q_{net}$. The heat transfer to the water is represented by area *a*-2-3-*b*-*a*. Thus, the thermal efficiency η of the Rankine cycle is

$$\eta = \frac{\text{area 1-2-3-4-1}}{\text{area } a\text{-2-3-}b\text{-}a} \tag{6.1}$$

that is, the desired output divided by the energy input (the purchased energy). Obviously, the thermal efficiency can be improved by increasing the numerator or by decreasing the denominator. This can be done by increasing the pump outlet pressure P_2, increasing the boiler outlet temperature T_3, or decreasing the turbine outlet pressure P_4.

Note that the efficiency of the Rankine cycle is less than that of a Carnot cycle operating between the high temperature T_3 and the low temperature T_1 since most of the heat transfer from a high-temperature reservoir (the boiler) occurs across large temperature differences, a situation that occurs in most heat exchangers (recall that heat transfer across a finite temperature difference is irreversible; i.e., a source of loss).

Let us now find expressions for the work and the heat transfer of the four devices of Fig. 6.1*a*. Section 4.8 provides the following expressions for the pump, boiler, turbine, and condenser:

$$w_P = v_1(P_2 - P_1) \qquad q_B = h_3 - h_2 \qquad w_T = h_3 - h_4 \qquad q_C = h_4 - h_1 \tag{6.2}$$

where w_P and q_C are expressed as positive quantities. In terms of the above, the thermal efficiency is

$$\eta = \frac{w_T - w_P}{q_B} \tag{6.3}$$

The pump work is usually quite small compared to the turbine work and can most often be neglected. With this approximation there results

$$\eta = \frac{w_T}{q_B} \tag{6.4}$$

This is the relation most often used for the thermal efficiency of the Rankine cycle.

EXAMPLE 6.1

A steam power plant is proposed to operate between the pressures of 10 kPa and 2 MPa with a maximum temperature of 400°C, as shown. Determine the maximum efficiency possible from the power cycle.

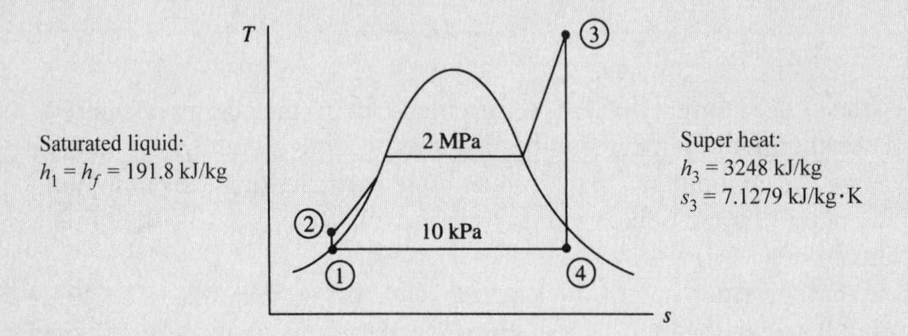

Solution

Let us include the pump work in the calculation and show that it is negligible. Also, we will assume a unit mass of working fluid since we are only interested in the efficiency. The pump work is [see Eq. (4.56)]

$$w_P = \frac{P_2 - P_1}{\rho} = \frac{2000 - 10}{1000} = 1.99 \text{ kJ/kg}$$

From the steam tables we find the values listed above. Using Eq. (4.52) we find that

$$h_2 = h_1 + w_P = 191.8 + 1.99 = 193.8 \text{ kJ/kg}$$

The heat input is found to be

$$q_B = h_3 - h_2 = 3248 - 193.8 = 3054 \text{ kJ/kg}$$

To locate state 4 we recognize that $s_4 = s_3 = 7.128$. Hence,

$$s_4 = s_f + x_4 s_{fg} \qquad \therefore 7.128 = 0.6491 + 7.5019 x_4$$

giving the quality of state 4 as $x_4 = 0.8636$. This allows us to find h_4 to be

$$h_4 = h_f + x_4 h_{fg} = 192 + 0.8636 \times 2393 = 2259 \text{ kJ/kg}$$

The work output from the turbine is

$$w_T = h_3 - h_4 = 3248 - 2259 = 989 \text{ kJ/kg}$$

Consequently, the efficiency is

$$\eta = \frac{w_T - w_P}{q_B} = \frac{989 - 1.99}{3054} = 0.323 \quad \text{or} \quad 32.3\%$$

Obviously, the work required in the pumping process is negligible, being only 0.2 percent of the turbine work. In engineering applications we often neglect quantities that have an influence of less than 3 percent, since invariably there is some quantity in the calculations that is known to only ±3 percent, for example, the mass flow rate, the dimensions of a pipe, or the density of the fluid.

6.2 Rankine Cycle Efficiency

The efficiency of the Rankine cycle can be improved by increasing the boiler pressure while maintaining the maximum temperature and the minimum pressure. The net increase in work output is the crosshatched area minus the shaded area of Fig. 6.2*a*; the added heat decreases by the shaded area minus the crosshatched area of Fig. 6.2*b*. This leads to a significant increase in efficiency, as Example 6.2 will illustrate. The disadvantage of raising the boiler pressure is that the quality of the steam exiting the turbine may become too low (less than 95 percent), resulting in

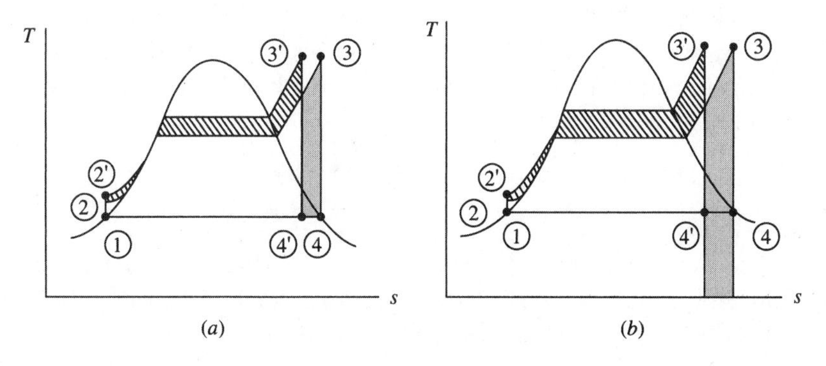

Figure 6.2 Effect of increased pressure on the Rankine cycle.

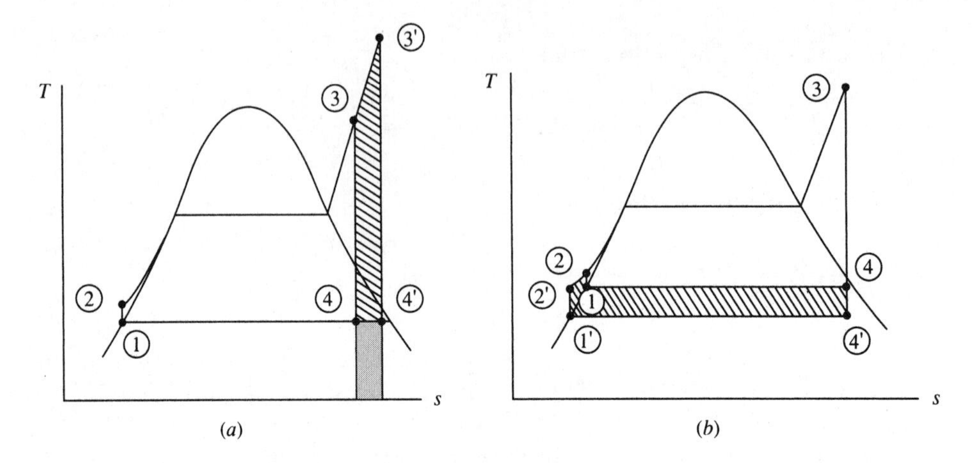

Figure 6.3 Effect of (*a*) increased maximum temperature and (*b*) decreased condenser pressure.

severe water droplet damage to the turbine blades and impaired turbine efficiency. However, this can be overcome, as we shall see.

Increasing the maximum temperature also results in an improvement in thermal efficiency of the Rankine cycle. In Fig. 6.3*a* the net work is increased by the crosshatched area and the heat input is increased by the sum of the crosshatched area and the shaded area, a smaller percentage increase than the work increase. Since the numerator of Eq. (6.4) realizes a larger percentage increase than the denominator, there will be a resulting increase in efficiency. This will be illustrated in Example 6.3. Another advantage of raising the boiler temperature is that the quality of state 4 is obviously increased, thereby reducing water droplet formation in the turbine.

A decrease in condenser pressure, illustrated in Fig. 6.3*b*, will also result in increased Rankine cycle efficiency. The net work will increase a significant amount, represented by the crosshatched area, and the heat input will increase a slight amount; the net result will be an increase in the Rankine cycle efficiency, as demonstrated in Example 6.4.

EXAMPLE 6.2

Increase the boiler pressure of Example 6.1 to 4 MPa while maintaining the maximum temperature and the minimum pressure. Neglect the work of the pump and calculate the percentage increase in the thermal efficiency.

Solution

The enthalpy h_2 is essentially equal to $h_1 \cong h_2 = 192$ kJ/kg. At 400°C and 4 MPa the enthalpy and entropy are $h_3 = 3214$ kJ/kg and $s_3 = 6.769$ kJ/kg·K. State 4 is in the quality region. Using $s_4 = s_3$, the quality is found to be

$$x_4 = \frac{s_4 - s_f}{s_{fg}} = \frac{6.769 - 0.6491}{7.5019} = 0.8158$$

Observe that the moisture content has increased to 18.4 ($x_4 = 81.6\%$) percent, an undesirable result. The enthalpy of state 4 is then

$$h_4 = h_f + x_4 h_{fg} = 192 + 0.8158 \times 2393 = 2144 \text{ kJ/kg}$$

The heat addition is

$$q_B = h_3 - h_2 = 3214 - 192 = 3022 \text{ kJ/kg}$$

and the turbine work output is

$$w_T = h_3 - h_4 = 3214 - 2144 = 1070 \text{ kJ/kg}$$

Finally, the thermal efficiency is

$$\eta = \frac{w_T}{q_B} = \frac{1070}{3022} = 0.354$$

The percentage increase in efficiency from that of Example 6.1 is

$$\% \text{ increase } = \frac{0.354 - 0.323}{0.323} \times 100 = 9.6\%$$

EXAMPLE 6.3

Increase the maximum temperature in the cycle of Example 6.1 to 600°C, while maintaining the boiler pressure and condenser pressure, and determine the percentage increase in thermal efficiency.

Solution

At 600°C and 2 MPa the enthalpy and entropy are $h_3 = 3690$ kJ/kg and $s_3 = 7.702$ kJ/ kg·K. State 4 remains in the quality region and, using $s_4 = s_3$, we have

$$x_4 = \frac{s_4 - s_f}{s_{fg}} = \frac{7.702 - 0.6491}{7.5019} = 0.940$$

Note here that the moisture content has been decreased to 6.0 percent, a desirable result. The enthalpy of state 4 is now found to be $h_4 = 192 + (0.940)(2393) = 2441$ kJ/kg. This allows us to calculate the thermal efficiency as

$$\eta = \frac{w_T}{q_B} = \frac{h_3 - h_4}{h_3 - h_2} = \frac{3690 - 2441}{3690 - 192} = 0.357$$

where h_2 is taken from Example 6.1. The percentage increase is

$$\% \text{ increase } = \frac{0.357 - 0.323}{0.323} \times 100 = 10.5\%$$

In addition to a significant increase in efficiency, the quality of the steam exiting the turbine is 94 percent, a much-improved value.

EXAMPLE 6.4

Decrease the condenser pressure of Example 6.1 to 4 kPa while maintaining the boiler pressure and maximum temperature, and determine the percentage increase in thermal efficiency.

Solution

The enthalpies $h_2 = 192$ kJ/kg and $h_3 = 3248$ kJ/kg remain as stated in Example 6.1. Using $s_3 = s_4 = 7.128$, with $P_4 = 4$ kPa, we find the quality to be

$$x_4 = \frac{s_4 - s_f}{s_{fg}} = \frac{7.128 - 0.4225}{8.0529} = 0.8327$$

Note that the moisture content of 16.7 percent has increased from that of Example 6.1, as expected. The enthalpy of state 4 is $h_4 = 121 + (0.8327)(2433) = 2147$ kJ/kg. The thermal efficiency is then

$$\eta = \frac{h_3 - h_4}{h_3 - h_2} = \frac{3248 - 2147}{3248 - 192} = 0.3603$$

The percentage increase is found to be

$$\% \text{ increase} = \frac{0.3603 - 0.323}{0.323} \times 100 = 11.5\%$$

This is the largest percentage increase which shows why the lowest possible condenser pressure is used.

6.3 The Reheat Cycle

It is apparent from the previous section that when operating a Rankine cycle with a high boiler pressure or a low condenser pressure it is difficult to prevent liquid droplets from forming in the low-pressure portion of the turbine. Since most metals cannot withstand temperatures above about 600°C, the *reheat cycle* is often used to prevent liquid droplet formation: the steam passing through the turbine is reheated at some intermediate pressure, thereby raising the temperature to state 5 in the *T*-s diagram of Fig. 6.4. The steam then passes through the low-pressure section of the turbine and enters the condenser at state 6. This controls or completely eliminates the moisture problem in the turbine. The reheat cycle does not significantly influence the thermal efficiency of the cycle, but it does result in a significant additional work output.

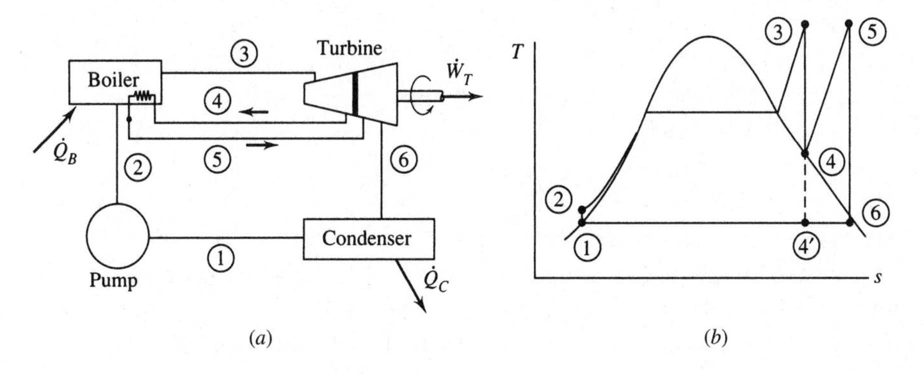

Figure 6.4 The reheat cycle.

EXAMPLE 6.5

High-pressure steam enters a turbine at 2 MPa and 400°C. It is reheated at a pressure of 800 kPa to 400°C and then expanded to 10 kPa. Determine the cycle efficiency.

Solution

At 10 kPa saturated water has an enthalpy of $h_1 = h_2 = 192$ kJ/kg. From Table C.3 we find $h_3 = 3248$ kJ/kg and $s_3 = 7.128$ kJ/kg·K. Setting $s_4 = s_3$ we interpolate, obtaining

$$h_4 = \frac{7.128 - 7.038}{7.233 - 7.038}(3056 - 2950) + 2950 = 2999 \text{ kJ/kg}$$

At 800 kPa and 400°C we have

$$h_5 = 3267 \text{ kJ/kg} \qquad \text{and} \qquad s_5 = 7.572 \text{ kJ/kg} \cdot \text{K}$$

In the quality region use $s_6 = s_5$ and find

$$x_6 = \frac{s_6 - s_f}{s_{fg}} = \frac{7.572 - 0.6491}{7.5019} = 0.923$$

Thus, $h_6 = 192 + (0.923)(2393) = 2401$ kJ/kg. The energy input and output are

$$q_B = (h_5 - h_4) + (h_3 - h_2) = 3267 - 2999 + 3248 - 192 = 3324 \text{ kJ/kg}$$

$$w_T = (h_5 - h_6) + (h_3 - h_4) = 3267 - 2401 + 3248 - 2999 = 1115 \text{ kJ/kg}$$

The thermal efficiency is then calculated to be

$$\eta = \frac{w_T}{q_B} = \frac{1115}{3324} = 0.335 \qquad \text{or} \qquad 33.5\%$$

Note: The fact that state 6 is in the mixture region is not of concern since x_6 is quite close to the saturated vapor state. As we will see in Sec. 6.7, turbine losses are able to move state 6 into the superheat region.

6.4 The Regenerative Cycle

In the conventional Rankine cycle, as well as in the reheat cycle, a considerable percentage of the total energy input is used to heat the high-pressure water from T_2 to its saturation temperature. The crosshatched area in Fig. 6.5*a* represents this necessary energy. To reduce this energy, the water is preheated before it enters the boiler by intercepting some of the steam as it expands in the turbine (for example, at state 5 of Fig. 6.5*b*) and mixing it with the water as it exits the first of the pumps, thereby preheating the water from T_2 to T_6. This would avoid the necessity of condensing all the steam, thereby reducing the amount of energy lost from the condenser. A cycle that utilizes this type of heating is a *regenerative cycle*, and the process is referred to as *regeneration.* A schematic representation of the major elements of such a cycle is shown in Fig. 6.6. The water entering the boiler is often referred to as *feedwater*, and the device used to mix the extracted steam and the condenser water is called a *feedwater heater.* When the condensate is mixed directly with the steam, it is done so in an *open feedwater heater*, as sketched in Fig. 6.6.

When analyzing a regenerative cycle we must consider a control volume surrounding the feedwater heater; see Fig. 6.7. A mass balance would result in

$$\dot{m}_6 = \dot{m}_5 + \dot{m}_2 \tag{6.5}$$

An energy balance, assuming an insulated heater, neglecting kinetic and potential energy changes, gives

$$\dot{m}_6 h_6 = \dot{m}_5 h_5 + \dot{m}_2 h_2 \tag{6.6}$$

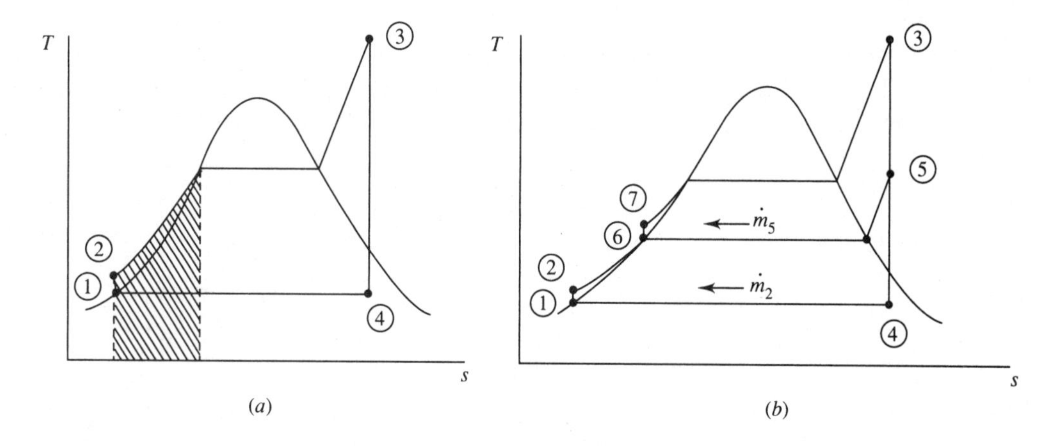

Figure 6.5 The regenerative cycle.

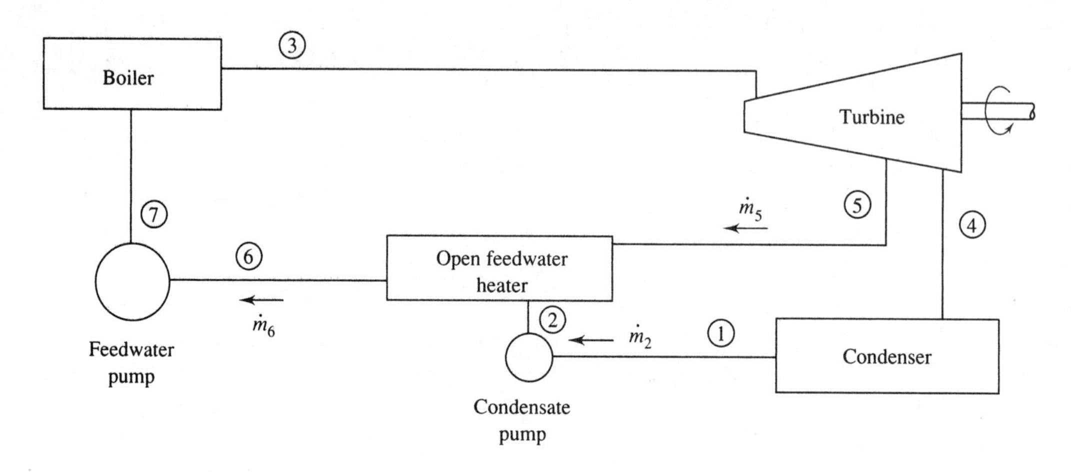

Figure 6.6 The major elements of the regenerative cycle.

Combining the above two equations gives the mass flux of intercepted steam:

$$\dot{m}_5 = \frac{h_6 - h_2}{h_5 - h_2} \times \dot{m}_6 \tag{6.7}$$

A *closed feedwater heater* that can be designed into a system using only one main pump is also a possibility, as sketched in Fig. 6.8. The closed feedwater heater is a heat exchanger in which the water passes through in tubes and the steam surrounds the tubes, condensing on the outer surfaces. The condensate thus formed, at temperature T_6, is pumped with a small condensate pump into the main feedwater line, as shown. Mass and energy balances are also required when analyzing a closed feedwater heater; if pump energy requirement is neglected in the analysis, the same relationship [see Eq. (6.7)] results.

The pressure at which the steam should be extracted from the turbine is approximated as follows. For one heater the steam should be extracted at the point that

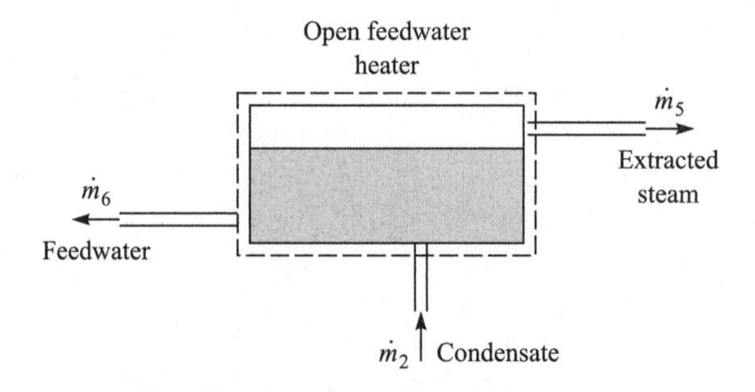

Figure 6.7 An open feedwater heater.

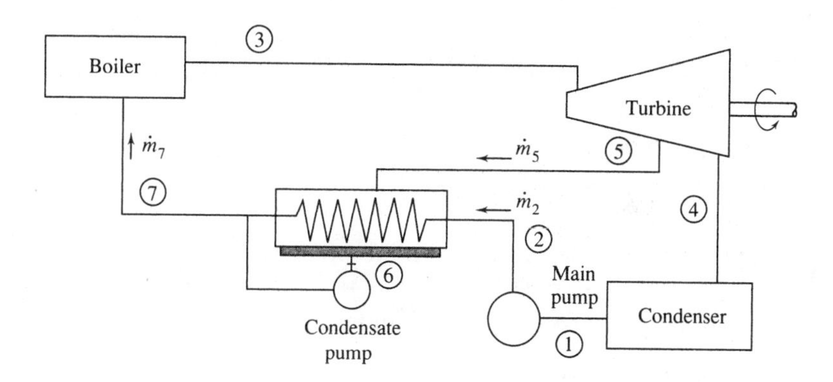

Figure 6.8 A closed feedwater heater.

allows the exiting feedwater temperature T_6 to be midway between the saturated steam temperature in the boiler and the condenser temperature. For several heaters the temperature difference should be divided as equally as possible.

It is not uncommon to combine a reheat cycle and a regenerative cycle, thereby avoiding the moisture problem and increasing the thermal efficiency. Ideal efficiencies significantly higher than for nonregenerative cycles can be realized with such a combination cycle.

There is an additional technique to effectively increase the "efficiency" of a power plant. There are special situations where a power plant can be located strategically so that the rejected steam from the condenser can be utilized to heat or cool (absorption refrigeration) buildings or it can be used in various industrial processes. This is often referred to as *cogeneration.* Often one-half of the rejected heat can be effectively used, almost doubling the "efficiency" of a power plant. The power plant must be located close to an industrial site or a densely populated area. A college campus is an obvious candidate for cogeneration, as are most large industrial plants. Several major cities are using rejected heat in downtown areas for cooling as well as heating.

EXAMPLE 6.6

The high-temperature cycle of Example 6.3 is to be modified by inserting an open feedwater heater with an extraction pressure of 200 kPa. Determine the percentage increase in thermal efficiency.

Solution

We have from Example 6.3 and the steam tables

$$h_1 \cong h_2 = 192 \text{ kJ/kg} \quad h_6 \cong h_7 = 505 \text{ kJ/kg} \quad h_3 = 3690 \text{ kJ/kg} \quad h_4 = 2442 \text{ kJ/kg}$$

Now, locate state 5. Using $s_5 = s_3 = 7.702$ kJ/kg · K, we interpolate and find, at 200 kPa,

$$h_5 = \frac{7.702 - 7.509}{7.709 - 7.509}(2971 - 2870) + 2870 = 2967 \text{ kJ/kg}$$

We now apply conservation of mass and the first law to a control volume surrounding the feedwater heater. For every kilogram of steam passing through the boiler, i.e., $m_6 = 1$ kg, we use Eq. (6.7) to find m_5:

$$m_5 = \frac{h_6 - h_2}{h_5 - h_2} \times m_6 = \frac{505 - 192}{2967 - 192} \times 1 = 0.1128 \text{ kg} \qquad \text{and} \qquad m_2 = 0.8872 \text{ kg}$$

The work output from the turbine is

$$w_T = h_3 - h_5 + (h_5 - h_4)m_2 = 3690 - 2967 + (2967 - 2442) \times 0.8872 = 1190 \text{ kJ}$$

The energy input to the boiler is $q_B = h_3 - h_7 = 3690 - 505 = 3185$ kJ/kg. The thermal efficiency is calculated to be

$$\eta = \frac{w_T}{q_B} = \frac{1190}{3185} = 0.374$$

The increase in efficiency is

$$\% \text{ increase} = \frac{0.374 - 0.357}{0.357} \times 100 = 4.8\%$$

EXAMPLE 6.7

The open feedwater heater shown is added to the reheat cycle of Example 6.5. Steam is extracted where the reheater interrupts the turbine flow. Determine the efficiency of this reheat-regeneration cycle.

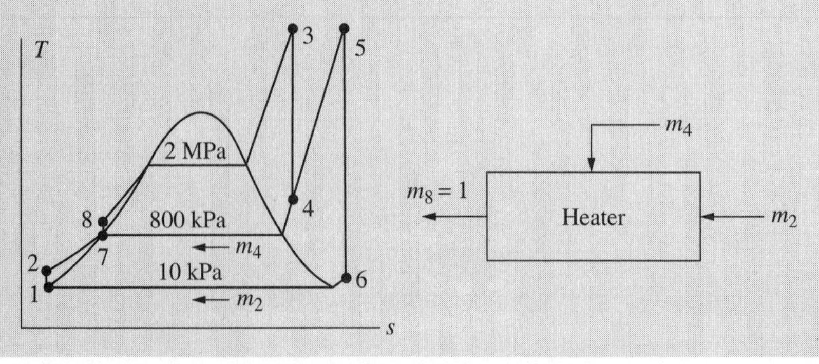

Solution

From the steam tables or from Example 6.5,

$$h_1 \cong h_2 = 192 \text{ kJ/kg} \qquad h_7 \cong h_8 = 721 \text{ kJ/kg} \qquad h_3 = 3248 \text{ kJ/kg}$$
$$h_5 = 3267 \text{ kJ/kg} \qquad h_6 = 2401 \text{ kJ/kg} \qquad h_4 = 2999 \text{ kJ/kg}$$

Continuity, using $m_8 = 1$ kg and the first law applied to the heater, gives

$$m_4 = \frac{h_8 - h_2}{h_4 - h_2} \times m_8 = \frac{721 - 192}{2999 - 192} \times 1 = 0.188 \text{ kg} \qquad \text{and} \qquad m_2 = 0.812 \text{ kg}$$

The turbine work output is then

$$w_T = h_3 - h_4 + (h_5 - h_6) \times m_2 = 3248 - 2999 + (3267 - 2401) \times 0.812 = 952 \text{ kJ/kg}$$

The energy input is

$$q_B = h_3 - h_8 + (h_5 - h_4)m_2 = 3248 - 721 + (3267 - 2999) \times 0.812 = 2745 \text{ kJ/kg}$$

The efficiency is calculated to be

$$\eta = \frac{w_T}{q_B} = \frac{952}{2745} = 0.347 \qquad \text{or} \qquad 34.7\%$$

Note the improvement of 7.4% in cycle efficiency over that of the basic Rankine cycle of Example 6.1.

6.5 Effect of Losses on Power Cycle Efficiency

The preceding sections dealt with ideal cycles assuming no pressure drop through the pipes in the boiler, no losses as the superheated steam passes over the blades in the turbine, no subcooling of the water leaving the condenser, and no pump losses during the compression process. The losses in the combustion process and the inefficiencies in the subsequent heat transfer to the fluid in the pipes of the boiler are not included here; those losses, which are in the neighborhood of 15 percent of the energy contained in the coal or oil, would be included in the overall plant efficiency.

There is actually only one substantial loss that must be accounted for when we calculate the actual cycle efficiency: the loss that occurs when the steam is expanded through the rows of turbine blades in the turbine. As the steam passes over a turbine blade, there is friction on the blade and the steam may separate from the rear portion of the blade. These losses result in a turbine efficiency of 80 to 89 percent. Turbine efficiency is defined as

$$\eta_T = \frac{w_a}{w_s} \tag{6.8}$$

where w_a is the actual work and w_s is the isentropic work. The definition of pump efficiency is

$$\eta_P = \frac{w_s}{w_a} \tag{6.9}$$

where the isentropic work input is obviously less than the actual input.

There is a substantial loss in pressure, probably 10 to 20 percent, as the fluid flows from the pump exit through the boiler to the turbine inlet. The loss can be overcome by simply increasing the exit pressure from the pump. This does require more pump work, but the pump work is still less than 1 percent of the turbine output and is thus negligible. Consequently, we ignore the boiler pipe losses. The condenser can be designed to operate such that the exiting water is very close to the saturated liquid condition. This will minimize the condenser losses so that they can also be neglected. The resulting actual Rankine cycle is shown on the *T-s* diagram in Fig. 6.9; the only significant loss is the turbine loss. Note the increase in entropy of state 4 as compared to state 3. Also, note the desirable effect of the decreased moisture content of state 4; in fact, state 4 may even move into the superheated region, as shown.

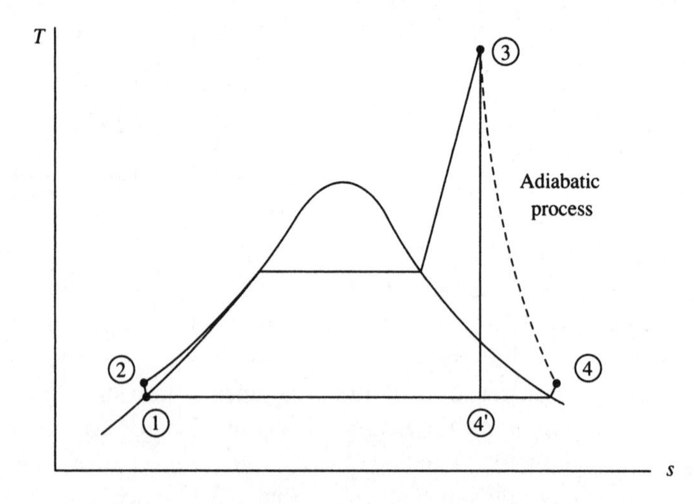

Figure 6.9 The Rankine cycle with turbine losses.

EXAMPLE 6.8

A Rankine cycle operates between pressures of 2 MPa and 10 kPa with a maximum temperature of 600°C. If the insulated turbine has an efficiency of 80 percent, calculate the cycle efficiency and the temperature of the steam at the turbine outlet.

Solution

From the steam tables we find $h_1 \simeq h_2 = 192$ kJ/kg, $h_3 = 3690$ kJ/kg, and $s_3 = 7.702$ kJ/kg·K. Setting $s_{4'} = s_3$ we find the quality and enthalpy of state 4′ (see Fig. 6.9) to be

$$x_{4'} = \frac{s_{4'} - s_f}{s_{fg}} = \frac{7.702 - 0.6491}{7.5019} = 0.940$$

$$\therefore h_{4'} = h_f + x_{4'} h_{fg} = 192 + 0.940 \times 2393 = 2440 \text{ kJ/kg}$$

From the definition of turbine efficiency,

$$\eta_T = \frac{w_a}{w_s} \qquad \text{or} \qquad 0.8 = \frac{w_a}{3690 - 2440} \qquad \therefore w_a = 1000 \text{ kJ/kg}$$

The cycle efficiency is then

$$\eta = \frac{w_T}{q_B} = \frac{1000}{3690 - 192} = 0.286 \qquad \text{or} \qquad 28.6\%$$

Note the substantial reduction from the ideal cycle efficiency of 35.7 percent, as calculated in Example 6.3.

The adiabatic process from state 3 to state 4 allows us to write

$$w_a = h_3 - h_4 \qquad \text{or} \qquad 1000 = 3690 - h_4 \qquad \therefore h_4 = 2690 \text{ kJ/kg}$$

At 10 kPa we find state 4 in the superheated region. The temperature is interpolated in the superheat region at 10 kPa to be

$$T_4 = \frac{2690 - 2688}{2783 - 2688}(150 - 100) + 100 = 101°\text{C}$$

Obviously, the moisture problem has been eliminated by the losses in the turbine; the losses tend to act as a small reheater.

6.6 The Vapor Refrigeration Cycle

It is possible to extract heat from a space by operating a vapor cycle, similar to the Rankine cycle, in reverse. Work input is required in the operation of such a cycle, as shown in Fig. 6.10*a*. The work is input by a compressor that increases the pressure, and thereby the temperature, through an isentropic compression in the ideal cycle. The working fluid (often, R134a) then enters a condenser in which heat is extracted, resulting in saturated liquid. The pressure is then reduced in an expansion process so that the fluid can be evaporated with the addition of heat from the refrigerated space.

The most efficient cycle, a Carnot cycle, is shown in Fig. 6.10*b*. There are, however, two major drawbacks when an attempt is made to put such a cycle into actual operation. First, it is not advisable to compress the mixture of liquid and vapor as

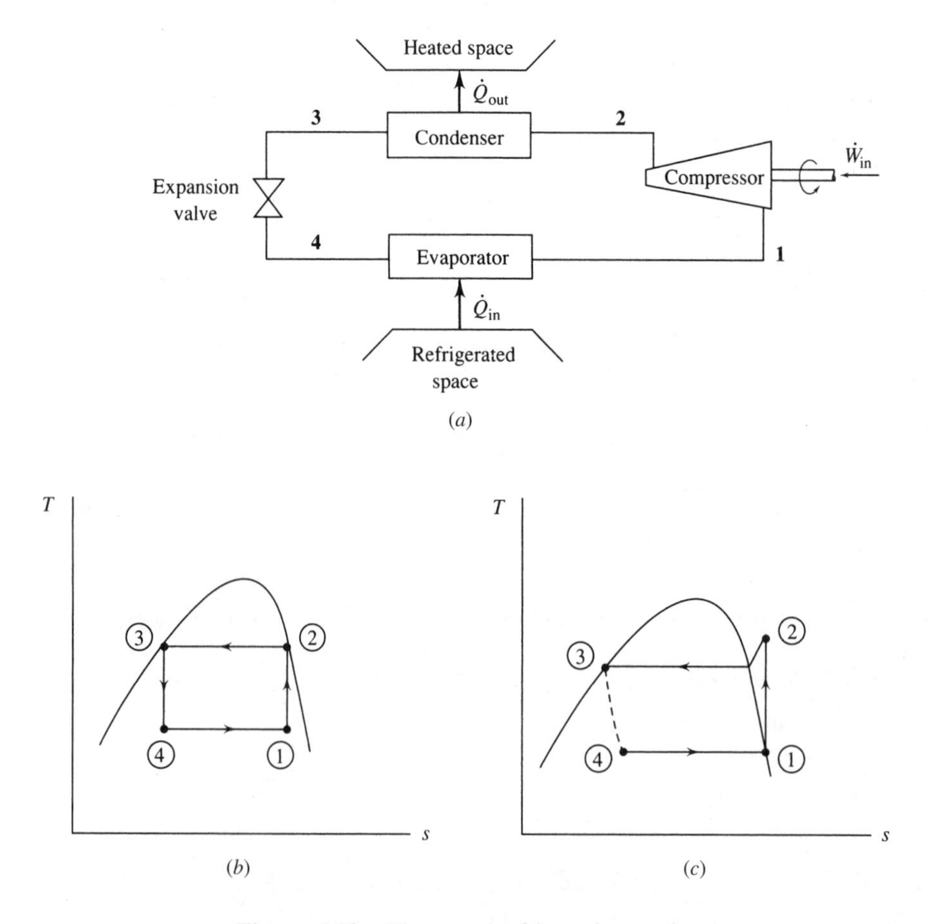

Figure 6.10 The vapor refrigeration cycle.

represented by state 1 in Fig. 6.10*b* since the liquid droplets would cause damage. Second, it would be quite expensive to construct a device to be used in the expansion process that would be nearly isentropic (no losses allowed). It is much simpler to reduce the pressure irreversibly by using an expansion valve in which enthalpy remains constant, as shown by the dotted line in Fig. 6.10*c*. Even though this expansion process is characterized by losses, it is considered to be part of the "ideal" vapor refrigeration cycle. Because the expansion process is a nonequilibrium process, the area under the *T-s* diagram does not represent the net work input.

The performance of the refrigeration cycle, when used as a refrigerator, is measured by

$$\text{COP} = \frac{\dot{Q}_{\text{in}}}{\dot{W}_{\text{in}}} \tag{6.10}$$

When the cycle is used as a heat pump, the performance is measured by

$$\text{COP} = \frac{\dot{Q}_{\text{out}}}{\dot{W}_{\text{in}}} \tag{6.11}$$

The coefficient of performance (COP) can attain values of perhaps 5 for properly designed heat pumps and 4 for refrigerators.

The condensation and evaporation temperatures, and hence the pressures, are established by the particular design of the refrigeration unit. For example, in a refrigerator that is designed to cool the freezer space to about −18°C (0°F), it is necessary to design the evaporator to operate at approximately −25°C to allow for effective heat transfer between the space and the cooling coils. The refrigerant condenses by transferring heat to air maintained at about 20°C; consequently, to allow for effective heat transfer from the coils that transport the refrigerant, the refrigerant must be maintained at a temperature of at least 27°C. This is shown in Fig. 6.11.

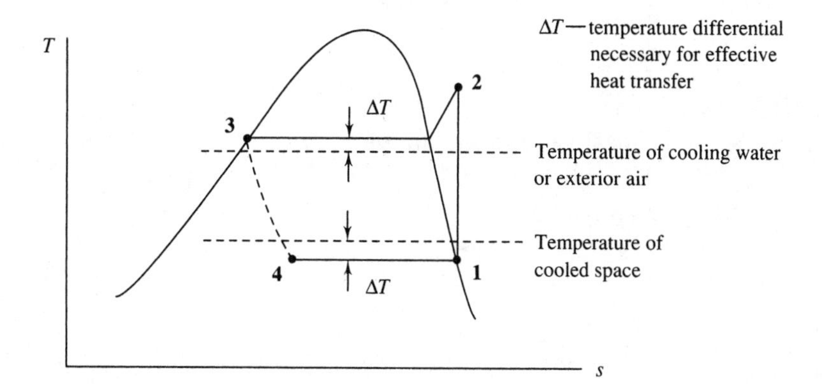

Figure 6.11 The refrigeration cycle showing design temperatures.

To accomplish refrigeration for most spaces, it is necessary that the evaporation temperature be quite low, in the neighborhood of −25°C, perhaps. This, of course, rules out water as a possible refrigerant. Two common refrigerants are ammonia (NH_3) and R134a. The thermodynamic properties of R134a are presented in App. D. The selection of a refrigerant depends on the high and low design temperatures. For example, temperatures well below −100°C are required to liquefy many gases. Neither ammonia nor R134a may be used at such low temperatures since they do not exist in a liquid form below −100°C. For most applications the refrigerant should be nontoxic, stable, environmentally benign, and relatively inexpensive.

Deviations from the ideal vapor refrigeration cycle include

- Pressure drops due to friction in connecting pipes.
- Heat transfer occurs from or to the refrigerant through the pipes connecting the components.
- Pressure drops occur through the condenser and evaporator tubes.
- Heat transfer occurs from the compressor.
- Frictional effects and flow separation occur on the compressor blades.
- The vapor entering the compressor may be slightly superheated.
- The temperature of the liquid exiting the condenser may be below the saturation temperature.

Quite often the size of an air conditioner is rated in tons, a rather interesting unit of rate of heat transfer. A "ton" of refrigeration is supposedly the heat rate necessary to melt a ton of ice in 24 hours. By definition, 1 ton of refrigeration equals 3.52 kW. Also, the performance of an air conditioner is often expressed as the EER (*the Energy Efficiency Rating*): the output in Btu's divided by the energy purchased in kWh. We multiply the COP by 3.412 to obtain the EER.

EXAMPLE 6.9

R134a is used in an ideal vapor refrigeration cycle operating between saturation temperatures of −24°C in the evaporator and 39.39°C in the condenser. Calculate the rate of refrigeration, the coefficient of performance, and the rating in horsepower per ton if the refrigerant flows at 0.6 kg/s.

Solution

Refer to Fig. 6.10*c*. The enthalpy of each state is needed. From Table D.1 we find that $h_1 = 232.8$ kJ/kg and $s_1 = 0.9370$ kJ/kg·K, and from Table D.2 we find

$h_3 = h_4 = 105.3$ kJ/kg. Using $s_1 = s_2$, we interpolate in Table D.3 at a pressure of 1.0 MPa (the pressure associated with the saturation temperature of 39.39°C), and find that

$$h_2 = \frac{0.937 - 0.9066}{0.9428 - 0.9066}(280.19 - 268.68) + 268.68 = 278.3 \text{ kJ/kg}$$

The rate of refrigeration is measured by the heat transfer rate needed in the evaporation process, namely,

$$\dot{Q}_E = \dot{m}(h_1 - h_4) = 0.6(232.8 - 105.3) = 76.5 \text{ kJ/s}$$

The power needed to operate the compressor is

$$\dot{W}_C = \dot{m}(h_2 - h_1) = 0.6(278.3 - 232.8) = 27.3 \text{ kW}$$

The coefficient of performance is then calculated to be

$$\text{COP} = \frac{\dot{Q}_{\text{in}}}{\dot{W}_C} = \frac{76.5}{27.3} = 2.80$$

The horsepower per ton of refrigeration is determined, with the appropriate conversion of units, as follows:

$$\text{hp/ton} = \frac{27.3/0.746}{76.5/3.52} = 1.68$$

EXAMPLE 6.10

The ideal refrigeration cycle of Example 6.9 is used in the operation of an actual refrigerator. It experiences the following real effects:

- The refrigerant leaving the evaporator is superheated to –20°C.
- The refrigerant leaving the condenser is subcooled to 40°C.
- The compressor is 80 percent efficient.

Calculate the actual rate of refrigeration and the coefficient of performance.

Solution

From Table D.1 we find, using $T_3 = 40°\text{C}$, that $h_3 = h_4 = 106.2$ kJ/kg. Also, from Table D.l we observe that $P_1 = 0.102$ MPa. From Table D.3, at $P_1 = 0.10$ MPa (close enough) and $T_1 = -20°\text{C}$,

$$h_1 = 236.5 \text{ kJ/kg} \qquad s_1 = 0.960 \text{ kJ/kg} \cdot \text{K}$$

If the compressor were isentropic, then, with $s_{2'} = s_1$ and $P_2 = 1.0$ MPa,

$$h_{2'} = \frac{0.960 - 0.9428}{0.9768 - 0.9428}(291.36 - 280.19) + 180.19 = 285.8 \text{ kJ/kg}$$

From the definition of efficiency, $\eta = w_s / w_a$, we have

$$0.8 = \frac{h_{2'} - h_1}{h_2 - h_1} = \frac{285.8 - 236.5}{h_2 - 236.5} \qquad \therefore h_2 = 298.1 \text{ kJ/kg}$$

The rate of refrigeration is

$$\dot{Q}_E = \dot{m}(h_1 - h_4) = 0.6 \times (236.5 - 106.2) = 78.2 \text{ kJ/s}$$

Note that the real effects have actually increased the capability to refrigerate a space. The coefficient of performance becomes

$$\text{COP} = \frac{q_E}{w_C} = \frac{78.2}{298.1 - 236.5} = 2.12$$

The 42 percent decrease in the COP occurs because the power input to the compressor has increased substantially. But, observe that the refrigeration has increased.

6.7 The Heat Pump

The heat pump utilizes the vapor refrigeration cycle discussed in Sec. 6.6. It can be used to heat a house in cool weather or cool a house in warm weather, as shown schematically in Fig. 6.12. Note that in the heating mode the house gains heat from the condenser, whereas in the cooling mode the house loses heat to the evaporator.

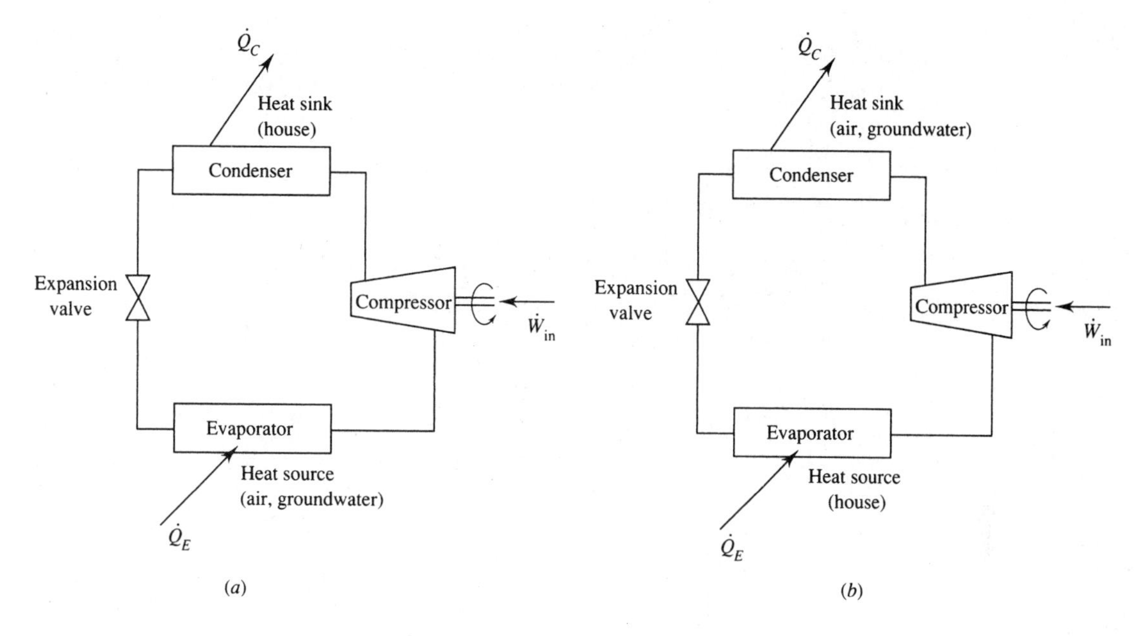

Figure 6.12 The heat pump. (*a*) Heating. (*b*) Cooling.

This is possible since the evaporator and the condenser are similar heat exchangers and valves are used to perform the desired switching.

A heat pump system is sized to meet the heating load or the cooling load, whichever is greater. In southern areas where the cooling loads are extremely large, the system may be oversized for the small heating demand of a chilly night; an air conditioner with an auxiliary heating system may be advisable in those cases. In a northern area where the large heating load demands a relatively large heat pump, the cooling load on a warm day may be too low for effective use of the heat pump; the large cooling capacity would quickly reduce the temperature of the house without a simultaneous reduction in the humidity, a necessary feature of any cooling system. In that case, a furnace that provides the heating with an auxiliary cooling system is usually advisable; or, the heat pump could be designed based on the cooling load, with an auxiliary heater for times of heavy heating demands.

EXAMPLE 6.11

A heat pump using R134a is proposed for heating a home that requires a maximum heating load of 300 kW. The evaporator operates at −12°C and the condenser at 800 kPa. Assume an ideal cycle.

(a) Determine the COP.

(b) Determine the cost of electricity at $0.07/kWh at peak load.

(c) Compare the R134a system with the cost of operating a furnace at peak load using natural gas at \$0.50/therm if there are 100 000 kJ/therm of natural gas.

Solution

(a) Refer to the *T-s* diagram of Fig. 6.10*c*. From Table D.1 we find $h_1 = 240$ kJ/kg, $s_1 = s_2 = 0.927$ kJ/kg·K, and $h_3 = h_4 = 93.4$ kJ/kg. Extrapolating at state 2 ($P_2 = 0.8$ MPa), there results

$$h_2 = 273.7 - \frac{0.9374 - 0.927}{0.9711 - 0.9374}(284.4 - 273.7) = 270.4 \text{ kJ/kg}$$

The heat rejected by the condenser is

$$\dot{Q}_C = \dot{m}(h_2 - h_3) \quad 300 = \dot{m}(270.4 - 93.4) \qquad \therefore \dot{m} = 1.695 \text{ kg/s}$$

The required power by the compressor is then

$$\dot{W}_{\text{Comp}} = \dot{m}(h_2 - h_1) = 1.695 \times (270.4 - 240) = 51.5 \text{ kW}$$

This results in a coefficient of performance of

$$\text{COP} = \frac{\dot{Q}_C}{\dot{W}_{\text{in}}} = \frac{300}{51.5} = 5.82$$

(b) Cost of electricity = 51.5 kW × \$0.07/kWh = \$3.60/h

(c) Assuming the gas furnace to be ideal, that is, it converts all of the energy of the gas into usable heat, we have

$$\text{Cost of gas} = \frac{300 \times 3600}{100\,000} \times 0.50 = \$5.40/\text{h}$$

The above gas furnace uses a heat pump that is assumed to operate ideally. A heat pump operates significantly below the ideal COP whereas a gas furnace can operate very near its ideal. The two systems are often quite competitive depending on the cost of electricity compared with the price of gas.

Quiz No. 1

1. The component of the Rankine cycle that leads to a relatively low cycle efficiency is
   (A) the pump
   (B) the boiler
   (C) the turbine
   (D) the condenser
2. The pump in the Rankine cycle of Fig. 6.1 increases the pressure from 20 kPa to 6 MPa. The boiler exit temperature is 600°C and the quality of the steam at the turbine exit is 100%. Find q_{boiler}.
   (A) 3410 kJ/kg
   (B) 3070 kJ/kg
   (C) 1050 kJ/kg
   (D) 860 kJ/kg
3. Find w_{turbine} for the Rankine cycle of Prob. 2.
   (A) 3410 kJ/kg
   (B) 3070 kJ/kg
   (C) 1050 kJ/kg
   (D) 860 kJ/kg
4. Find w_{pump} for the Rankine cycle of Prob. 2.
   (A) 4 kJ/kg
   (B) 6 kJ/kg
   (C) 8 kJ/kg
   (D) 10 kJ/kg
5. The turbine efficiency of the turbine in Prob. 2 is nearest
   (A) 64%
   (B) 72%
   (C) 76%
   (D) 81%

6. The cooling water in the condenser of Prob. 2 is allowed a 10°C increase. If $\dot{m}_{\text{steam}} = 2$ kg/s, determine $\dot{m}_{\text{cooling water}}$ if saturated liquid leaves the condenser.
   (A) 80 kg/s
   (B) 91 kg/s
   (C) 102 kg/s
   (D) 113 kg/s
7. The steam passing through the turbine of the power cycle of Prob. 2 is reheated at 800 kPa to 400°C. The thermal efficiency of the resulting cycle is nearest
   (A) 33%
   (B) 37%
   (C) 40%
   (D) 43%
8. If the turbine of Prob. 2 were 80 percent efficient, estimate the temperature that would exist at the turbine outlet.
   (A) 60°C
   (B) 65°C
   (C) 70°C
   (D) 75°C
9. An open feedwater heater is inserted into the cycle of Prob. 2 at 800 kPa. Calculate the mass m_5 in Fig. 6.5 if $m_6 = 1$ kg.
   (A) 0.170 kg
   (B) 0.192 kg
   (C) 0.221 kg
   (D) 0.274 kg
10. An open feed water heater accepts superheated steam at 400 kPa and 200°C and mixes the steam with condensate at 20 kPa from an ideal condenser. If 20 kg/s flows out of the condenser, what is the mass flow rate of the superheated steam?
    (A) 2.26 kg/s
    (B) 2.71 kg/s
    (C) 3.25 kg/s
    (D) 3.86 kg/s

11. R134a is used in an ideal refrigeration cycle between pressures of 120 kPa and 1000 kPa. If the compressor requires 10 hp, calculate the rate of refrigeration.
    (A) 15.9 kJ/s
    (B) 17.2 kJ/s
    (C) 21.8 kJ/s
    (D) 24.4 kJ/s
12. The coefficient of performance for the refrigerator of Prob. 11 is nearest
    (A) 2.31
    (B) 2.76
    (C) 2.93
    (D) 3.23
13. A heat pump using R134a as the refrigerant provides 80 MJ/h to a building. The cycle operates between pressures of 1000 and 200 kPa. Assuming an ideal cycle the compressor horsepower is nearest
    (A) 3.28
    (B) 3.87
    (C) 4.21
    (D) 4.97
14. The volume flow rate into the compressor of the ideal heat pump of Prob. 13 is nearest
    (A) 0.031 m^3/s
    (B) 0.026 m^3/s
    (C) 0.019 m^3/s
    (D) 0.013 m^3/s

Quiz No. 2

1. The Rankine power cycle is idealized. Which of the following is not one of the idealizations?
    (A) Friction is absent.
    (B) Heat transfer does not occur across a finite temperature difference.
    (C) Pressure drops in pipes are neglected.
    (D) Pipes connecting components are insulated.

2. A steam power plant is designed to operate on a Rankine cycle with a condenser pressure of 100 kPa and boiler outlet temperature of 500°C. If the pump outlet pressure is 2 MPa, calculate the ideal turbine work output.
   (A) 764 kJ/kg
   (B) 723 kJ/kg
   (C) 669 kJ/kg
   (D) 621 kJ/kg
3. The boiler requirement of the cycle of Prob. 2 is nearest
   (A) 2280 kJ/kg
   (B) 2750 kJ/kg
   (C) 2990 kJ/kg
   (D) 3050 kJ/kg
4. The maximum possible thermal efficiency of the cycle of Prob. 2 is nearest
   (A) 31%
   (B) 28%
   (C) 25%
   (D) 23%
5. If $\dot{m} = 2$ kg/s for the cycle of Prob. 2, the pump horsepower would be nearest
   (A) 4.0 hp
   (B) 5.1 hp
   (C) 6.2 hp
   (D) 7.0 hp
6. The steam passing through the turbine of the power cycle of Prob. 2 is reheated at 800 kPa to 500°C. The thermal efficiency of the resulting reheat Rankine cycle is nearest
   (A) 26.2%
   (B) 28.0%
   (C) 28.8%
   (D) 29.2%

7. An open feedwater heater is inserted into the Rankine cycle of Prob. 2 at 800 kPa. Calculate the mass m_5 in Fig. 6.5 if $m_6 = 1$ kg.

   (A) 0.14 kg

   (B) 0.13 kg

   (C) 0.12 kg

   (D) 0.11 kg

8. The Rankine cycle of Prob. 2 with the feedwater heater of Prob. 7 (refer to Fig. 6.5) has a maximum efficiency nearest

   (A) 24%

   (B) 26%

   (C) 28%

   (D) 30%

9. If the turbine of Prob. 2 is 90 percent efficient, estimate the temperature at the turbine outlet.

   (A) 152°C

   (B) 154°C

   (C) 156°C

   (D) 158°C

10. Refrigerant flows through the coil on the back of most refrigerators. This coil is the

    (A) evaporator

    (B) intercooler

    (C) reheater

    (D) condenser

11. An ideal vapor refrigeration cycle utilizes R134a as the working fluid between saturation temperatures of –28 and 40°C. For a flow of 0.6 kg/s, determine the rate of refrigeration.

    (A) 68.3 kJ/s

    (B) 70.2 kJ/s

    (C) 74.5 kJ/s

    (D) 76.9 kJ/s

12. The coefficient of performance for the refrigerator of Prob. 11 is nearest
    (A) 3.52
    (B) 3.15
    (C) 2.74
    (D) 2.52
13. The compressor of a refrigeration cycle accepts R134a as saturated vapor at 200 kPa and compresses it to 1200 kPa. The R134a leaves the condenser at 40°C. Determine the mass flow rate of R134a for 10 tons of refrigeration.
    (A) 0.261 kg/s
    (B) 0.349 kg/s
    (C) 0.432 kg/s
    (D) 0.566 kg/s
14. Calculate the minimum compressor horsepower for the ideal heat pump of Prob. 13.
    (A) 10.1 hp
    (B) 11.8 hp
    (C) 13.3 hp
    (D) 15.2 hp
15. A heat pump uses groundwater at 12°C as an energy source. If the energy delivered by the heat pump is to be 60 MJ/h, estimate the minimum mass flow rate of groundwater if the compressor operates with R134a between pressures of 100 kPa and 1.0 MPa.
    (A) 0.421 kg/s
    (B) 0.332 kg/s
    (C) 0.303 kg/s
    (D) 0.287 kg/s

CHAPTER 7

Power and Refrigeration Gas Cycles

Several cycles utilize a gas as the working substance, the most common being the Otto cycle and the diesel cycle, used in internal combustion engines. The word "cycle" used in reference to an internal combustion engine is technically incorrect since the working fluid does not undergo a thermodynamic cycle; air enters the engine, mixes with a fuel, undergoes combustion, and exits the engine as exhaust gases. This is often referred to as an *open cycle*, but we should keep in mind that a thermodynamic cycle does not really occur; the engine itself operates in what we could call a *mechanical cycle*. We do, however, analyze an internal combustion engine as though the working fluid operated on a cycle; it is an approximation that allows us to predict influences of engine design on such quantities as efficiency and fuel consumption. It also allows us to compare the features of the various cycles.

7.1 The Air-Standard Cycle

In this section we introduce engines that utilize a gas as the working fluid. Spark-ignition engines that burn gasoline and compression-ignition (diesel) engines that burn fuel oil are the two most common engines of this type.

The operation of a gas engine can be analyzed by assuming that the working fluid does indeed go through a complete thermodynamic cycle. The cycle is often called an *air-standard cycle.* All the air-standard cycles we will consider have certain features in common:

- Air is the working fluid throughout the entire cycle. The mass of the small quantity of injected fuel is negligible.
- There is no inlet process or exhaust process.
- A heat transfer process replaces the combustion process with energy transferred from an external source.
- The exhaust process, used to restore the air to its original state, is replaced with a constant-volume process transferring heat to the surroundings so that no work is accomplished, as in an actual cycle.
- All processes are assumed to be in quasiequilibrium.
- The air is assumed to be an ideal gas with constant specific heats.

A number of the engines we will consider make use of a closed system with a piston-cylinder arrangement, as shown in Fig. 7.1. The cycle shown on the P-v and T-s diagrams in the figure is representative. The diameter of the piston is called the *bore*, and the distance the piston travels in one direction is the *stroke.* When the piston is at top dead center (TDC), the volume occupied by the air in the cylinder is at a minimum; this volume is the *clearance volume.* When the piston moves to bottom dead center (BDC), the air occupies the maximum volume. The difference between the maximum volume and the clearance volume is the *displacement volume.* The clearance volume is often implicitly presented as the *percent clearance c*, the ratio of the clearance volume to the displacement volume. The *compression ratio r* is defined to be the ratio of the volume in the cylinder with the piston at BDC to the clearance volume, that is, referring to Fig. 7.1,

$$r = \frac{V_1}{V_2} \tag{7.1}$$

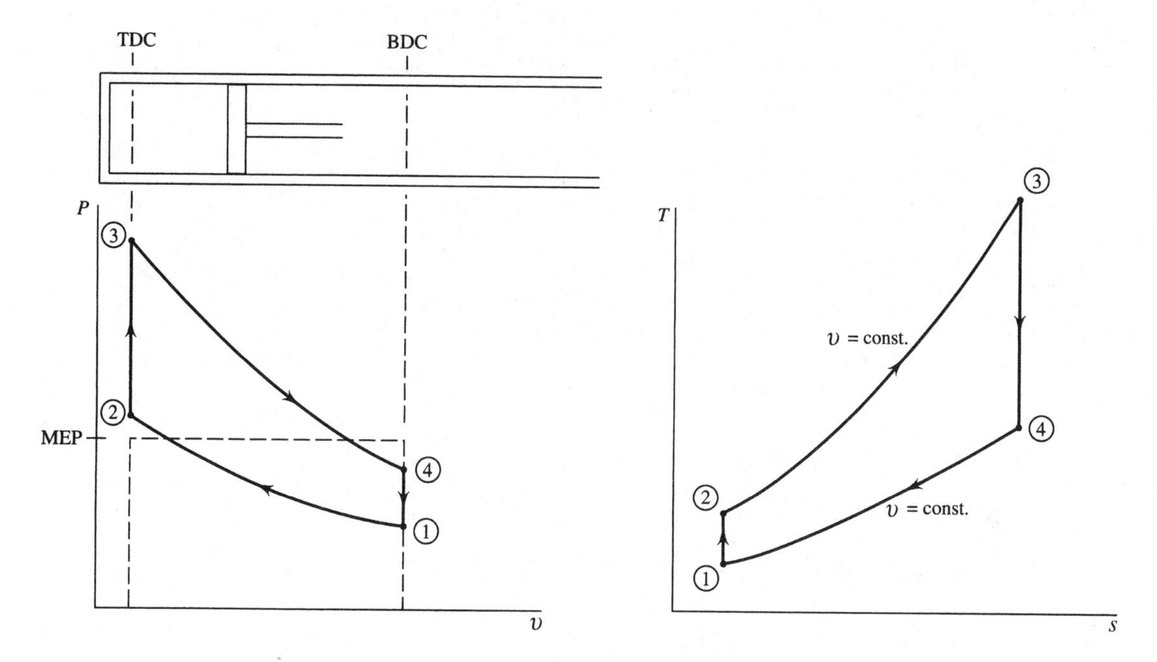

Figure 7.1 The cycle of a piston-cylinder gasoline engine.

The *mean effective pressure* (MEP) is another quantity that is often used when rating piston-cylinder engines; it is the pressure that, if acting on the piston during the power stroke, would produce an amount of work equal to that actually done during the entire cycle. Thus,

$$W_{\text{cycle}} = \text{MEP} \times (V_{\text{BDC}} - V_{\text{TDC}}) \tag{7.2}$$

In Fig. 7.1 this means that the enclosed area of the actual cycle is equal to the area under the MEP dotted line.

EXAMPLE 7.1

An engine operates with air on the cycle shown in Fig.7.1 with isentropic processes $1 \to 2$ and $3 \to 4$. If the compression ratio is 12, the minimum pressure is 200 kPa, and the maximum pressure is 10 MPa, determine the percent clearance and the MEP, assuming constant specific heats.

Solution

The percent clearance is given by

$$c = \frac{V_2}{V_1 - V_2} \times 100$$

But the compression ratio is $r = V_1/V_2 = 12$. Thus,

$$c = \frac{V_2}{12V_2 - V_2} \times 100 = \frac{1}{11} \times 100 = 9.09\%$$

To determine the MEP we must calculate the area under the *P-v* diagram (the work), a rather difficult task. The work from 3 → 4 is, using $Pv^k = C$,

$$w_{3\text{-}4} = \int P dv = C \int_{v_3}^{v_4} \frac{dv}{v^k} = \frac{C}{1-k}(v_4^{1-k} - v_3^{1-k}) = \frac{P_4 v_4 - P_3 v_3}{1-k}$$

where $C = P_4 v_4^k = P_3 v_3^k$. But we know that $v_4/v_3 = 12$, so $w_{3\text{-}4} = \frac{v_3}{1-k}(12P_4 - P_3)$.

Likewise, the work from 1 → 2 is $w_{1\text{-}2} = \frac{v_2}{1-k}(P_2 - 12P_1)$. Since no work occurs in the two constant-volume processes, we find, using $v_2 = v_3$,

$$w_{\text{cycle}} = \frac{v_2}{1-k}(12P_4 - P_3 + P_2 - 12P_1)$$

The pressures P_2 and P_4 are found as follows:

$$P_2 = P_1 \left(\frac{v_1}{v_2}\right)^k = 200 \times 12^{1.4} = 1665 \text{ kPa}$$

$$P_4 = P_3 \left(\frac{v_3}{v_4}\right)^k = 10\,000 \times \left(\frac{1}{12}\right)^{1.4} = 308 \text{ kPa}$$

so that

$$w_{\text{cycle}} = \frac{v_2}{-0.4}(12 \times 308 - 10\,000 + 1665 - 12 \times 200) = 20\,070 v_2$$

But, by definition,

$$w_{\text{cycle}} = \text{MEP} \times (v_1 - v_2) = \text{MEP} \times (12v_2 - v_2)$$

Equating the two expressions yields

$$\text{MEP} = \frac{20\,070}{11} = 1824 \text{ kPa}$$

7.2 The Carnot Cycle

This ideal cycle was treated in detail in Chap. 5. Recall that the thermal efficiency of a Carnot engine,

$$\eta_{\text{Carnot}} = 1 - \frac{T_L}{T_H} \tag{7.3}$$

exceeds that of any real engine operating between the given temperatures. We will use this efficiency as an upper limit for all engines operating between T_L and T_H.

7.3 The Otto Cycle

The four processes that form the *Otto cycle* are displayed in the *T-s* and *P-v* diagrams of Fig. 7.2. The piston starts at state 1 at BDC and compresses the air until it reaches TDC at state 2. Instantaneous combustion then occurs, due to spark ignition, resulting in a sudden jump in pressure to state 3 while the volume remains constant (a quasiequilibrium heat addition process). The process that follows is the power stroke as the air (the combustion products) expands isentropically to state 4. In the final process, heat transfer to the surroundings occurs and the cycle is completed.

The thermal efficiency of the Otto cycle is found from

$$\eta = \frac{\dot{W}_{\text{net}}}{\dot{Q}_{\text{in}}} = \frac{\dot{Q}_{\text{in}} - \dot{Q}_{\text{out}}}{\dot{Q}_{\text{in}}} = 1 - \frac{\dot{Q}_{\text{out}}}{\dot{Q}_{\text{in}}} \tag{7.4}$$

Noting that the two heat transfer processes occur during constant-volume processes, for which the work is zero, there results

$$\dot{Q}_{\text{in}} = \dot{m}(u_3 - u_2) = \dot{m}C_v(T_3 - T_2) \qquad \dot{Q}_{\text{out}} = \dot{m}(u_4 - u_1) = \dot{m}C_v(T_4 - T_1) \tag{7.5}$$

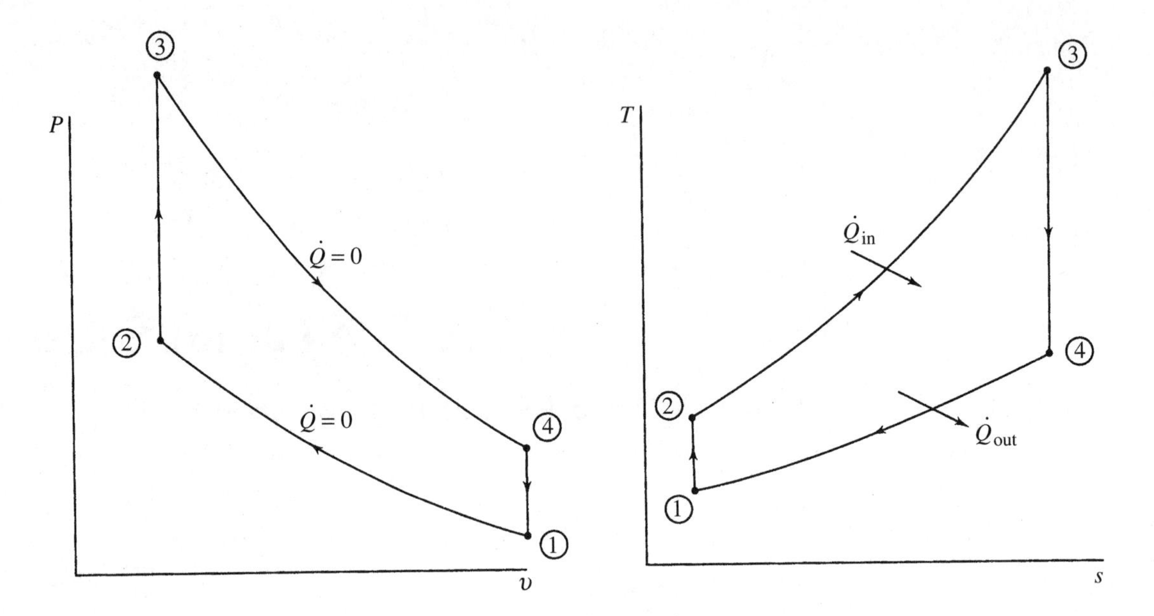

Figure 7.2 The Otto cycle.

where we have assumed each quantity to be positive. Then

$$\eta = 1 - \frac{T_4 - T_1}{T_3 - T_2} \tag{7.6}$$

This can be written as

$$\eta = 1 - \frac{T_1}{T_2}\left(\frac{T_4/T_1 - 1}{T_3/T_2 - 1}\right) \tag{7.7}$$

For the isentropic processes we have

$$\frac{T_2}{T_1} = \left(\frac{V_1}{V_2}\right)^{k-1} \quad \text{and} \quad \frac{T_3}{T_4} = \left(\frac{V_4}{V_3}\right)^{k-1} \tag{7.8}$$

But, using $V_1 = V_4$ and $V_3 = V_2$, we see that

$$\frac{T_2}{T_1} = \frac{T_3}{T_4} \tag{7.9}$$

Thus, Eq. (7.7) gives the thermal efficiency as

$$\eta = 1 - \frac{T_1}{T_2} = 1 - \left(\frac{V_2}{V_1}\right)^{k-1} = 1 - \frac{1}{r^{k-1}} \tag{7.10}$$

We see that the thermal efficiency in this Otto cycle is dependent only on the compression ratio r: higher the compression ratio, higher the thermal efficiency.

EXAMPLE 7.2

An Otto cycle is proposed to have a compression ratio of 10 while operating with a low air temperature of 227°C and a low pressure of 200 kPa. If the work output is to be 1000 kJ/kg, calculate the maximum possible thermal efficiency and compare with that of a Carnot cycle. Also, calculate the MEP. Assume constant specific heats.

Solution

The Otto cycle provides the model for this engine. The maximum possible thermal efficiency for the engine would be

$$\eta = 1 - \frac{1}{r^{k-1}} = 1 - \frac{1}{10^{0.4}} = 0.602 \qquad \text{or} \qquad 60.2\%$$

Since process $1 \rightarrow 2$ is isentropic, we find that

$$T_2 = T_1\left(\frac{v_1}{v_2}\right)^{k-1} = 500 \times 10^{0.4} = 1256 \text{ K}$$

The net work for the cycle is given by

$$\begin{aligned} w_{\text{net}} &= w_{1\text{-}2} + \cancel{w}_{2\text{-}3} + w_{3\text{-}4} + \cancel{w}_{4\text{-}1} \\ &= C_v(T_1 - T_2) + C_v(T_3 - T_4) \\ 1000 &= 0.717 \times (500 - 1256 + T_3 - T_4) \end{aligned}$$

But, for the isentropic process $3 \rightarrow 4$,

$$T_3 = T_4\left(\frac{v_4}{v_3}\right)^{k-1} = T_4 \times 10^{0.4} = 2.512 T_4$$

Solving the last two equations: $T_3 = 3573$ K and $T_4 = 1422$ K, so that

$$\eta_{\text{Carnot}} = 1 - \frac{T_L}{T_H} = 1 - \frac{500}{3573} = 0.860 \quad \text{or} \quad 86.0\%$$

The Otto cycle efficiency is less than that of a Carnot cycle because the heat transfer processes in the Otto cycle are highly irreversible.

The MEP is found by using the equation

$$w_{\text{net}} = \text{MEP} \times (v_1 - v_2)$$

The ideal-gas law and the compression ratio provides

$$v_1 = \frac{RT_1}{P_1} = \frac{0.287 \times 500}{200} = 0.7175 \text{ m}^3/\text{kg} \quad \text{and} \quad v_2 = \frac{v_1}{10}$$

and $$\text{MEP} = \frac{w_{\text{net}}}{v_1 - v_2} = \frac{1000}{0.7175 - 0.07175} = 1549 \text{ kPa}$$

7.4 The Diesel Cycle

If the compression ratio is large enough, the temperature of the air in the cylinder when the piston approaches TDC will exceed the ignition temperature of diesel fuel. This will occur if the compression ratio is about 14 or greater. No external spark is needed; the diesel fuel is simply injected into the cylinder and combustion occurs because of the high temperature of the compressed air. This type of engine is referred to as a *compression-ignition engine.* The ideal cycle used to model the engine is the *diesel cycle*, shown in Fig. 7.3. The difference between this cycle and the Otto cycle is that, in the diesel cycle, the heat is added during a constant-pressure process.

The cycle begins with the piston at BDC, state 1; compression of the air occurs isentropically to state 2 at TDC; heat addition takes place (this models the injection and combustion of fuel) at constant pressure until state 3 is reached; expansion occurs isentropically to state 4 at BDC; constant-volume heat rejection completes the cycle and returns the air to the original state. Note that the power stroke includes the heat addition process and the expansion process.

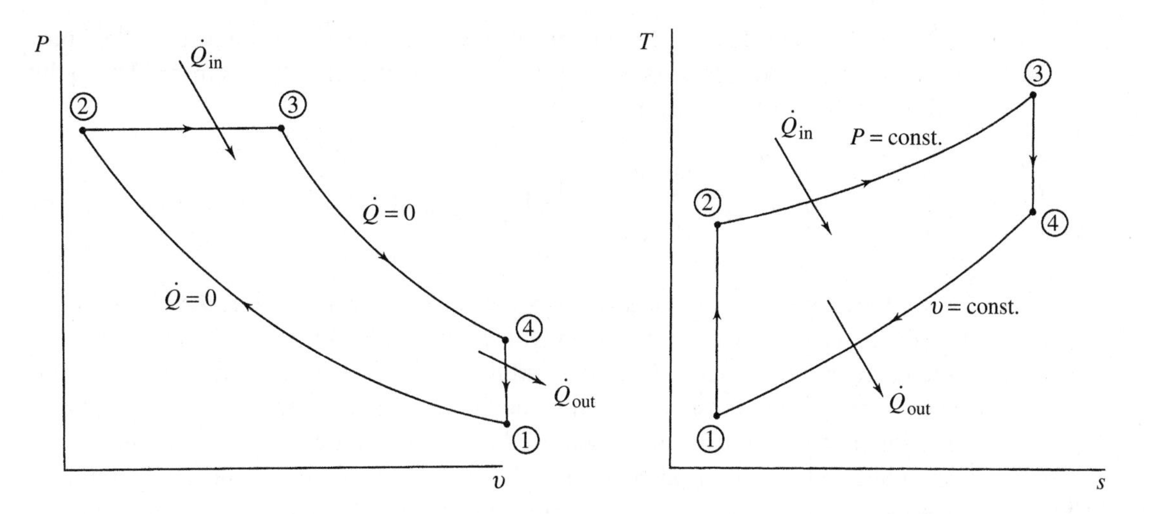

Figure 7.3 The diesel cycle.

The thermal efficiency of the diesel cycle is expressed as

$$\eta = \frac{\dot{W}_{net}}{\dot{Q}_{in}} = 1 - \frac{\dot{Q}_{out}}{\dot{Q}_{in}} \tag{7.11}$$

For the constant-volume process and the constant-pressure process

$$\dot{Q}_{out} = \dot{m}C_v(T_4 - T_1) \qquad \dot{Q}_{in} = \dot{m}C_p(T_3 - T_2) \tag{7.12}$$

The efficiency is then

$$\eta = 1 - \frac{C_v(T_4 - T_1)}{C_p(T_3 - T_2)} = 1 - \frac{T_4 - T_1}{k(T_3 - T_2)} \tag{7.13}$$

This can be put in the form

$$\eta = 1 - \frac{T_1}{kT_2}\left(\frac{T_4/T_1 - 1}{T_3/T_2 - 1}\right) \tag{7.14}$$

This expression for the thermal efficiency is often written in terms of the compression ratio r and the *cutoff ratio* r_c which is the ratio of volumes from TDC to the end of combustion. We then have

$$\eta = 1 - \frac{1}{r^{k-1}}\left(\frac{r_c^k - 1}{k(r_c - 1)}\right) \qquad \text{where} \qquad r_c = V_3/V_2 \tag{7.15}$$

From this expression we see that, for a given compression ratio r, the efficiency of the diesel cycle is less than that of an Otto cycle. For example, if $r = 8$ and $r_c = 2$, the Otto cycle efficiency is 56.3 percent and the diesel cycle efficiency is 49.0 percent. As r_c increases, the diesel cycle efficiency decreases. In practice, however, a compression ratio of 20 or so can be achieved in a diesel engine; using $r = 20$ and $r_c = 2$, we would find $\eta = 64.7$ percent. Thus, because of the higher compression ratios, a diesel engine typically operates at a higher efficiency than a gasoline engine.

EXAMPLE 7.3

A diesel cycle, with a compression ratio of 18, operates on air with a low pressure of 200 kPa and a low temperature 200°C. If the high temperature is limited to 2000 K, determine the thermal efficiency and the MEP. Assume constant specific heats.

Solution

The cutoff ratio r_c is found first. We have

$$v_1 = \frac{RT_1}{P_1} = \frac{0.287 \times 473}{200} = 0.6788 \text{ m}^3\text{/kg} \quad \text{and} \quad v_2 = v_1/18 = 0.03771 \text{ m}^3\text{/kg}$$

Since process $1 \rightarrow 2$ is isentropic and $2 \rightarrow 3$ is constant pressure, we find

$$T_2 = T_1\left(\frac{v_1}{v_2}\right)^{k-1} = 473 \times 18^{0.4} = 1503 \text{ K}$$

$$P_3 = P_2 = P_1\left(\frac{v_1}{v_2}\right)^{k} = 200 \times 18^{1.4} = 11\,440 \text{ kPa}$$

The cutoff ratio results:

$$v_3 = \frac{RT_3}{P_3} = \frac{0.287 \times 2000}{11440} = 0.05017 \text{ m}^3\text{/kg} \qquad \text{so that}$$

$$r_c = \frac{v_3}{v_2} = \frac{0.05017}{0.03771} = 1.3305$$

The thermal efficiency is now calculated as

$$\eta = 1 - \frac{1}{r^{k-1}} \frac{r_c^k - 1}{k(r_c - 1)} = 1 - \frac{1}{18^{0.4}} \frac{1.3305^{1.4} - 1}{1.4(1.3305 - 1)} = 0.666 \quad \text{or} \quad 66.6\%$$

The temperature of state 4 is found to be, using $v_1 = v_4$,

$$T_4 = T_3 \left(\frac{v_3}{v_4} \right)^{k-1} = 2000 \times \left(\frac{0.05017}{0.6788} \right)^{0.4} = 705.5 \text{ K}$$

The first law then allows us to calculate the MEP:

$$q_{\text{in}} = C_p (T_3 - T_2) = 1.0 \times (2000 - 1503) = 497 \text{ kJ/kg}$$

$$q_{\text{out}} = C_v (T_4 - T_1) = 0.717 \times (705.5 - 473) = 166.7 \text{ kJ/kg}$$

$$\therefore \text{MEP} = \frac{w_{\text{net}}}{v_1 - v_2} = \frac{q_{\text{in}} - q_{\text{out}}}{v_1 - v_2} = \frac{497 - 166.7}{0.6788 - 0.03771} = 515 \text{ kPa}$$

EXAMPLE 7.4

Rework Example 7.3 using Table E.1; do not assume constant specific heats. Calculate the thermal efficiency, the MEP, and the errors in both quantities as calculated in Example 7.3.

Solution

Equation (7.10) that gives the thermal efficiency of the Otto cycle was obtained assuming constant specific heats; so it cannot be used. We must calculate the work output and the heat added. Refering to Table E.1:

$$T_1 = 473 \text{ K}$$

$$\therefore v_1 = \frac{0.287 \times 473}{200} = 0.6788 \text{ m}^3/\text{kg} \qquad u_1 = 340 \text{ kJ/kg} \qquad v_{r1} = 198.1$$

The process from state 1 to state 2 is isentropic:

$$v_{r2} = v_{r1} \frac{v_2}{v_1} = 198.1 \times \frac{1}{18} = 11.01$$

$$\therefore T_2 = 1310 \text{ K} \qquad P_{r2} = 342 \qquad \text{and} \qquad h_2 = 1408 \text{ kJ/kg}$$

The specific volume and pressure at state 2 are

$$v_2 = \frac{v_1}{18} = \frac{0.6788}{18} = 0.03771 \text{ m}^3/\text{kg} \quad \text{and} \quad P_3 = P_2 = \frac{0.287 \times 1310}{0.03771} = 9970 \text{ kPa}$$

At state 3 $T_3 = 2000$ K, $h_3 = 2252.1$ kJ/kg. Then

$$v_3 = \frac{RT_3}{p_3} = \frac{0.287 \times 2000}{9970} = 0.05757 \text{ m}^3/\text{kg} \qquad \text{and} \qquad v_{r3} = 2.776$$

$$\therefore\ v_{r4} = v_{r3} \times \frac{v_4}{v_3} = 2.776 \times \frac{0.6788}{0.05757} = 32.73$$

Hence, at state 4 we find from Table E.1 that $T_4 = 915$ K and $u_4 = 687.5$ kJ/kg. The first law results in

$$q_{\text{in}} = h_3 - h_2 = 2252 - 1408 = 844 \text{ kJ/kg}$$

$$q_{\text{out}} = u_4 - u_1 = 687.5 - 339 = 348.5 \text{ kJ/kg}$$

Finally,

$$\eta = \frac{w_{\text{net}}}{q_{\text{in}}} = \frac{844 - 348.5}{844} = 0.587 \qquad \text{or} \qquad 58.7\%$$

$$\text{MEP} = \frac{w_{\text{net}}}{v_1 - v_2} = \frac{844 - 348.5}{0.6788 - 0.03771} = 773 \text{ kPa}$$

The percent errors in the answers to Example 7.3 are

$$\%\ \text{error in}\ \eta = \frac{66.6 - 58.7}{58.7} \times 100 = 13.5\%$$

$$\%\ \text{error in MEP} = \frac{515 - 773}{773} \times 100 = -33.3\%$$

The errors are substantial primarily due to large temperature differences experienced in the diesel cycle. However, the constant specific heat analyses are of interest particularly in parametric studies where various designs are considered.

7.5 The Brayton Cycle

The gas turbine is another mechanical system that produces power. It may operate on an open cycle when used as a truck engine, or on a closed cycle when used in a power plant. In open cycle operation, air enters the compressor, passes through a constant-pressure

combustion chamber, then through a turbine, and finally exits as products of combustion to the atmosphere, as shown in Fig. 7.4*a*. In closed cycle operation the combustion chamber is replaced with a heat exchanger in which energy enters the cycle from some exterior source; an additional heat exchanger transfers heat from the cycle so that the air is returned to its initial state, as shown in Fig. 7.4*b*.

The ideal cycle used to model the gas turbine is the *Brayton cycle*. It utilizes isentropic compression and expansion, as indicated in Fig. 7.5. The efficiency of such a cycle is given by

$$\eta = 1 - \frac{\dot{Q}_{\text{out}}}{\dot{Q}_{\text{in}}} = 1 - \frac{C_p(T_4 - T_1)}{C_p(T_3 - T_2)} = 1 - \frac{T_1}{T_2}\left(\frac{T_4/T_1 - 1}{T_3/T_2 - 1}\right) \tag{7.16}$$

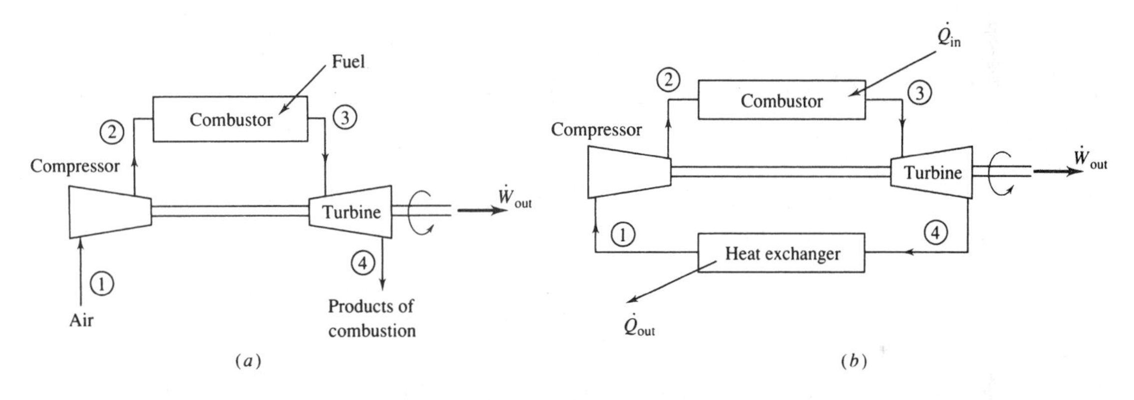

Figure 7.4 The Brayton cycle components. (*a*) Open cycle. (*b*) Closed cycle.

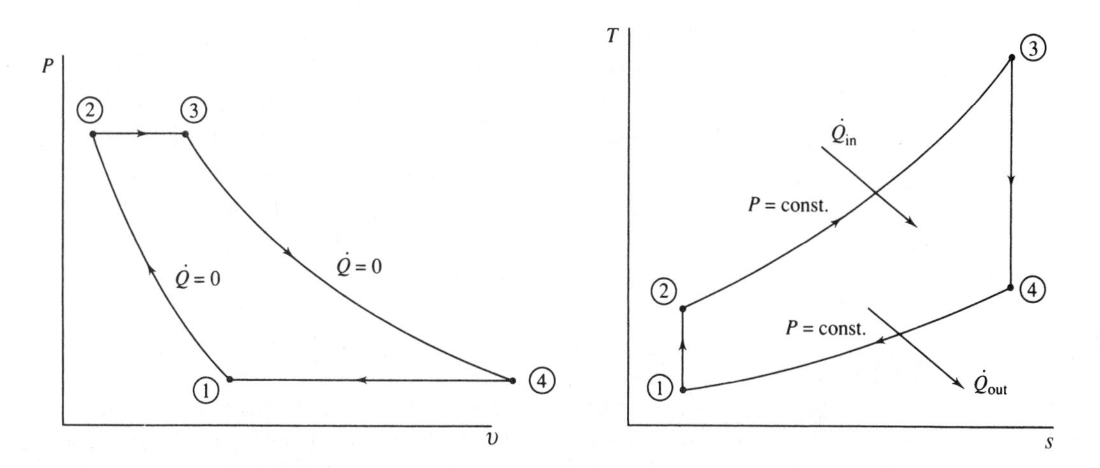

Figure 7.5 The Brayton cycle.

Using the isentropic relations

$$\frac{P_2}{P_1} = \left(\frac{T_2}{T_1}\right)^{k/(k-1)} \qquad \frac{P_3}{P_4} = \left(\frac{T_3}{T_4}\right)^{k/(k-1)} \tag{7.17}$$

and observing that $P_2 = P_3$ and $P_1 = P_4$, we see that

$$\frac{T_2}{T_1} = \frac{T_3}{T_4} \quad \text{or} \quad \frac{T_4}{T_1} = \frac{T_3}{T_2} \tag{7.18}$$

Hence, the thermal efficiency can be written as

$$\eta = 1 - \frac{T_1}{T_2} = 1 - \left(\frac{P_1}{P_2}\right)^{(k-1)/k} \tag{7.19}$$

In terms of the pressure ratio $r_p = P_2/P_1$, the thermal efficiency is

$$\eta = 1 - r_p^{(1-k)/k} \tag{7.20}$$

This expression for thermal efficiency was obtained using constant specific heats. For more accurate calculations the gas tables from App. E should be used.

In an actual gas turbine the compressor and the turbine are not isentropic; some losses do occur. These losses, usually in the neighborhood of 15 percent, significantly reduce the efficiency of the gas-turbine engine.

Another important feature of the gas turbine that seriously limits thermal efficiency is the high work requirement of the compressor, measured by the *back work ratio*, BWR = $\dot{W}_{comp} / \dot{W}_{turb}$. The compressor may require up to 80 percent of the turbine's output (a back work ratio of 0.8), leaving only 20 percent for net work output. This relatively high limit is experienced when the efficiencies of the compressor and turbine are too low. Solved problems illustrate this point.

EXAMPLE 7.5

Air enters the compressor of a gas turbine at 100 kPa and 25°C. For a pressure ratio of 5 and a maximum temperature of 850°C, determine the back work ratio (BWR) and the thermal efficiency for this Brayton cycle. Assume constant specific heat.

Solution

To find the back work ratio we observe that

$$\text{BWR} = \frac{w_{\text{comp}}}{w_{\text{turb}}} = \frac{C_p(T_2 - T_1)}{C_p(T_3 - T_4)} = \frac{T_2 - T_1}{T_3 - T_4}$$

The temperatures are $T_1 = 298$ K, $T_3 = 1123$ K, and

$$T_2 = T_1\left(\frac{P_2}{P_1}\right)^{(k-1)/k} = 298 \times 5^{0.2857} = 472$$

$$T_4 = T_3\left(\frac{P_4}{P_3}\right)^{(k-1)/k} = 1123 \times 5^{-0.2857} = 709 \text{ K}$$

The back work ratio is then

$$\text{BWR} = \frac{w_{\text{comp}}}{w_{\text{turb}}} = \frac{472.0 - 298}{1123 - 709.1} = 0.420 \qquad \text{or} \qquad 42.0\%$$

The thermal efficiency is

$$\eta = 1 - r^{(1-k)/k} = 1 - 5^{-0.2857} = 0.369 \qquad \text{or} \qquad 36.9\%$$

EXAMPLE 7.6

Assume the compressor and the gas turbine in Example 7.5 each have an efficiency of 75 percent. Determine the back work ratio (BWR) and the thermal efficiency for the Brayton cycle, assuming constant specific heats.

Solution

We can calculate the desired quantities if we determine w_{comp}, w_{turb}, and q_{in}. The compressor work is

$$w_{\text{comp}} = \frac{w_{s,\text{comp}}}{\eta_{\text{comp}}} = \frac{C_p(T_{2'} - T_1)}{\eta_{\text{comp}}}$$

where $w_{s,\text{comp}}$ is the isentropic work and $T_{2'}$ is the temperature of state 2′ assuming an isentropic process where state 2 is the actual state. We then have, using $T_{2'} = T_2$ from Example 7.5,

$$w_{\text{comp}} = \frac{1.00}{0.75}(472 - 298) = 232 \text{ kJ/kg}$$

Likewise, there results

$$w_{turb} = \eta_{turb} w_{s,turb} = \eta_{turb} C_p (T_3 - T_{4'})$$
$$= 0.75 \times 1.0 \times (1123 - 709.1) = 310.4 \text{ kJ/kg}$$

where $T_{4'}$ is T_4 as calculated in Example 7.5. State 4 is the actual state and state 4′ is the isentropic state. The back work ratio is then

$$\text{BWR} = \frac{w_{comp}}{w_{turb}} = \frac{232}{310.4} = 0.747 \qquad \text{or} \qquad 74.7\%$$

The heat transfer input necessary in this cycle is $q_{in} = h_3 - h_2 = C_p(T_3 - T_2)$, where T_2 is the actual temperature of the air leaving the compressor. It is found by returning to the compressor:

$$w_{comp} = C_p (T_2 - T_1) \qquad 232 = 1.00 \times (T_2 - 298) \qquad \therefore T_2 = 530 \text{ K}$$

Thus,

$$q_{in} = 1.00 \times (1123 - 530) = 593 \text{ kJ/kg}$$

The thermal efficiency of the cycle can then be written as

$$\eta = \frac{w_{net}}{q_{in}} = \frac{w_{turb} - w_{comp}}{q_{in}} = \frac{310.4 - 232}{593} = 0.132 \qquad \text{or} \qquad 13.2\%$$

It is obvious why the Brayton cycle cannot operate efficiently with relatively low efficiencies of the compressor and turbine. Efficiencies had to be raised into the 90's before the Brayton cycle became a cycle that could effectively be used to produce power.

7.6 The Regenerative Brayton Cycle

The heat transfer from the simple gas-turbine cycle of the previous section is simply lost to the surroundings—either directly, with the products of combustion—or from a heat exchanger. Some of this exit energy can be utilized since the temperature

of the flow exiting the turbine is greater than the temperature of the flow entering the compressor. A counterflow heat exchanger, a *regenerator*, is used to transfer some of this energy to the air leaving the compressor, as shown in Fig. 7.6. For an ideal regenerator the exit temperature T_3 would equal the entering temperature T_5; and, similarly, T_2 would equal T_6. Since less energy is rejected from the cycle, the thermal efficiency is expected to increase. It is given by

$$\eta = \frac{w_{\text{turb}} - w_{\text{comp}}}{q_{\text{in}}} \tag{7.21}$$

Using the first law, expressions for q_{in} and w_{turb} are found to be

$$q_{\text{in}} = C_p(T_4 - T_3) \qquad w_{\text{turb}} = C_p(T_4 - T_5) \tag{7.22}$$

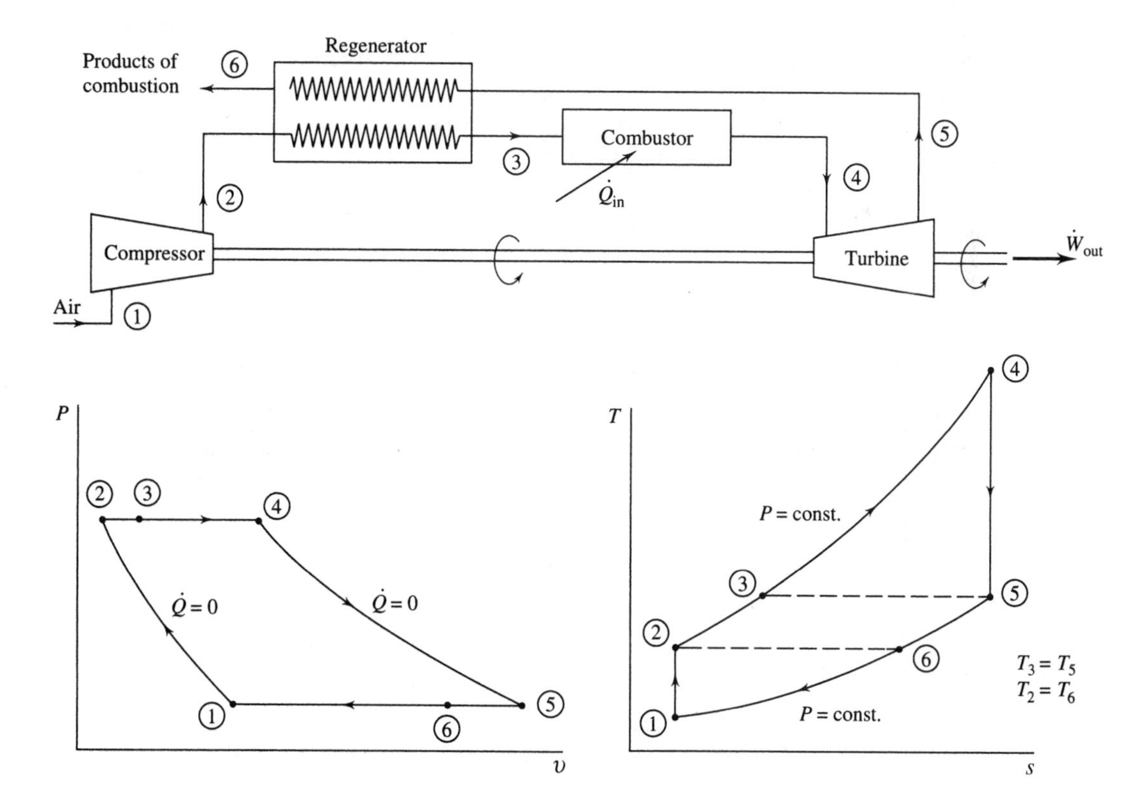

Figure 7.6 The regenerative Brayton cycle.

Hence, for the ideal regenerator in which $T_3 = T_5$ and $q_{in} = w_{turb}$, the thermal efficiency can be written as

$$\eta = 1 - \frac{w_{comp}}{w_{turb}} = 1 - \frac{T_2 - T_1}{T_4 - T_5} = 1 - \frac{T_1}{T_4}\left(\frac{T_2/T_1 - 1}{1 - T_5/T_4}\right) \tag{7.23}$$

Using the appropriate isentropic relation, after some algebra, this can be written in the form

$$\eta = 1 - \frac{T_1}{T_4} r_p^{(k-1)/k} \tag{7.24}$$

Note that this expression for thermal efficiency is quite different from that for the Brayton cycle. For a given pressure ratio, the efficiency increases as the ratio of minimum to maximum temperature decreases. But, perhaps more surprisingly, as the pressure ratio increases the efficiency decreases, an effect opposite to that of the Brayton cycle. Hence it is not surprising that for a given regenerative cycle temperature ratio, there is a particular pressure ratio for which the efficiency of the Brayton cycle will equal the efficiency of the regenerative cycle. This is shown for a temperature ratio of 0.25 in Fig. 7.7.

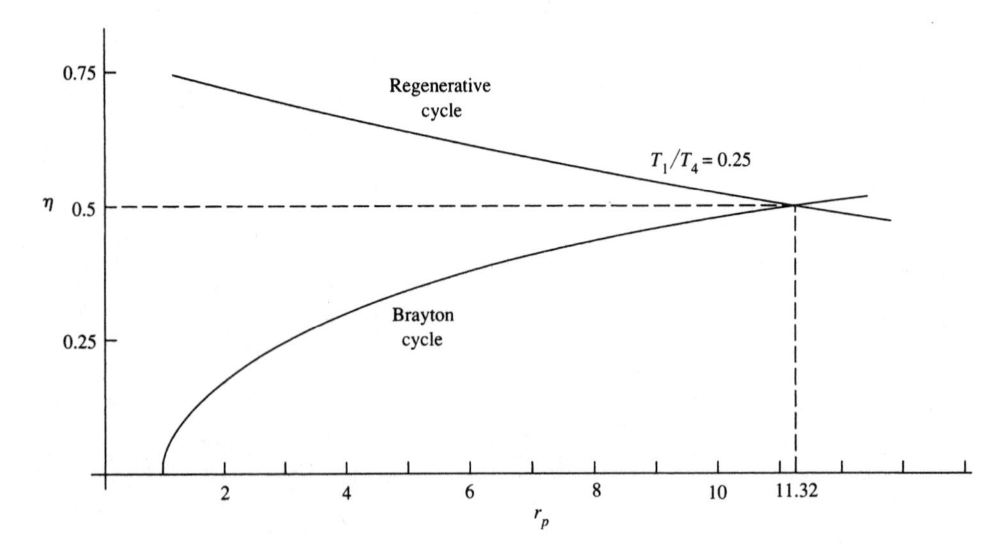

Figure 7.7 Efficiencies of the Brayton and regenerative cycles.

In practice the temperature of the air leaving the regenerator at state 3 must be less than the temperature of the air entering at state 5. Also, $T_6 > T_2$. The effectiveness, or efficiency, of a regenerator is measured by

$$\eta_{\text{reg}} = \frac{h_3 - h_2}{h_5 - h_2} \tag{7.25}$$

This is equivalent to

$$\eta_{\text{reg}} = \frac{T_3 - T_2}{T_5 - T_2} \tag{7.26}$$

if we assume an ideal gas with constant specific heats. For the ideal regenerator $T_3 = T_5$ and $\eta_{\text{reg}} = 1$. Regenerator efficiencies exceeding 80 percent are common.

EXAMPLE 7.7

Add an ideal regenerator to the gas-turbine cycle of Example 7.5 and calculate the thermal efficiency and the back work ratio, assuming constant specific heats.

Solution

The thermal efficiency is found using Eq. (7.24):

$$\eta = 1 - \frac{T_1}{T_4}\left(\frac{P_2}{P_1}\right)^{(k-1)/k} = 1 - \frac{298}{1123} \times 5^{0.2857} = 0.58 \qquad \text{or} \qquad 58\%$$

This represents a 57 percent increase in efficiency, a rather large effect. Note that, for the information given, the back work ratio does not change; hence, $\text{BWR} = w_{\text{comp}}/w_{\text{turb}} = 0.420$.

7.7 The Combined Cycle

The Brayton cycle efficiency is quite low primarily because a substantial amount of the energy input is exhausted to the surroundings. This exhausted energy is usually at a relatively high temperature and thus it can be used effectively to produce power. One possible application is the *combined Brayton-Rankine cycle* in which the high-temperature exhaust gases exiting the gas turbine are used to supply energy to the boiler of the Rankine cycle, as illustrated in Fig. 7.8. Note that the temperature T_9 of the Brayton cycle gases exiting the boiler is less than the temperature T_3 of the

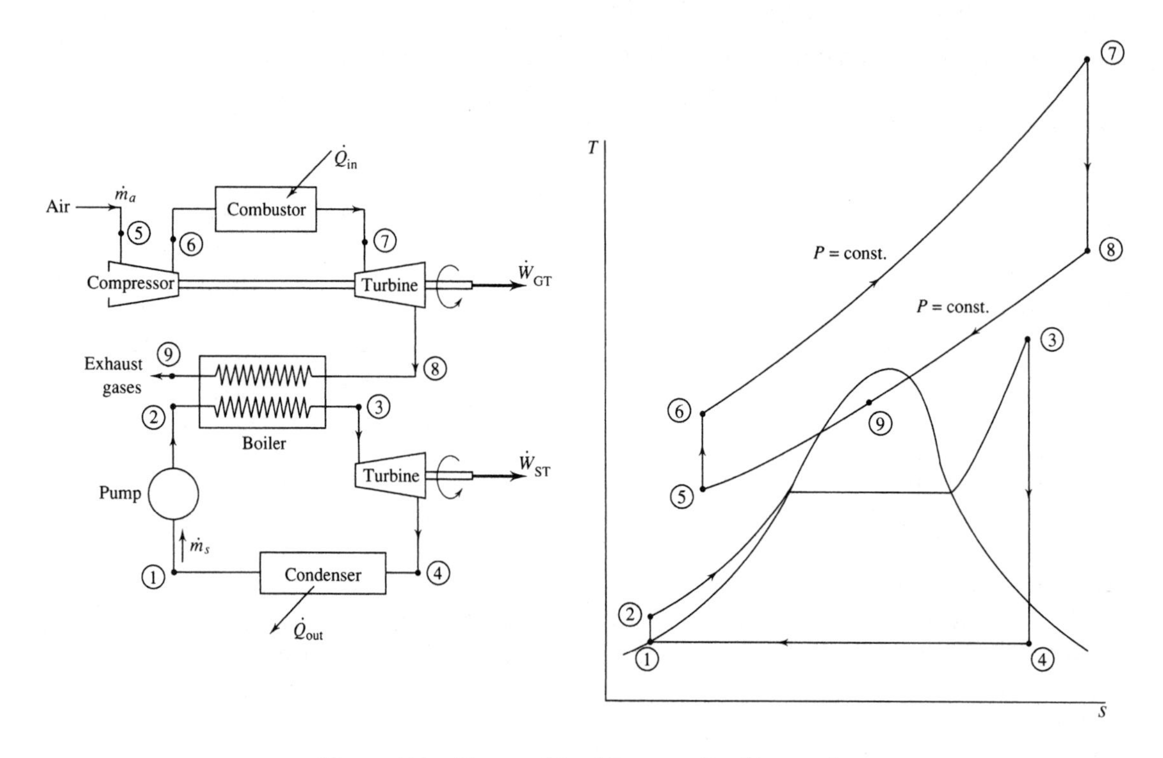

Figure 7.8 The combined Brayton-Rankine cycle.

Rankine cycle steam exiting the boiler; this is possible in the counterflow heat exchanger, the boiler.

To relate the air mass flow rate $\dot{m}_a$ of the Brayton cycle to the steam mass flow rate $\dot{m}_s$ of the Rankine cycle, we use an energy balance in the boiler; it gives (see Fig. 7.8),

$$\dot{m}_a(h_8 - h_9) = \dot{m}_s(h_3 - h_2) \tag{7.27}$$

assuming no additional energy addition in the boiler, which would be possible with an oil burner, for example. The cycle efficiency would be found by considering the purchased energy as $\dot{Q}_{in}$, the energy input in the combustor. The output is the sum of the net output $\dot{W}_{GT}$ from the gas turbine and the output $\dot{W}_{ST}$ from the steam turbine. The combined cycle efficiency is thus given by

$$\eta = \frac{\dot{W}_{GT} + \dot{W}_{ST}}{\dot{Q}_{in}} \tag{7.28}$$

An example will illustrate the increase in efficiency of such a combined cycle.

EXAMPLE 7.8

A simple steam power plant operates between pressures of 10 kPa and 4 MPa with a maximum temperature of 400°C. The power output from the steam turbine is 100 MW. A gas turbine provides the energy to the boiler; it accepts air at 100 kPa and 25°C, and has a pressure ratio of 5 and a maximum temperature of 850°C. The exhaust gases exit the boiler at 350 K. Determine the thermal efficiency of the combined Brayton-Rankine cycle. Refer to Fig. 7.8.

Solution

If we neglect the work of the pump, the enthalpy remains unchanged across the pump. Hence, $h_2 = h_1 = 192$ kJ/kg. At 400°C and 4 MPa we have $h_3 = 3214$ kJ/kg and $s_3 = 6.769$ kJ/kg · K. State 4 is located by noting that $s_4 = s_3$ so that the quality is

$$x_4 = \frac{s_4 - s_f}{s_{fg}} = \frac{6.768 - 0.6491}{7.5019} = 0.8156$$

$$\therefore h_4 = h_f + x_4 h_{fg} = 192 + 0.8156 \times 2393 = 2144 \text{ kJ/kg}$$

The steam mass flow rate is found using the turbine output as follows:

$$\dot{W}_{\text{ST}} = \dot{m}_s (h_3 - h_4) \qquad 100\,000 = \dot{m}_s (3214 - 2144)$$

$$\therefore \dot{m}_s = 93.46 \text{ kg/s}$$

Considering the gas-turbine cycle,

$$T_6 = T_5 \left(\frac{P_6}{P_5} \right)^{(k-1)/k} = 298 \times 5^{0.2857} = 472.0 \text{ K}$$

$$T_8 = T_7 \left(\frac{P_8}{P_7} \right)^{(k-1)/k} = 1123 \times 5^{-0.2857} = 709.1 \text{ K}$$

Thus, we have, for the boiler,

$$\dot{m}_s (h_3 - h_2) = \dot{m}_a c_p (T_8 - T_9)$$

$$93.46(3214 - 192) = \dot{m}_a \times 1.00(709.1 - 350)$$

$$\therefore \dot{m}_a = 786.5 \text{ kg/s}$$

The power produced by the gas turbine is (note that this is not $\dot{W}_{\text{GT}}$)

$$\dot{W}_{\text{turb}} = \dot{m}_a C_p (T_7 - T_8) = 786.5 \times 1.00(1123 - 709.1) = 325\,500 \text{ kW}$$

The energy needed by the compressor is

$$\dot{W}_{\text{comp}} = \dot{m}_a C_p (T_6 - T_5) = 786.5 \times 1.00(472 - 298) = 136\,900 \text{ kW}$$

Hence, the net gas-turbine output is

$$\dot{W}_{\text{GT}} = \dot{W}_{\text{turb}} - \dot{W}_{\text{comp}} = 325.5 - 136.9 = 188.6 \text{ MW}$$

The energy input by the combustor is

$$\dot{Q}_{\text{in}} = \dot{m}_a C_p (T_7 - T_6) = 786.5 \times 1.00(1123 - 472) = 512\,000 \text{ kW}$$

The above calculations allow us to determine the combined cycle efficiency as

$$\eta = \frac{\dot{W}_{\text{ST}} + \dot{W}_{\text{GT}}}{\dot{Q}_{\text{in}}} = \frac{100 + 188.6}{512} = 0.564 \quad \text{or} \quad 56.4\%$$

Note that this efficiency is 59 percent higher than the Rankine cycle (see Example 6.2) and 53 percent higher than the Brayton cycle (see Example 7.5). Using steam reheaters, steam regenerators, gas intercoolers, and gas reheaters could increase cycle efficiency even more.

7.8 The Gas Refrigeration Cycle

If the flow of the gas is reversed in the Brayton cycle of Fig. 7.5, the gas undergoes an isentropic expansion process as it flows through the turbine, resulting in a substantial reduction in temperature, as shown in Fig. 7.9. The gas with low turbine exit temperature can be used to refrigerate a space to temperature T_2 by extracting heat at rate $\dot{Q}_{\text{in}}$ from the refrigerated space.

Figure 7.9 illustrates a *closed* refrigeration cycle. (An *open* cycle system is used in aircraft; air is extracted from the atmosphere at state 2 and inserted into the passenger compartment at state 1. This provides both fresh air and cooling.) An additional heat exchanger may be used, like the regenerator of the Brayton power cycle, to provide an even lower turbine exit temperature, as illustrated in Example 7.10. The gas does not enter the expansion process (the turbine) at state 5; rather, it passes through an *internal* heat exchanger (it does not exchange heat with the surroundings). This allows the temperature of the gas entering the

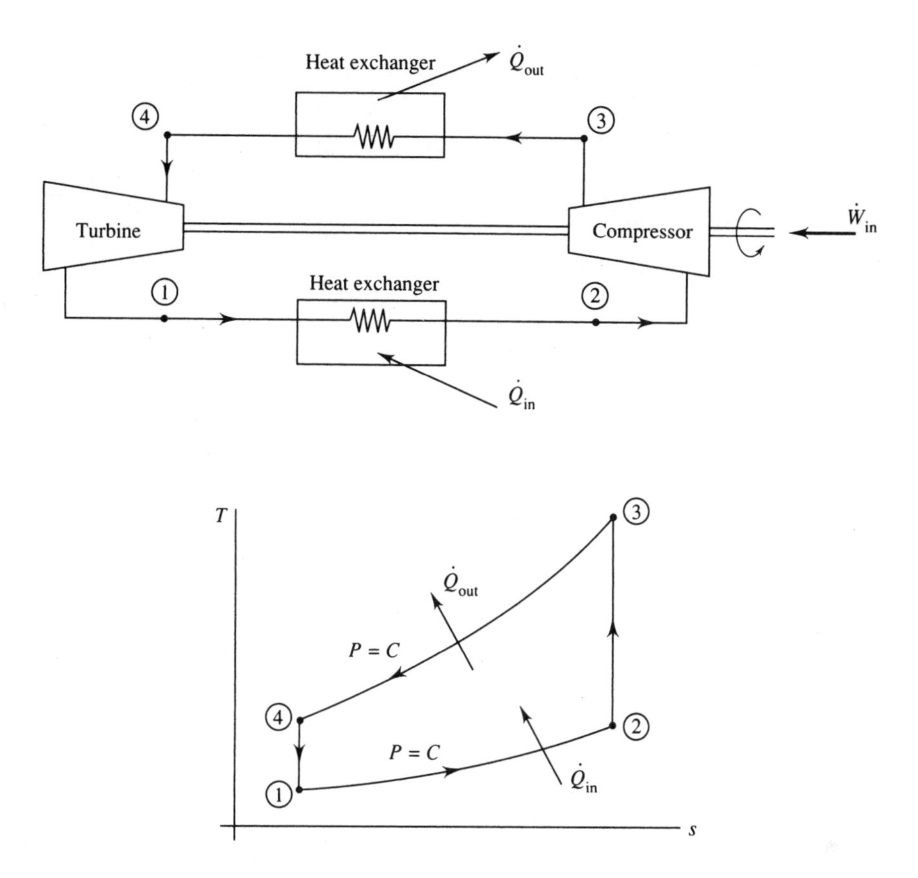

Figure 7.9 The gas refrigeration cycle.

turbine to be much lower than that of Fig. 7.9. The temperature T_1 after the expansion is so low that gas liquefaction is possible. It should be noted, however, that the coefficient of performance is actually reduced by the inclusion of an internal heat exchanger.

A reminder: when the purpose of a thermodynamic cycle is to cool or heat a space, we do not define a cycle's efficiency; rather, we define its coefficient of performance:

$$\text{COP} = \frac{\text{desired effect}}{\text{energy purchased}} = \frac{\dot{Q}_{in}}{\dot{W}_{in}} \tag{7.29}$$

where $\dot{W}_{in} = \dot{m}(h_{comp} - h_{turb})$.

EXAMPLE 7.9

Air enters the compressor of a simple gas refrigeration cycle at −10°C and 100 kPa. For a compression ratio of 10 and a turbine inlet temperature of 30°C, calculate the minimum cycle temperature and the coefficient of performance using the ideal-gas equations of this section.

Solution

Assuming isentropic compression and expansion processes we find

$$T_3 = T_2\left(\frac{P_3}{P_2}\right)^{(k-1)/k} = 263\times 10^{0.2857} = 508\text{ K}$$

$$T_1 = T_4\left(\frac{P_1}{P_4}\right)^{(k-1)/k} = 303\times 10^{-0.2857} = 157\text{ K}$$

$$= -116°\text{C}$$

The COP is now calculated as follows:

$$q_{\text{in}} = C_p(T_2 - T_1) = 1.00(263-157) = 106\text{ kJ/kg}$$

$$w_{\text{comp}} = C_p(T_3 - T_2) = 1.00(508-263) = 245\text{ kJ/kg}$$

$$w_{\text{turb}} = C_p(T_4 - T_1) = 1.00(303-157) = 146\text{ kJ/kg}$$

$$\therefore \text{COP} = \frac{q_{\text{in}}}{w_{\text{comp}} - w_{\text{turb}}} = \frac{106}{245-146} = 1.07$$

This coefficient of performance is quite low when compared with that of a vapor refrigeration cycle. Thus, gas refrigeration cycles are used only for special applications.

EXAMPLE 7.10

Use the given information for the compressor of the refrigeration cycle of Example 7.9 but add an ideal internal heat exchanger, a regenerator, as illustrated below, so that the air temperature entering the turbine is –40°C. Calculate the minimum cycle temperature and the coefficient of performance.

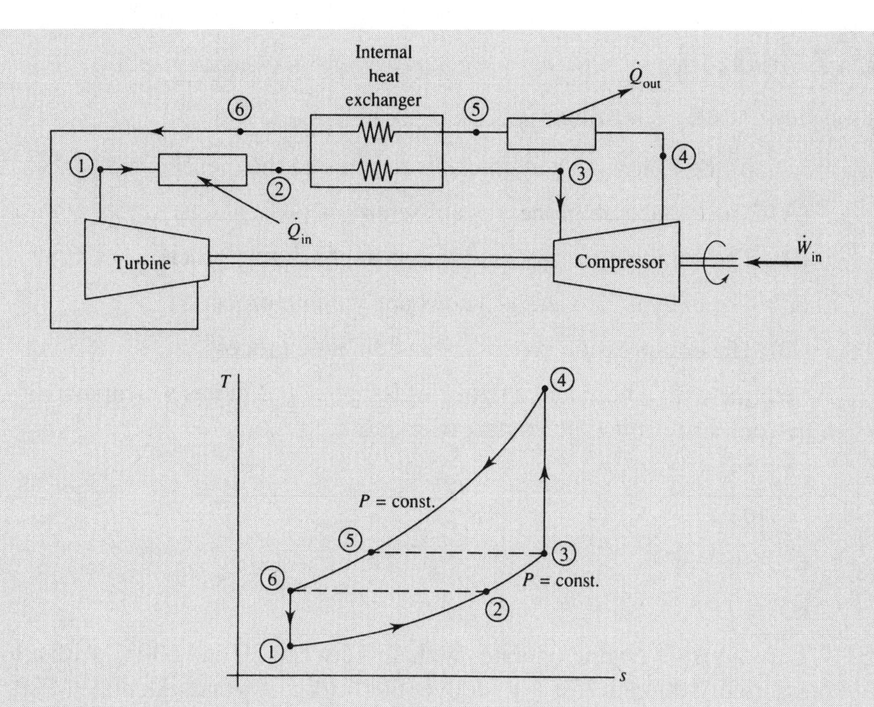

Solution

Assuming isentropic compression we again have

$$T_4 = T_3\left(\frac{P_4}{P_3}\right)^{(k-1)/k} = 263\times 10^{0.2857} = 508\ \text{K}$$

For an ideal internal heat exchanger we would have $T_5 = T_3 = 263$ K and $T_6 = T_2 = 233$ K. The minimum cycle temperature is

$$T_1 = T_6\left(\frac{P_1}{P_6}\right)^{(k-1)/k} = 233\times 10^{-0.2857} = 121\ \text{K} = -152°\text{C}$$

The COP follows:

$$q_{in} = C_p(T_2 - T_1) = 1.00(233-121) = 112\ \text{kJ/kg}$$

$$w_{comp} = C_p(T_4 - T_3) = 1.00(508-263) = 245\ \text{kJ/kg}$$

$$w_{turb} = C_p(T_6 - T_1) = 1.00(233-121) = 112\ \text{kJ/kg}$$

$$\therefore \text{COP} = \frac{q_{in}}{w_{comp} - w_{turb}} = \frac{112}{245-112} = 0.842$$

Obviously, the COP is quite low, but so is the refrigeration temperature.

Quiz No. 1

(Assume constant specific heats)

1. Which of the following statements is not true of the diesel cycle?
   (A) The expansion process is an isentropic process.
   (B) The combustion process is a constant-volume process.
   (C) The exhaust process is a constant-volume process.
   (D) The compression process is an adiabatic process.

2. An engine with a bore and a stroke of 0.2 m × 0.2 m has a clearance of 5 percent. Determine the compression ratio.
   (A) 23
   (B) 21
   (C) 19
   (D) 17

3. A Carnot piston engine operates with air between 20 and 600°C with a low pressure of 100 kPa. If it is to deliver 800 kJ/kg of work, calculate MEP. (See Figs. 5.5 and 5.6.)
   (A) 484 kPa
   (B) 374 kPa
   (C) 299 kPa
   (D) 243 kPa

4. The heat rejected in the Otto cycle of Fig. 7.2 is
   (A) $C_p(T_4 - T_1)$
   (B) $C_p(T_3 - T_4)$
   (C) $C_v(T_4 - T_1)$
   (D) $C_v(T_3 - T_4)$

5. The maximum allowable pressure in an Otto cycle is 8 MPa. Conditions at the beginning of the air compression are 85 kPa and 22°C. Calculate the required heat addition if the compression ratio is 8.
   (A) 2000 kJ/kg
   (B) 2400 kJ/kg
   (C) 2800 kJ/kg
   (D) 3200 kJ/kg

6. The MEP for the Otto cycle of Prob. 5 is nearest
   (A) 900 kPa
   (B) 1100 kPa
   (C) 1300 kPa
   (D) 1500 kPa
7. A diesel cycle operates on air which enters the compression process at 85 kPa and 30°C. If the compression ratio is 16 and the maximum temperature is 2000°C, the cutoff ratio is nearest
   (A) 2.47
   (B) 2.29
   (C) 2.04
   (D) 1.98
8. If the power output of the diesel cycle of Prob. 7 is 500 hp, the mass flow rate of air is nearest
   (A) 0.532 kg/s
   (B) 0.467 kg/s
   (C) 0.431 kg/s
   (D) 0.386 kg/s
9. Air enters the compressor of a gas turbine at 85 kPa and 0°C. If the pressure ratio is 6 and the maximum temperature is 1000°C, using the ideal-gas equations, the thermal efficiency is nearest
   (A) 52%
   (B) 48%
   (C) 44%
   (D) 40%
10. Using the ideal-gas law, the back work ratio for the Brayton cycle of Prob. 9 is nearest
    (A) 0.305
    (B) 0.329
    (C) 0.358
    (D) 0.394

11. For the ideal-gas turbine with regenerator shown below, $\dot{W}_{out}$ is nearest

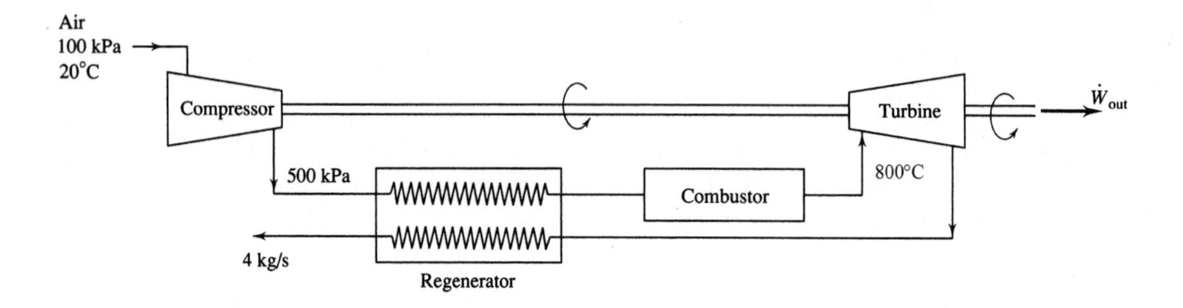

(A) 950 kW
(B) 900 kW
(C) 850 kW
(D) 800 kW

12. The back work ratio for the cycle of Prob. 11 is nearest
(A) 0.432
(B) 0.418
(C) 0.393
(D) 0.341

13. Air enters the compressor of an ideal-gas refrigeration cycle at 10°C and 80 kPa. If the maximum and minimum temperatures are 250 and –50°C, the compressor work is nearest
(A) 170 kJ/kg
(B) 190 kJ/kg
(C) 220 kJ/kg
(D) 240 kJ/kg

14. The pressure ratio across the compressor in the cycle of Prob. 13 is nearest
(A) 7.6
(B) 8.2
(C) 8.6
(D) 8.9

15. The COP for the refrigeration cycle of Prob. 13 is nearest
    (A) 1.04
    (B) 1.18
    (C) 1.22
    (D) 1.49

Quiz No. 2

(Assume constant specific heats)

1. The exhaust process in the Otto and diesel cycles is replaced with a constant-volume process for what primary reason?
    (A) To simulate the zero work of the actual exhaust process.
    (B) To simulate the zero heat transfer of the actual process.
    (C) To restore the air to its original state.
    (D) To ensure that the first law is satisfied.
2. An air-standard cycle operates in a piston-cylinder arrangement with the following four processes: $1 \to 2$: isentropic compression from 100 kPa and 15°C to 2 MPa; $2 \to 3$: constant-pressure heat addition to 1200°C; $3 \to 4$: isentropic expansion; and $4 \to 1$: constant-volume heat rejection. Calculate the heat addition.
    (A) 433 kJ/kg
    (B) 487 kJ/kg
    (C) 506 kJ/kg
    (D) 522 kJ/kg
3. A Carnot piston engine operates on air between high and low pressures of 3 MPa and 100 kPa with a low temperature of 20°C. For a compression ratio of 15, calculate the thermal efficiency. (See Figs. 5.5 and 5.6.)
    (A) 40%
    (B) 50%
    (C) 60%
    (D) 70%

4. A spark-ignition engine operates on an Otto cycle with a compression ratio of 9 and temperature limits of 30 and 1000°C. If the power output is 500 kW, the thermal efficiency is nearest

   (A) 50%

   (B) 54%

   (C) 58%

   (D) 64%

5. The mass flow rate of air required for the Otto cycle of Prob. 4 is nearest

   (A) 1.6 kg/s

   (B) 1.8 kg/s

   (C) 2.2 kg/s

   (D) 2.0 kg/s

6. A diesel engine is designed to operate with a compression ratio of 16 and air entering the compression stroke at 110 kPa and 20°C. If the energy added during combustion is 1800 kJ/kg, calculate the cutoff ratio.

   (A) 3.2

   (B) 3.0

   (C) 2.8

   (D) 2.6

7. The MEP of the diesel cycle of Prob. 6 is nearest

   (A) 1430 kPa

   (B) 1290 kPa

   (C) 1120 kPa

   (D) 1080 kPa

8. Air enters the compressor of a Brayton cycle at 80 kPa and 30°C and compresses it to 500 kPa. If 1800 kJ/kg of energy is added in the combustor, calculate the compressor work requirement.

   (A) 304 kJ/kg

   (B) 286 kJ/kg

   (C) 232 kJ/kg

   (D) 208 kJ/kg

9. The new output of the turbine of Prob. 8 is nearest
   (A) 874 kJ/kg
   (B) 826 kJ/kg
   (C) 776 kJ/kg
   (D) 734 kJ/kg
10. The BWR of the Brayton cycle of Prob. 8 is nearest
    (A) 0.22
    (B) 0.24
    (C) 0.26
    (D) 0.28
11. A regenerator is installed in the gas turbine of Prob. 8. Determine the cycle efficiency if its effectiveness is 100 percent.
    (A) 88%
    (B) 85%
    (C) 82%
    (D) 79%
12. The engines on a commercial jet aircraft operate on which of the basic cycles?
    (A) Otto
    (B) Diesel
    (C) Carnot
    (D) Brayton
13. Air flows at the rate of 2.0 kg/s through the compressor of an ideal-gas refrigeration cycle where the pressure increases to 500 kPa from 100 kPa. The maximum and minimum cycle temperatures are 300 and −20°C, respectively. Calculate the power needed to drive the compressor fluid using the ideal-gas equations.
    (A) 198 kW
    (B) 145 kW
    (C) 126 kW
    (D) 108 kW

14. The COP of the refrigeration cycle of Prob. 13 is nearest
    (A) 2.73
    (B) 2.43
    (C) 1.96
    (D) 1.73

CHAPTER 8

Psychrometrics

Thus far in our study of thermodynamics we have considered only single-component systems. In this chapter we will study the two-component system of air and water vapor, generally called *psychrometrics*. The more general presentation in which several components make up the system is not considered in this text. The two-component mixture of air and water vapor must be considered when designing air-conditioning systems, whether they are for heating, for cooling, or simply for dehumidification. A cooling tower used to remove heat from a power plant is an example of an air-conditioning system as is the usual heating and cooling system in an office building.

8.1 Gas-Vapor Mixtures

Air is a mixture of nitrogen, oxygen, and argon plus traces of some other gases. When water vapor is not included, we refer to it as *dry air*. If water vapor is included, as in *atmospheric air*, we must be careful to properly account for it. At the relatively low atmospheric temperature we can treat dry air as an ideal gas with constant specific heats. It is also possible to treat the water vapor in the air as an ideal gas, even though the water vapor may be at the saturation state. Consequently, we can consider atmospheric air to be a mixture of two ideal gases.

Two models are used to study a mixture of ideal gases. The *Amagat model* treats each component as though it exists separately at the same pressure and temperature of the mixture; the total volume is the sum of the volumes of the components. In this chapter we use the *Dalton model* in which each component occupies the same volume and has the same temperature as the mixture; the total pressure is the sum of the component pressures, termed the *partial pressures*. For the Dalton model

$$P = P_a + P_v \tag{8.1}$$

where the total pressure P is the sum of the partial pressure P_a of the dry air and the partial pressure P_v of the water vapor (called the *vapor pressure*). We will assume the total pressure to be 100 kPa, unless stated otherwise.

To show that the water vapor in the atmosphere can be treated as an ideal gas, consider an atmosphere at 20°C. From Table C.1 at 20°C we see that $P_v = 2.338$ kPa, the pressure needed to vaporize water at 20°C. The ideal-gas law gives

$$v = \frac{RT}{P_v} = \frac{0.4165 \times 293}{2.338} = 57.8 \text{ m}^3\text{/kg} \tag{8.2}$$

Observe that this is essentially the same value found in Table C.1 for v_g at 20°C. Hence, we conclude that water vapor found in the atmosphere can be treated as an ideal gas.

Since we assume that the water vapor is an ideal gas, its enthalpy is dependent on temperature only. Hence we use the enthalpy of the water vapor to be the enthalpy of saturated water vapor at the temperature of the air, expressed as

$$h_v(T) = h_g(T) \tag{8.3}$$

In Fig. 8.1 this means that $h_1 = h_2$, where $h_2 = h_g$ from the steam tables at $T = T_1$.

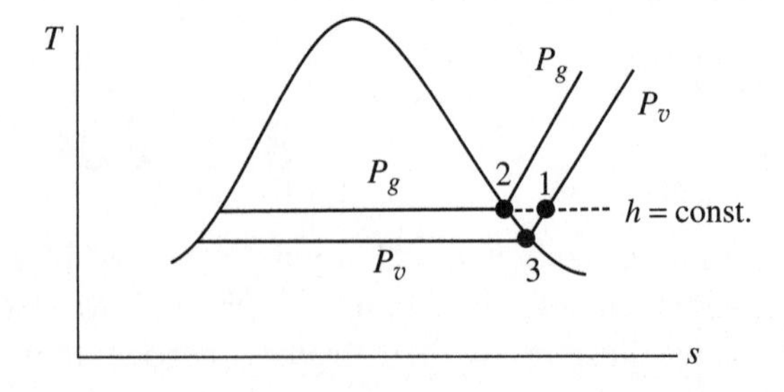

Figure 8.1 *T-s* diagram for the water vapor component.

This is acceptable for situations in which the total pressure is relatively low (near atmospheric pressure) and the temperature is below about 60°C.

The amount of water vapor in the air of state 1 in Fig. 8.1 is related to the relative humidity and the humidity ratio. The *relative humidity* ϕ is defined as the ratio of the mass of the water vapor m_v to the maximum amount of water vapor m_g the air can hold at the same temperature:

$$\phi = \frac{m_v}{m_g} \tag{8.4}$$

Using the ideal-gas law we find

$$\phi = \frac{P_v V/RT}{P_g V/RT} = \frac{P_v}{P_g} \tag{8.5}$$

where the constant-pressure lines for P_v and P_g are shown in Fig. 8.1.

The *humidity ratio* ω (also referred to as *specific humidity*) is the ratio of the mass of water vapor to the mass of dry air:

$$\omega = \frac{m_v}{m_a} \tag{8.6}$$

Using the ideal-gas law for air and water vapor, this becomes

$$\omega = \frac{P_v V/R_v T}{P_a V/R_a T} = \frac{P_v/R_v}{P_g/R_a} = \frac{P_v/0.4615}{P_a/0.287} = 0.622\frac{P_v}{P_a} \tag{8.7}$$

Combining Eqs. (8.5) and (8.7), we relate ϕ and ω by

$$\omega = 0.622\frac{P_g}{P_a}\phi \qquad \phi = 1.608\frac{P_a}{P_g}\omega \tag{8.8}$$

Note that at state 3 in Fig. 8.1 the relative humidity is 1.0 (100%). Also note that for a given mass of water vapor in the air, ω remains constant but ϕ varies depending on the temperature.

When making an energy balance in an air-conditioning system we must use enthalpy, as with all control volumes involving gases. The total enthalpy is the sum of the enthalpy of the air and the enthalpy of the water vapor:

$$\begin{aligned} H &= H_a + H_v \\ &= m_a h_a + m_v h_v \end{aligned} \tag{8.9}$$

Divide by m_a and obtain

$$h = h_a + \omega h_v \tag{8.10}$$

where h is the specific enthalpy of the mixture per unit mass of dry air.

The temperature of the air as measured by a conventional thermometer is referred to as the *dry-bulb temperature* T (T_1 in Fig. 8.1). The temperature at which condensation begins if air is cooled at constant pressure is the *dew-point temperature* T_{dp} (T_3 in Fig. 8.1). If the temperature falls below the dew-point temperature, condensation occurs and the amount of water vapor in the air decreases. This may occur on a cool evening; it also may occur on the cool coils of an air conditioner, or the inside of a pane of glass when it's cold outside, or on the outside of your glass filled with ice water sitting on a table.

EXAMPLE 8.1
The air at 25°C and 100 kPa in a 150-m³ room has a relative humidity of 60%. Calculate the humidity ratio, the dew point, the mass of water vapor in the air, and the volume percentage of the room that is water vapor using the equations in this section.

Solution
By Eq. (8.5), the partial pressure of the water vapor is

$$P_v = P_g \phi = 3.29 \times 0.6 = 1.98 \text{ kPa}$$

where P_g is the saturation vapor pressure at 25°C interpolated (a straight-line method) in Table C.1. The partial pressure of the air is then

$$P_a = P - P_v = 100 - 1.98 = 98.0$$

where we have used the total pressure of the air in the room to be at 100 kPa. The humidity ratio is then

$$\omega = 0.622 \frac{P_v}{P_a} = 0.622 \times \frac{1.98}{98.0} = 0.0126 \text{ kg water/kg dry air}$$

The dew point is the temperature T_3 of Fig. 8.1 associated with the partial pressure P_v. It is found by interpolation in Table C.1 or Table C.2, whichever appears to be easier: interpolation gives $T_{\text{dp}} = 17.4$°C.

From the definition of the humidity ratio the mass of water vapor is found to be

$$m_v = \omega m_a = \omega \frac{P_a V}{R_a T} = 0.0126\left(\frac{98.0 \times 150}{0.287 \times 298}\right) = 2.17 \text{ kg}$$

We have assumed that both the dry air and the water vapor fill the entire 150 m^3. To find the volume percentage of the water vapor we must compute the moles N of each component:

$$N_v = \frac{m_v}{M_v} = \frac{2.17}{18} = 0.1203 \text{ mol}$$

$$N_a = \frac{m_a}{M_a} = \frac{98.0 \times 150 / 0.287 \times 298}{28.97} = 5.93 \text{ mol}$$

The fraction of the volume occupied by the water vapor is

$$\frac{N_v}{N_a + N_v} = \frac{0.1203}{5.93 + 0.1203} = 0.0199 \qquad \text{or} \qquad 1.99\%$$

This demonstrates that air with 60% humidity is about 2% water vapor by volume. We usually ignore this when analyzing air and consider air to be dry air as we have done in the problems up to this point. Ignoring the water vapor does not lead to significant errors in most engineering applications. It must be included, however, when considering problems involving, for example, combustion and air-conditioning.

EXAMPLE 8.2

The air in Example 8.1 is cooled below the dew point to 10°C. Estimate the percentage of water vapor that will condense. Reheat the air to 25°C and calculate the relative humidity.

Solution

At 10°C the air is saturated, with $\phi = 100\%$. In Fig. 8.1 we are at a state on the saturation line that lies below state 3. At 10°C we find from Table C.1 that $P_v =$ 1.228 kPa, so that

$$P_a = P - P_v = 100 - 1.228 = 98.77 \text{ kPa}$$

The humidity ratio is then

$$\omega = 0.622\frac{P_v}{P_a} = 0.622\frac{1.228}{98.77} = 0.00773 \text{ kg water/kg dry air}$$

The difference in the humidity ratio just calculated and the humidity ratio of Example 8.1 is $\Delta\omega = 0.0126 - 0.00773 = 0.00487$ kg water/kg dry air. The mass of water vapor removed (condensed) is found to be

$$\Delta m_v = \Delta\omega \times m_a = 0.00487\left(\frac{98.0 \times 150}{0.287 \times 298}\right) = 0.837 \text{ kg water}$$

where we have used the initial mass of dry air. The percentage that condenses is $(0.837/2.17) \times 100 = 38.6\%$.

As we reheat the air back to 25°C, the ω remains constant at 0.00773. Using Eq. (8.8), the relative humidity is then reduced to

$$\phi = 1.608\frac{\omega P_a}{P_g} = 1.608\frac{0.00773 \times 98.77}{3.29} = 0.373 \quad \text{or} \quad 37.3\%$$

where P_g is used as the saturation pressure at 25°C interpolated in Table C.l.

8.2 Adiabatic Saturation and Wet-Bulb Temperatures

It is quite difficult to measure the relative humidity ϕ and the humidity ratio ω directly, at least with any degree of accuracy. This section presents two indirect methods for accurately determining these quantities.

Consider a relatively long insulated channel, shown in Fig. 8.2; air with an unknown relative humidity enters, moisture is added to the air by the sufficiently long pool of water, and saturated air exits. This process involves no heat transfer because the channel is insulated and hence the exiting temperature is called the *adiabatic saturation temperature.* Let us find an expression for the desired humidity ratio ω_1. Consider that the liquid water added is at temperature T_2. An energy balance on this control volume, neglecting kinetic and potential energy changes, is done considering the dry air and the water-vapor components. With $Q = W = 0$ we have

$$\dot{m}_{v1}h_{v1} + \dot{m}_{a1}h_{a1} + \dot{m}_f h_{f2} = \dot{m}_{a2}h_{a2} + \dot{m}_{v2}h_{v2} \tag{8.11}$$

But, from conservation of mass for both the dry air and the water vapor,

$$\dot{m}_{a1} = \dot{m}_{a2} = \dot{m}_a \qquad \dot{m}_{v1} + \dot{m}_f = \dot{m}_{v2} \tag{8.12}$$

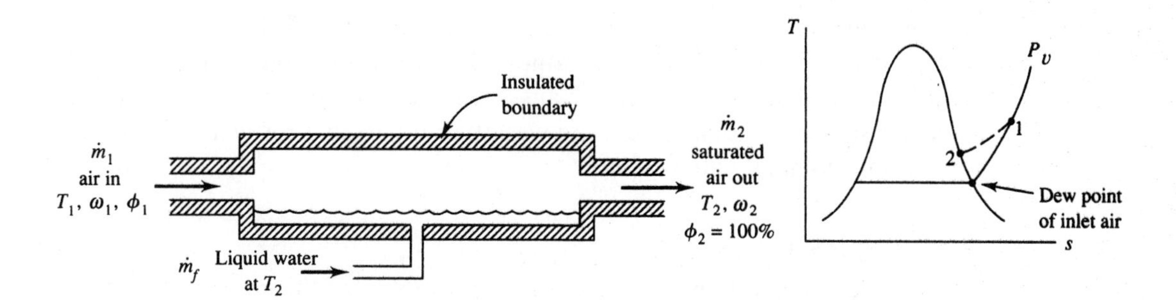

Figure 8.2 Setup used to find ω of air.

Using the definition of ω in Eq. (8.6), the above equations allow us to write

$$\dot{m}_a\omega_1 + \dot{m}_f = \omega_2\dot{m}_a \tag{8.13}$$

Substituting this into Eq. (8.11) for $\dot{m}_f$ there results, using $h_v = h_g$,

$$\dot{m}_a\omega_1 h_{g1} + \dot{m}_a h_{a1} + (\omega_2 - \omega_1)\dot{m}_a h_{f2} = \dot{m}_a h_{a2} + \omega_2\dot{m}_a h_{g2} \tag{8.14}$$

At state 2 we know that $\phi_2 = 1.0$ and, using Eq. (8.8),

$$\omega_2 = 0.622\frac{P_{g2}}{P - P_{g2}} \tag{8.15}$$

Thus, Eq. (8.14) becomes

$$\omega_1 = \frac{\omega_2 h_{fg2} + C_p(T_2 - T_1)}{h_{g1} - h_{f2}} \tag{8.16}$$

where $h_{a2} - h_{a1} = C_p(T_2 - T_1)$ for the dry air and $h_{fg2} = h_{g2} - h_{f2}$. Consequently, if we measure the temperatures T_2 and T_1 and the total pressure P, we can find ω_2 from Eq. (8.15) and then ω_1 from Eq. (8.16) since the remaining quantities are given in App. C.

Because T_2 is significantly less than T_1, the apparatus shown in Fig. 8.2 can be used to cool an air stream. This is done in relatively dry climates so that T_2 is reduced but usually not to the saturation temperature. Such a device is often referred to as a "swamp cooler." A fan blowing air through a series of wicks that stand in water is quite effective at cooling low-humidity air.

Using the device of Fig. 8.2 to obtain the adiabatic saturation temperature is a rather involved process. A much simpler approach is to wrap the bulb of a thermometer with a cotton wick saturated with water, and then either to blow air over the wick or to swing the thermometer overhead through the air until the temperature reaches a steady-state value. This *wet-bulb temperature* T_{wb} and the adiabatic saturation temperature are essentially the same for water if the pressure is approximately atmospheric.

EXAMPLE 8.3

The dry-bulb and wet-bulb temperatures of a 100-kPa air stream are measured to be 40 and 20°C, respectively. Determine the humidity ratio, the relative humidity, and the specific enthalpy of the air.

Solution

We use Eq. (8.16) to find ω_1. But first ω_2 is found using Eq. (8.15):

$$\omega_2 = 0.622\frac{P_{g2}}{P - P_{g2}} = 0.622\frac{2.338}{100 - 2.338} = 0.01489$$

where P_{g2} is the saturation pressure at 20°C. Now Eq. (8.16) gives

$$\omega_1 = \frac{\omega_2 h_{fg2} + C_p(T_2 - T_1)}{h_{g1} - h_{f2}} = \frac{0.01489 \times 2454 + 1.0(20 - 40)}{2574 - 83.9}$$
$$= 0.00664 \text{ kg water/kg dry air}$$

The partial pressure of the water vapor is found using Eq. (8.7):

$$\omega_1 = 0.622\frac{P_{v1}}{P_{a1}} \qquad 0.00664 = 0.622\frac{P_{v1}}{100 - P_{v1}} \qquad \therefore P_{v1} = 1.0567 \text{ kPa}$$

The relative humidity is obtained from Eq. (8.5):

$$\phi = \frac{P_{v1}}{P_{g1}} = \frac{1.0567}{7.383} = 0.143 \quad \text{or} \quad 14.3\%$$

The specific enthalpy is found by assuming a zero value for air at $T = 0$°C (note that Tables C.1 and C.2 select $h = 0$ at $T = 0$°C so that establishes the datum). Using Eq. (8.10) the enthalpy for the mixture is

$$h = h_a + \omega h_v = C_p T + \omega h_g = 1.0 \times 40 + 0.00664 \times 2574 = 57.1 \text{ kJ/kg dry air}$$

where we have used $h_v = h_g$ at 40°C (see Fig. 8.1). The enthalpy is always expressed per mass unit of dry air.

8.3 The Psychrometric Chart

A convenient way of relating the various properties associated with a water vapor-air mixture is to plot these quantities on a *psychrometric chart*, displayed in App. F, and shown in Fig. 8.3. Any two of the properties $(T, \phi, \omega, T_{dp}, T_{wb})$ establish a state from which the other properties are determined. As an example, consider a state A that is located by specifying the dry-bulb temperature (T_{db} or, simply, T) and the relative humidity. The wet-bulb temperature would be read at 1, the dew-point temperature at 2, the enthalpy at 3, and the humidity ratio at 4. Referring to App. F, a dry-bulb temperature of 30°C and a relative humidity of 80% would provide the following: T_{dp} = 26°C, T_{wb} = 27°C, h = 85 kJ/(kg dry air), and ω = 0.0215 kg water/(kg dry air). The chart provides us with a quick, relatively accurate method (large charts are used in the industry) for finding the quantities of interest. If the pressure is significantly different from 100 kPa, the equations presented in the preceding sections must be used.

EXAMPLE 8.4

Using App. F, rework Example 8.3 (T_{db} = 40°C, T_{wb} = 20°C) to find ω, ϕ, and h.

Solution

Using the chart, the intersection of T_{db} = 40°C and T_{wb} = 20°C gives

$$\omega = 0.0066 \text{ kg water/kg dry air} \qquad \phi = 14\% \qquad h = 57.5 \text{ kJ/kg dry air}$$

These values are less accurate than those calculated in Example 8.3, but certainly are acceptable. The simplicity of using the psychrometric chart is obvious.

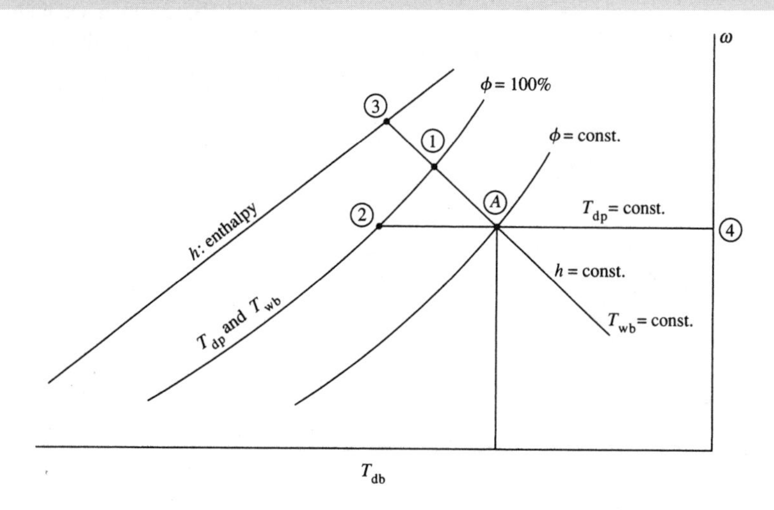

Figure 8.3 The psychrometric chart.

8.4 Air-Conditioning Processes

Generally, people feel most comfortable when the air is in the "comfort zone": the temperature is between 22°C (72°F) and 27°C (80°F) and the relative humidity is between 40 and 60%. In Fig. 8.4, the area enclosed by the dotted lines represents the comfort zone. There are several situations in which air must be conditioned to put it in the comfort zone:

- The air is too cold or too hot. Heat is simply added or extracted. This is represented by *A*-C and *B-C* in Fig. 8.4.
- The air is too cold and the humidity is too low. The air can first be heated, and then moisture added, as in *D-E-C*.
- The temperature is acceptable but the humidity is too high. The air is first cooled from *F* to *G*. Moisture is removed from *G* to *H*. Heat is added from *H* to *I*.
- The air is too hot and the humidity is low. Moisture is added, and the process represented by *J-K* results. (This is a "swamp cooler.")

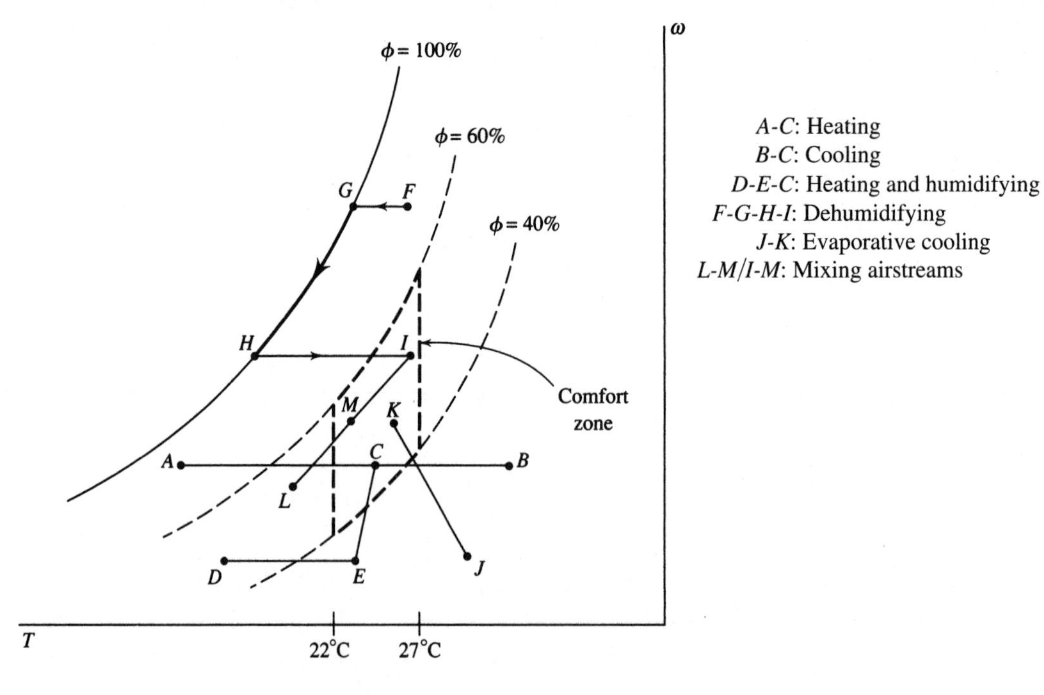

Figure 8.4 The conditioning of air.

- The air is too hot. An air stream from the outside is mixed with an air stream from the inside to provide natural cooling or fresh air. Process *I-M* represents the warmer inside air mixed with the outside air represented by *L-M*. State *M* represents the mixed air and lies along a line connecting *L* and *I*.

Several of these situations will be considered in the following examples. The first law will be used to predict the heating or cooling needed or to establish the final state.

Often the *volume flow rate* is used in air-conditioning problems; it is the average velocity in a conduit times the area of the conduit. It is customary to use the symbol Q_f to represent the volume flow rate. Referring to Eq. (4.45), we observe that

$$Q_f = AV \qquad \text{and} \qquad \dot{m} = \rho Q_f \tag{8.17}$$

The units on Q_f are m³/s. (The English units are often stated as cfs for ft³/sec or cfm for ft³/min.)

EXAMPLE 8.5

Hot, dry air at 40°C and 10% relative humidity passes through an evaporative cooler. Water is added as the air passes through a series of wicks and the mixture exits at 27°C. Find the outlet relative humidity, the amount of water added, and the lowest temperature that could be realized.

Solution

The heat transfer is negligible in an evaporative cooler, so that $h_2 = h_1$. A constant enthalpy line is shown in Fig. 8.4 and is represented by *J-K*. At 27°C on the psychrometric chart we find that

$$\phi_2 = 45\%$$

The added water is found to be

$$\omega_2 - \omega_1 = 0.010 - 0.0046 = 0.0054 \text{ kg water/kg dry air}$$

The lowest possible temperature occurs when $\phi = 100\%$: $T_{\text{min}} = 18.5°\text{C}$.

EXAMPLE 8.6

Outside air at 5°C and 70% relative humidity is heated to 25°C. Calculate the rate of heat transfer needed if the incoming volume flow rate is 50 m³/min. Also, find the final relative humidity.

Solution
First, let's find the mass flow rate of air. The density of dry air is found using the partial pressure P_{a1} in the ideal-gas law:

$$P_{a1} = P - P_{v1} = P - \phi P_{g1} = 100 - 0.7 \times 0.872 = 99.4 \text{ kPa}$$

$$\therefore \rho_{a1} = \frac{P_{a1}}{R_a T_1} = \frac{99.4}{0.287 \times 278} = 1.246 \text{ kg/m}^3$$

The mass flow rate of dry air is then

$$\dot{m}_a = Q_f \rho_a = \frac{50 \text{ m}^3/\text{min}}{60 \text{ s/min}} \times 1.246 \ \frac{\text{kg}}{\text{m}^3} = 1.038 \text{ kg/s}$$

Using the psychrometric chart at state 1 ($T_1 = 5°\text{C}$, $\phi_1 = 70\%$), we find $h_1 = 14$ kJ/kg air. Since ω remains constant (no moisture is added or removed), we follow curve *A-C* in Fig. 8.4; at state 2 we find that $h_2 = 35$ kJ/kg air. Hence,

$$\dot{Q} = \dot{m}_a (h_2 - h_1) = 1.038(35 - 14) = 21.8 \text{ kJ/s}$$

At state 2 we also note from the chart that $\phi_2 = 19\%$, a relatively low value.

EXAMPLE 8.7
Outside air at 5°C and 40% relative humidity is heated to 25°C and the final relative humidity is raised to 40% while the temperature remains constant by introducing steam at 400 kPa into the air stream. Find the needed rate of heat transfer if the incoming volume flow rate of air is 60 m³/min. Calculate the rate of steam supplied and the state of the steam introduced.

Solution
The process we must follow is first simple heating and then humidification, shown as *D-E-C* in Fig. 8.4. The partial pressure of dry air is

$$P_{a1} = P - P_{v1} = P - \phi P_{g1} = 100 - 0.4 \times 0.872 = 99.7 \text{ kPa}$$

where we have assumed standard atmospheric pressure. The dry air density is

$$\rho_{a1} = \frac{P_{a1}}{R_a T_1} = \frac{99.7}{0.287 \times 278} = 1.25 \text{ kg/m}^3$$

so that the mass flow rate of dry air is

$$\dot{m}_a = Q_f \rho_a = \frac{60 \text{ m}^3/\text{min}}{60 \text{ s/min}} \times 1.25 \frac{\text{kg}}{\text{m}^3} = 1.25 \text{ kg/s}$$

The rate of heat addition is found using h_1 and h_2 from the psychrometric chart:

$$\dot{Q} = \dot{m}_a (h_2 - h_1) = 1.25 \times (31 - 10) = 26 \text{ kJ/s}$$

We assume that all the heating is done in the *D-E* process and that humidification takes place in a process in which the steam is mixed with the air flow. Assuming a constant temperature in the mixing process, conservation of mass demands that

$$\dot{m}_s = (\omega_3 - \omega_2)\dot{m}_a = (0.008 - 0.0021) \times 1.25 = 0.0074 \text{ kg/s}$$

where the air enters the humidifier at state 2 and leaves at state 3.

An energy balance around the humidifier provides $h_s \dot{m}_s = (h_3 - h_2)\dot{m}_a$. Hence,

$$h_s = \frac{\dot{m}_a}{\dot{m}_s}(h_3 - h_2) = \frac{1.25}{0.0074} \times (45 - 33) = 2030 \text{ kJ/kg}$$

This is less than h_g at 400 kPa. Consequently, the temperature of the steam is 144°C and the quality is

$$x_s = \frac{2030 - 604.7}{2133.8} = 0.67$$

Only two significant figures are used because of inaccuracy using the relatively small psychrometric chart provided.

EXAMPLE 8.8

Outside air at 30°C and 90% relative humidity is conditioned so that it enters a building at 23°C and 40% relative humidity. Estimate the amount of moisture removed, the heat removed, and the necessary added heat.

Solution

The overall process is sketched as *F-G-H-I* in Fig. 8.4. Heat is removed during the *F-H* process, moisture is removed during the *G-H* process, and heat is added during the *H-I* process. Using the psychrometric chart, we find the moisture removed to be

$$\Delta\omega = \omega_3 - \omega_2 = 0.007 - 0.0245 = -0.0175 \text{ kg water/kg dry air}$$

where states 2 and 3 are at G and H, respectively. (The negative sign implies moisture removal.)

The heat that must be removed to cause the air to follow the F-G-H process is

$$q_{\text{out}} = h_3 - h_1 = 26 - 93 = -67 \text{ kJ/kg dry air}$$

We neglected the small amount of heat due to the liquid state.

The heat that must be added to change the state of the air from the saturated state at H to the desired state at I is

$$q_{\text{in}} = h_4 - h_3 = 41 - 26 = 15 \text{ kJ/kg dry air}$$

If the air flow rate were given, the equipment could be sized.

EXAMPLE 8.9

Outside cool air at 15°C and 40% relative humidity (air stream 1) is mixed with inside air taken near the ceiling at 32°C and 70% relative humidity (air stream 2). Determine the relative humidity and temperature of the resultant air stream 3 if the outside flow rate is 40 m³/min and the inside flow rate is 20 m³/min.

Solution

An energy and mass balance of the mixing of air stream 1 with air stream 2 to produce air stream 3 would reveal the following facts relative to the psychrometric chart (see line L-M-I in Fig. 8.4):

- State 3 lies on a straight line connecting state 1 and state 2.
- The ratio of the distance 2-3 to the distance 3-1 is equal to $\dot{m}_{a1}/\dot{m}_{a2}$.

First, we determine $\dot{m}_{a1}$ and $\dot{m}_{a2}$:

$$P_{a1} = P - P_{v1} = 100 - 1.7 \times 0.4 = 99.3 \text{ kPa}$$
$$P_{a2} = P - P_{v2} = 100 - 4.9 \times 0.7 = 96.6 \text{ kPa}$$

$$\therefore \rho_{a1} = \frac{99.3}{0.287 \times 288} = 1.20 \text{ kg/m}^3 \qquad \rho_{a2} = \frac{95.2}{0.287 \times 305} = 1.09 \text{ kg/m}^3$$

$$\therefore \dot{m}_{a1} = Q_{f1}\rho_{a1} = 40 \times 1.20 = 48.0 \text{ kg/min} \qquad \dot{m}_{a2} = 20 \times 1.09 = 21.8 \text{ kg/min}$$

Then, locate states 1 and 2 on the psychrometric chart:

$$h_1 = 37 \text{ kJ/kg dry air} \qquad \omega_1 = 0.0073 \text{ kg water/kg dry air}$$
$$h_2 = 110 \text{ kJ/kg dry air} \qquad \omega_2 = 0.0302 \text{ kg water/kg dry air}$$

State 3 is located by the ratio

$$\frac{d_{2\text{-}3}}{d_{3\text{-}1}} = \frac{\dot{m}_{a1}}{\dot{m}_{a2}} = \frac{48.0}{21.8} = 2.20$$

where $d_{2\text{-}3}$ is the distance from state 2 to state 3. State 3 is positioned on the psychrometric chart, and we find

$$\phi_3 = 65\% \qquad \text{and} \qquad T_3 = 20°\text{C}$$

Note: State 3 is located nearer state 1 since the greater outside flow dominates.

EXAMPLE 8.10

Water is used to remove the heat from the condenser of a power plant; 10 000 kg/min of 40°C water enters a cooling tower, as shown. Water leaves at 25°C. Air enters at 20°C and leaves at 32°C. Estimate the volume flow rate of air into the cooling tower, and the mass flow rate of water that leaves the cooling tower from the bottom.

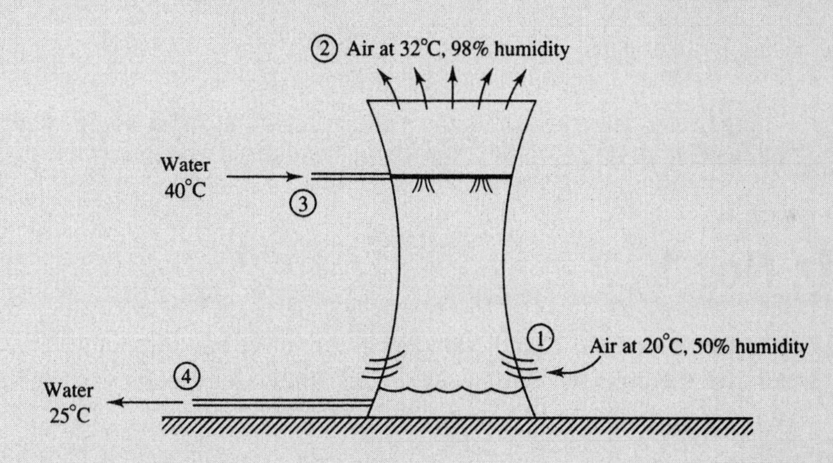

Solution

An energy balance for the cooling tower provides

$$\dot{m}_{a1}h_1 + \dot{m}_{w3}h_3 = \dot{m}_{a2}h_2 + \dot{m}_{w4}h_4$$

where $\dot{m}_{a1} = \dot{m}_{a2} = \dot{m}_a$ is the mass flow rate of dry air. From the psychrometric chart we find

$$h_1 = 37 \text{ kJ/kg dry air} \qquad \omega_1 = 0.0073 \text{ kg water/kg dry air}$$
$$h_2 = 110 \text{ kJ/kg dry air} \qquad \omega_2 = 0.030 \text{ kg water/kg dry air}$$

(Extrapolate to find h_2 and ω_2.) From the steam tables we use h_f at 40°C and find h_3 = 167.5 kJ/kg and h_4 = 104.9 kJ/kg. A mass balance on the water results in $\dot{m}_{w4} = \dot{m}_{w3} - (\omega_2 - \omega_1)\dot{m}_a$. Substituting this into the energy balance, with $\dot{m}_{a1} = \dot{m}_{a2} = \dot{m}_a$ results in

$$\dot{m}_a = \frac{\dot{m}_{w3}(h_4 - h_3)}{h_1 - h_2 + (\omega_2 - \omega_1)h_4} = \frac{10\,000 \times (104.9 - 167.5)}{37 - 110 + (0.03 - 0.0073)104.9} = 8870 \text{ kg/min}$$

From the psychrometric chart we find that $v_1 = 0.84$ m³/kg dry air. This allows us to find the volume flow rate:

$$Q_f = \dot{m}_a v_1 = 8870 \times 0.84 = 7450 \text{ m}^3\text{/min}$$

This air flow rate requires fans, although there is some "chimney effect" since the hotter air tends to rise. Returning to the mass balance provides

$$\dot{m}_4 = \dot{m}_{w3} - (\omega_2 - \omega_1)\dot{m}_a = 10\,000 - (0.03 - 0.0073) \times 8870 = 9800 \text{ kg/min}$$

If the exiting water is returned to the condenser, it must be augmented by 200 kg/min so that 10 000 kg/min is furnished. This is the *makeup water.*

Quiz No. 1

1. Atmospheric air at 30°C and 100 kPa has a relative humidity of 40%. Determine the humidity ratio using the equations (for accuracy).
   (A) 10.1024
   (B) 0.1074
   (C) 0.1102
   (D) 0.1163
2. The air in a 12 m × 15 m × 3 m room is at 20°C and 100 kPa, with a 50% relative humidity. Estimate the humidity ratio using the equations.
   (A) 0.00716
   (B) 0.00728
   (C) 0.00736
   (D) 0.00747

3. Using equations, estimate the mass of water vapor in the room of Prob. 2.
   (A) 4.67 kg
   (B) 4.93 kg
   (C) 5.23 kg
   (D) 5.81 kg
4. Using equations, estimate the total enthalpy ($h = 0$ at 0°C) in the room of Prob. 2.
   (A) 21 900 kJ
   (B) 22 300 kJ
   (C) 23 800 kJ
   (D) 24 600 kJ
5. Outside air at 25°C has a relative humidity of 60%. Using the psychrometric chart, the expected wet-bulb temperature is nearest
   (A) 19.3°C
   (B) 18.3°C
   (C) 17.5°C
   (D) 16.2°C
6. For which of the following situations would you expect condensation to not occur?
   (A) On a glass of ice water sitting on the kitchen table
   (B) On the grass on a cool evening in Tucson, Arizona
   (C) On your glasses when you come into a warm room on a cold winter day
   (D) On the inside of a windowpane in an apartment that has high air infiltration on a cold winter day
7. The dry bulb temperature is 35°C and the wet-bulb temperature is 27°C. What is the relative humidity?
   (A) 65%
   (B) 60%
   (C) 55%
   (D) 50%

8. The air in a conference room is to be conditioned from its present state of 18°C and 40% humidity. Select the appropriate conditioning strategy.
   (A) Heat and dehumidify
   (B) Cool, dehumidify, and then heat
   (C) Heat, dehumidify, and then cool
   (D) Heat and humidify
9. Air at 5°C and 80% humidity is heated to 25°C in an enclosed room. The final humidity is nearest
   (A) 36%
   (B) 32%
   (C) 26%
   (D) 22%
10. Air at 35°C and 70% humidity in a 3 m × 10 m × 20 m classroom is cooled to 25°C and 40% humidity. How much water is removed?
    (A) 4 kg
    (B) 6 kg
    (C) 11 kg
    (D) 16 kg
11. A rigid 2-m^3 tank contains air at 160°C and 400 kPa and a relative humidity of 20%. Heat is removed until the final temperature is 20°C. Determine the temperature at which condensation begins.
    (A) 101°C
    (B) 109°C
    (C) 116°C
    (D) 123°C
12. The mass of water condensed during the cooling process in Prob. 11 is nearest
    (A) 4 kg
    (B) 6 kg
    (C) 11 kg
    (D) 16 kg

13. The heat transfer necessary for the cooling of Prob. 11 to take place is nearest
    (A) 5900 kJ
    (B) 6300 kJ
    (C) 6700 kJ
    (D) 7100 kJ
14. Outside air at 10°C and 60% relative humidity mixes with 50 m^3/min of inside air at 28°C and 40% relative humidity. If the outside flow rate is 30 m^3/min, estimate the relative humidity using the psychrometric chart.
    (A) 43%
    (B) 46%
    (C) 49%
    (D) 52%
15. The mass flow rate of the exiting stream in Prob. 14 is nearest
    (A) 94 kg/min
    (B) 99 kg/min
    (C) 110 kg/min
    (D) 125 kg/min

Quiz No. 2

1. Outside air at 30°C and 100 kPa is observed to have a dew point of 20°C. Find the relative humidity using equations (for accuracy).
   (A) 53.8%
   (B) 54.0%
   (C) 54.4%
   (D) 55.1%
2. Atmospheric air has a dry-bulb temperature of 30°C and a wet-bulb temperature of 20°C. Calculate the humidity ratio using the equations.
   (A) 0.1054
   (B) 0.1074
   (C) 0.1092
   (D) 0.1113

3. Using equations, estimate the relative humidity of the air in Prob. 2.
   (A) 40.2%
   (B) 39.7%
   (C) 39.5%
   (D) 39.1%
4. Using equations, estimate the enthalpy per kg of dry air ($h = 0$ at 0°C) in Prob. 2.
   (A) 58.9 kJ/kg
   (B) 58.1 kJ/kg
   (C) 57.5 kJ/kg
   (D) 57.0 kJ/kg
5. Atmospheric air at 10°C and 60% relative humidity is heated to 21°C. Use the psychrometric chart to estimate the final humidity.
   (A) 17%
   (B) 20%
   (C) 23%
   (D) 26%
6. The rate of heat transfer needed in Prob. 5 if the mass flow rate of dry air is 50 kg/min is nearest
   (A) 26 kJ/s
   (B) 22 kJ/s
   (C) 18 kJ/s
   (D) 14 kJ/s
7. Which temperature is most different from the outside temperature as measured with a conventional thermometer?
   (A) Wet-bulb temperature
   (B) Dry-bulb temperature
   (C) Dew-point temperature
   (D) Ambient temperature

8. Estimate the liters of water in the air in a 3 m × 10 m × 20 m room if the temperature is 20°C and the humidity is 70%.
   (A) 7.9 L
   (B) 7.2 L
   (C) 6.1 L
   (D) 5.5 L
9. An evaporative cooler inlets air at 40°C and 20% humidity. If the exiting air is at 80% humidity, the exiting temperature is nearest
   (A) 30°C
   (B) 28°C
   (C) 25°C
   (D) 22°C
10. Air at 30°C and 80% humidity is to be conditioned to 20°C and 40% humidity. How much heating is required?
   (A) 30 kJ/kg
   (B) 25 kJ/kg
   (C) 20 kJ/kg
   (D) 15 kJ/kg
11. Outside air at 15°C and 40% humidity is mixed with inside air at 32°C and 70% humidity taken near the ceiling. Determine the relative humidity of the mixed stream if the outside flow rate is 40 m^3/min and the inside flow rate is 20 m^3/min.
   (A) 64%
   (B) 58%
   (C) 53%
   (C) 49%
12. The temperature of the mixed stream of Prob. 11 is nearest
   (A) 28°C
   (B) 25°C
   (C) 23°C
   (D) 20°C

13. One hundred m^3/min of outside air at 36°C and 80% relative humidity is conditioned for an office building by cooling and heating. Estimate the rate of cooling if the final state of the air is 25°C and 40% relative humidity.
    (A) 9920 kJ/min
    (B) 9130 kJ/min
    (C) 8830 kJ/min
    (D) 8560 kJ/min
14. The heating required to condition the air in Prob. 13 is nearest
    (A) 1610 kJ/min
    (B) 1930 kJ/min
    (C) 2240 kJ/min
    (D) 2870 kJ/min

CHAPTER 9

Combustion

Several of the devices we have considered in our study of thermodynamics have involved heat transfer from an external source. Most often, the source of the energy supplied by the heat transfer comes from a combustion process, such as in an internal combustion engine or in a power plant. In this chapter, we will consider the process of the combustion of one of the substances included in Table B.6. We will balance the combustion equation, present complete and incomplete combustion, apply the first law with an energy balance, and calculate the temperature of the products of combustion.

9.1 Combustion Equations

Combustion involves the burning of a fuel with oxygen or a substance containing oxygen such as air. A chemical-reaction equation relates the components before and after the chemical process takes place. Let us begin our study of this particular variety of chemical-reaction equations by considering the combustion of propane in a pure oxygen environment. The chemical reaction is represented by

$$C_3H_8 + 5O_2 \Rightarrow 3CO_2 + 4H_2O \qquad (9.1)$$

Note that the number of moles of the elements on the left-hand side may not equal the number of moles on the right-hand side. However, the number of atoms of an element must remain the same before, after, and during a chemical reaction; this demands that the mass of each element be conserved during combustion.

In writing the equation, we have demonstrated some knowledge of the products of the reaction. Unless otherwise stated we will assume *complete combustion:* the products of the combustion of a hydrocarbon fuel will include H_2O and CO_2 but no CO, OH, C, or H_2. *Incomplete combustion* results in products that contain H_2, CO, C, and/or OH.

For a simple chemical reaction, such as that of Eq. (9.1), we can immediately write down a balanced chemical equation. For more complex reactions the following systematic steps prove useful:

1. Set the number of moles of fuel equal to 1.
2. Balance CO_2 with number of C from the fuel.
3. Balance H_2O with H from the fuel.
4. Balance O_2 from CO_2 and H_2O.

For the combustion of propane in Eq. (9.1), we assumed that the process occurred in a pure oxygen environment. Actually, such a combustion process would normally occur in air. For our purposes we assume that air consists of 21% O_2 and 79% N_2 by volume (we ignore the 1% argon) so that for each mole of O_2 in a reaction we will have

$$\frac{79}{21} = 3.76 \ \frac{\text{mol N}_2}{\text{mol O}_2} \tag{9.2}$$

Thus, assuming that N_2 will not undergo a chemical reaction, Eq. (9.1) is replaced with

$$C_3H_8 + 5(O_2 + 3.76N_2) \Rightarrow 3CO_2 + 4H_2O + 18.8N_2 \tag{9.3}$$

The minimum amount of air that supplies sufficient O_2 for complete combustion of the fuel is called *theoretical air* or *stoichiometric air.* When complete combustion is achieved with theoretical air, the products contain no CO or O_2. In practice, it is found that if complete combustion is to occur, air must be supplied in an amount greater than theoretical air. This is due to the chemical kinetics and molecular activity of the reactants and products. Thus, we often speak in terms of *percent theoretical air* or *percent excess air*, where they are related by

$$\% \text{ theoretical air} = 100\% + \% \text{ excess air} \tag{9.4}$$

Slightly insufficient air results in CO being formed; some hydrocarbons may result from larger deficiencies.

The parameter that relates the amount of air used in a combustion process is the *air-fuel ratio* (*AF*), which is the ratio of the mass of air to the mass of fuel. The reciprocal is the *fuel-air ratio* (*FA*). Thus

$$AF = \frac{m_{\text{air}}}{m_{\text{fuel}}} \qquad FA = \frac{m_{\text{fuel}}}{m_{\text{air}}} \tag{9.5}$$

Again, considering propane combustion with theoretical air, as in Eq. 9.3, the air-fuel ratio is

$$AF = \frac{m_{\text{air}}}{m_{\text{fuel}}} = \frac{5 \times 4.76 \times 29}{1 \times 44} = 15.69 \ \frac{\text{kg air}}{\text{kg fuel}} \tag{9.6}$$

where we have used the molecular weight of air as 29 kg/kmol (close enough to 28.97) and that of propane as 44 kg/kmol. If, for the combustion of propane, $AF > 15.69$, a *lean mixture* occurs; if $AF < 15.69$, a *rich mixture* results.

The combustion of hydrocarbon fuels involves H_2O in the products of combustion. The calculation of the dew point of the products is often of interest; it is the saturation temperature at the partial pressure of the water vapor. If the temperature drops below the dew point, the water vapor begins to condense. The condensate usually contains corrosive elements, and thus it is often important to ensure that the temperature of the products does not fall below the dew point. This is especially necessary in the chimney of a house; the corrosive elements can eat through the building materials and cause substantial damage. For that reason mid-efficiency furnaces (80% or so) require chimney inserts that are able to resist the corrosive elements.

When analyzing a mixture of gases, either a gravimetric analysis based on mass (or weight) or a volumetric analysis based on moles (or volume) may be used. If the analysis is presented on a dry basis, the water vapor is not included. An Orsat analyzer and a gas chromatograph are two of the devices used to measure the products of combustion.

EXAMPLE 9.1

Butane is burned with dry air at an air-fuel ratio of 20. Calculate the percent excess air, the volume percentage of CO_2 in the products, and the dew-point temperature of the products.

Solution

The reaction equation for theoretical air is (see Table B.3 for the composition of butane)

$$C_4H_{10} + 6.5(O_2 + 3.76N_2) \Rightarrow 4CO_2 + 5H_2O + 24.44N_2$$

The air-fuel ratio for theoretical air is

$$AF_{\text{th}} = \frac{m_{\text{air}}}{m_{\text{fuel}}} = \frac{6.5 \times 4.76 \times 29}{1 \times 58} = 15.47 \ \frac{\text{kg air}}{\text{kg fuel}}$$

This represents 100% theoretical air. The actual air-fuel ratio is 20. The excess air is then

$$\%\ \text{excess air} = \frac{AF_{\text{actual}} - AF_{\text{th}}}{AF_{\text{th}}} \times 100 = \frac{20 - 15.47}{15.47} \times 100 = 29.28\%$$

The reaction equation with 129.28% theoretical air is

$$C_4H_{10} + 6.5 \times 1.2928 \times (O_2 + 3.76N_2) \Rightarrow 4CO_2 + 5H_2O + 1.903O_2 + 31.6N_2$$

The volume percentage is obtained using the total moles in the products of combustion. For CO_2 we have

$$\%\ \text{CO} = \frac{4}{42.5} \times 100 = 9.41\%$$

To find the dew-point temperature of the products we need the partial pressure P_v of the water vapor. We know that

$$P_v = \frac{N_v RT}{V} \qquad \text{and} \qquad P = \frac{NRT}{V}$$

where N is the number of moles. Then the ratio of the above two equations provides

$$P_v = P\frac{N_v}{N} = 100 \times \frac{5}{42.5} = 11.76 \text{ kPa} \qquad \text{or} \qquad 0.01176 \text{ MPa}$$

where we have assumed an atmospheric pressure of 100 kPa. Using Table C.2 we find the dew-point temperature to be $T_{\text{dp}} = 49°\text{C}$.

EXAMPLE 9.2

Butane is burned with 90% theoretical air. Calculate the volume percentage of CO in the products and the air-fuel ratio. Assume no hydrocarbons in the products.

Solution

For incomplete combustion we add CO to the products of combustion. Using the reaction equation from Example 9.1,

$$C_4H_{10} + 0.9 \times 6.5 \times (O_2 + 3.76N_2) \Rightarrow aCO_2 + 5H_2O + 22N_2 + bCO$$

With atomic balances on the carbon and oxygen we find:

$$\left.\begin{array}{ll} \text{C:} & 4 = a + b \\ \text{O:} & 11.7 = 2a + 5 + b \end{array}\right\} \quad \therefore a = 2.7 \qquad b = 1.3$$

The volume percentage of CO is then

$$\%CO = \frac{11.3}{31} \times 100 = 4.19\%$$

The air-fuel ratio is

$$AF = \frac{m_{\text{air}}}{m_{\text{fuel}}} = \frac{9 \times 6.5 \times 4.76 \times 29}{1 \times 58} = 13.92 \ \frac{\text{kg air}}{\text{kg fuel}}$$

EXAMPLE 9.3

Butane is burned with dry air, and volumetric analysis of the products on a dry basis (the water vapor is not measured) gives 11.0% CO_2, 1.0% CO, 3.5% O_2, and 84.6% N_2. Determine the percent theoretical air.

Solution

The problem is solved assuming that there are 100 moles of dry products. The chemical equation is

$$aC_4H_{10} + b(O_2 + 3.76N_2) \Rightarrow CO_2 + 1CO + 3.5O_2 + 84.6N_2 + cH_2O$$

We perform the following balances:

$$\left.\begin{array}{l} \text{C: } 4a = 11 + 1 \\ \text{H: } 10a = 2c \\ \text{O: } 2b = 22 + 1 + 7 + c \end{array}\right\} \quad \therefore a = 3 \qquad c = 15 \qquad b = 22.5$$

Dividing through the chemical equation by the value of a so that we have 1 mol of fuel, provides

$$C_4H_{10} + 7.5(O_2 + 3.76N_2) \Rightarrow 3.67CO_2 + 0.33CO + 1.17O_2 + 28.17N_2 + 5H_2O$$

Comparing this with the combustion equation of Example 9.1 using theoretical air, we find

$$\%\ \text{theoretical air} = \frac{7.5}{6.5} \times 100 = 115.4\%$$

EXAMPLE 9.4

Volumetric analysis of the products of combustion of an unknown hydrocarbon, measured on a dry basis, gives 10.4% CO_2, 1.2% CO, 2.8% O_2, and 85.6% N_2. Determine the composition of the hydrocarbon and the percent excess air.

Solution

The chemical equation for 100 mol dry products is

$$C_aH_b + c(O_2 + 3.76N_2) \Rightarrow 10.4CO_2 + 1.2CO + 2.8O_2 + 85.6N_2 + dH_2O$$

Balancing each element,

$$\left.\begin{aligned} &\text{C: } a = 10.4 + 1.2 \\ &\text{N: } 3.76c = 85.6 \\ &\text{O: } 2c = 20.8 + 1.2 + 5.6 + d \\ &\text{H: } b = 2d \end{aligned}\right\} \quad \begin{aligned} \therefore a &= 11.6 \qquad & c &= 22.8 \\ d &= 18.9 \qquad & b &= 37.9 \end{aligned}$$

The chemical formula for the fuel is $C_{11.6}H_{37.9}$. This could represent a mixture of hydrocarbons, but it is not any species listed in App. B, since the ratio of hydrogen atoms to carbon atoms is 37.9/11.6 = 3.27 = 13/4, i.e., C_4H_{13}.

To find the percent theoretical air we must have the chemical equation using 100% theoretical air. It would be

$$C_{11.6}H_{37.9} + 21.08(O_2 + 3.76N_2) \Rightarrow 11.6CO_2 + 18.95H_2O + 79.26N_2$$

Using the number of moles of air from the actual chemical equation, we find

$$\%\ \text{excess air} = \%\ \text{theoretical air} - 100 = \frac{22.8}{21.08} \times 100 - 100 = 8\%$$

9.2 Enthalpy of Formation, Enthalpy of Combustion, and the First Law

When a chemical reaction occurs, there may be considerable change in the chemical composition of a system. The problem this creates is that for a control volume the mixture that exits is different from the mixture that enters. Since different zeros are used for enthalpy, it is necessary to establish a standard reference state, chosen as 25°C (77°F) and 1 atm and denoted by the superscript " °", for example, h°.

Consider the combustion of H_2 with O_2, resulting in H_2O:

$$H_2 + \frac{1}{2}O_2 \Rightarrow H_2O(l) \tag{9.7}$$

where the symbol (l) after a chemical compound implies the liquid phase; the symbol (g) would imply the gaseous phase. If no symbol is given, a gas is implied. If H_2 and O_2 enter a combustion chamber at 25°C (77°F) and 1 atm and $H_2O(l)$ leaves the chamber at 25°C (77°F) and 1 atm, then the measured heat transfer will be −285 830 kJ for each kmol of $H_2O(l)$ formed. The negative sign on the heat transfer means energy has left the control volume, as shown schematically in Fig. 9.1.

The first law applied to a combustion process in a control volume with no shaft work or changes in kinetic and potential energy is

$$Q = H_P - H_R \tag{9.8}$$

where H_P is the enthalpy of the *products of combustion* that leave the combustion chamber and H_R is the enthalpy of the *reactants* that enter. If the reactants are stable elements, as in our example in Fig. 9.1, and the process is at constant temperature and constant pressure, then the enthalpy change is called the *enthalpy of formation*, denoted by h_f°. The enthalpies of formation of numerous compounds are listed in Table B.5. Note that some compounds have a positive h_f°, indicating that they

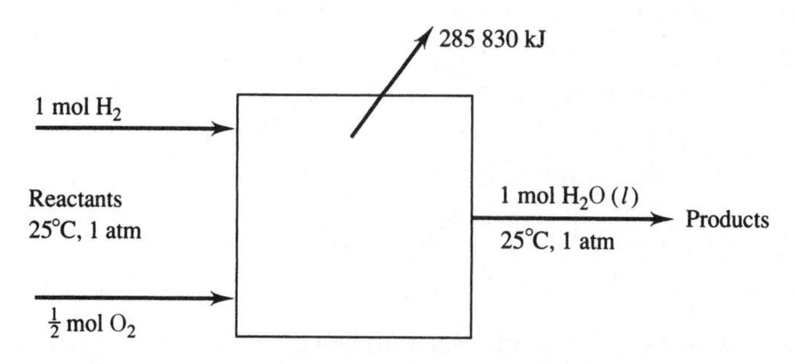

Figure 9.1 The control volume used during combustion of hydrogen and oxygen.

require energy to form (an *endothermic reaction*); others have a negative h_f°, indicating that they give off energy when they are formed (an *exothermic reaction*).

The enthalpy of formation is the enthalpy change when a compound is formed. The enthalpy change when a compound undergoes complete combustion at constant temperature and pressure is called the *enthalpy of combustion.* For example, the enthalpy of formation of H_2 is zero, and the enthalpy of combustion of H_2 is 285 830 kJ/kmol. Values are listed for several compounds in Table B.6. If the products contain liquid water, the enthalpy of combustion is the *higher heating value* (HHV); if the products contain water vapor, the enthalpy of combustion is the *lower heating value.* The difference between the higher heating value and the lower heating value is the heat of vaporization h_{fg} at standard conditions.

If the reactants and products consist of several components, the first law, allowing for shaft work but neglecting kinetic and potential energy changes, is

$$Q - W_S = \sum_{\text{prod}} N_i \left(\bar{h}_f^\circ + \bar{h} - \bar{h}^\circ\right)_i - \sum_{\text{react}} N_i \left(\bar{h}_f^\circ + \bar{h} - \bar{h}^\circ\right)_i \tag{9.9}$$

where N_i represents the number of moles of substance i. The work is often zero, but not in, for example, a combustion turbine.

If combustion occurs in a rigid chamber, like a bomb calorimeter, so that no work results, the first law is

$$\begin{aligned} Q &= U_{\text{products}} - U_{\text{reactants}} \\ &= \sum_{\text{prod}} N_i \left(\bar{h}_f^\circ + \bar{h} - \bar{h}^\circ - P_v\right)_i - \sum_{\text{react}} N_i \left(\bar{h}_f^\circ + \bar{h} - \bar{h}^\circ - P_v\right)_i \end{aligned} \tag{9.10}$$

where we have used enthalpy ($u = h - P_v$) since the h_i values are tabulated. Since the volume of any liquid or solid is negligible compared to the volume of the gases, we write Eq. (9.10) as

$$Q = \sum_{\text{prod}} N_i \left(\bar{h}_f^\circ + \bar{h} - \bar{h}^\circ - \bar{R}T\right)_i - \sum_{\text{react}} N_i \left(\bar{h}_f^\circ + \bar{h} - \bar{h}^\circ - \bar{R}T\right)_i \tag{9.11}$$

If $N_{\text{prod}} = N_{\text{react}}$ then Q for the rigid volume is equal to Q for the control volume for the isothermal process.

We employ one of the following methods to find $(\bar{h} - \bar{h}^\circ)$:

- For a solid or liquid: Use $\bar{h} - \bar{h}^\circ = \bar{C}_p \Delta T$.
- For gases:
  1. Assume an ideal gas with constant specific heat so that $\bar{h} - \bar{h}^\circ = \bar{C}_p \Delta T$.
  2. Assume an ideal gas and use tabulated values for $\bar{h}$.

Which method to use for gases is left to the judgment of the engineer. In our examples, we'll usually use method 2 for gases with Tables E.1 to E.6 since temperature changes for combustion processes are often quite large and method 1 introduces substantial error.

EXAMPLE 9.5

Calculate the enthalpy of combustion of gaseous propane and of liquid propane assuming the reactants and products to be at 25°C and 1 atm. Assume liquid water in the products exiting the steady-flow combustion chamber.

Solution

Assuming theoretical air (the use of excess air would not influence the result since the process is isothermal), the chemical equation is

$$C_3H_8 + 5(O_2 + 3.76N_2) \Rightarrow 3CO_2 + 4H_2O(l) + 18.8N_2$$

where, for the HHV, a liquid is assumed for H_2O. The first law becomes, for the isothermal process $h = h^\circ$,

$$Q = H_P - H_R = \sum_{\text{prod}} N_i (\bar{h}_f^\circ)_i - \sum_{\text{react}} N_i (\bar{h}_f^\circ)_i$$

$$= 3(-393\,520) + 4(-285\,830) - (-103\,850) = -2\,220\,000 \text{ kJ}$$

This is the enthalpy of combustion; it is stated with the negative sign. The sign is dropped for the HHV; for gaseous propane it is 2220 MJ for each kmol of fuel.

For liquid propane we find

$$Q = 3(-393\,520) + 4(-285\,830) - (-103\,850 - 15\,060) = -2\,205\,000 \text{ kJ}$$

This is slightly less than the HHV for gaseous propane, because some energy is needed to vaporize the liquid fuel.

EXAMPLE 9.6

Calculate the heat transfer required if propane and air enter a steady-flow combustion chamber at 25°C and 1 atm and the products leave at 600 K and 1 atm. Use theoretical air and one mole of fuel.

Solution

The combustion equation is written using H_2O in the vapor form due to the high exit temperature:

$$C_3H_8 + 5(O_2 + 3.76N_2) \Rightarrow 3C_2 + 4H_2O(g) + 18.8N_2$$

The first law takes the form [see Eq. (9.9)]

$$\begin{aligned} Q &= \sum_{\text{prod}} N_i(\bar{h}_f^\circ + \bar{h} - \bar{h}^\circ)_i - \sum_{\text{react}} N_i(\bar{h}_f^\circ + \bar{h} - \bar{h}^\circ)_i \\ &= 3(-393\,520) + 22\,280 - 9360 + 4(-241\,810 + 20\,400 - 9900) \\ &\quad + 18.8(17\,560 - 8670) - (-103\,850) \\ &= -1\,821\,900 \text{ kJ} \end{aligned}$$

where we have used method 2 listed for gases. This heat transfer is less than the enthalpy of combustion of propane, as it should be, since some energy is needed to heat the products to 600 K.

EXAMPLE 9.7

Liquid octane at 25°C fuels a jet engine. Air at 600 K enters the insulated combustion chamber and the products leave at 1000 K. The pressure is assumed constant at 1 atm. Estimate the exit velocity using theoretical air.

Solution

The equation is

$$C_8H_{18}(l) + 12.5(O_2 + 3.76N_2) \Rightarrow 8CO_2 + 9H_2O + 47N_2$$

The first law, with $Q = W_S = 0$ and including the kinetic energy change (neglect V_{inlet}) is

$$0 = H_P - H_R + \frac{V^2}{2} M_P \qquad \text{or} \qquad V^2 = \frac{2}{M_P}(H_R - H_P)$$

where M_p is the mass of the products (one mole of fuel). For the products,

$$\begin{aligned} H_P &= 8(-39\,3520 + 42\,770 - 9360) + 9(-241\,810 + 35\,880 - 9900) \\ &\quad + 47(30\,130 - 8670) = -3\,814\,700 \text{ kJ} \end{aligned}$$

For the reactants,

$$H_R = (-249\,910) + 12.5(17\,930 - 8680) + 47(17\,560 - 8670) = 283\,540 \text{ kJ}$$

The mass of the products is $M_P = 8 \times 44 + 9 \times 18 + 47 \times 28 = 1830$ kg, and so

$$V^2 = \frac{2}{1830}(0.28354 + 3.8147) \times 10^9 \qquad \therefore \ V = 2120 \text{ m/s}$$

EXAMPLE 9.8
Liquid octane is burned with 300% excess air. The octane and air enter the steady-flow combustion chamber at 25°C and 1 atm and the products exit at 1000 K and 1 atm. Estimate the heat transfer for one mole of fuel.

Solution
The reaction with theoretical air is

$$C_8H_{18} + 12.5(O_2 + 3.76N_2) \Rightarrow 8CO_2 + 9H_2O + 47N_2$$

For 300% excess air (400% theoretical air) the reaction is

$$C_8H_{18}(l) + 50(O_2 + 3.76N_2) \Rightarrow 8CO_2 + 9H_2O + 37.5O_2 + 188N_2$$

The first law applied to the combustion chamber is

$$Q = H_P - H_R = 8(-393\,520 + 42\,770 - 9360) + 9(-241\,810 + 35\,880 - 9900)$$
$$+ 37.5(31\,390 - 8680) + 188(30\,130 - 8670) - (-249\,910) = 312\,700 \text{ kJ}$$

In this situation heat must be added to obtain the exit temperature.

EXAMPLE 9.9
A constant-volume bomb calorimeter is surrounded by water at 20°C. One mole of liquid propane is burned with pure oxygen in the calorimeter. Determine the heat transfer to be expected.

Solution
The complete combustion of propane follows:

$$C_3H_8(l) + 5O_2 \Rightarrow 3CO_2 + 4H_2O(g)$$

The surrounding water sustains a constant-temperature process, so that Eq. (9.11) becomes

$$Q = \sum_{\text{prod}} N_i \left(\bar{h}_f^\circ\right)_i - \sum_{\text{react}} N_i \left(\bar{h}_f^\circ\right)_i + (N_R - N_P)\bar{R}T$$

$$= [3(-393\,520) + 4(-241\,820)] - (103\,850 + 15\,060) + (6 - 7) \times 8.314 \times 293$$

$$= -2.269 \times 10^6 \text{ kJ} \qquad \text{or} \qquad 2269 \text{ MJ}$$

9.3 Adiabatic Flame Temperature

If we consider a combustion process that takes place adiabatically, with no work or changes in kinetic and potential energy, then the temperature of the products is referred to as the *adiabatic flame temperature*. The maximum adiabatic flame temperature that can be achieved occurs at theoretical air. This fact allows us to control the adiabatic flame temperature by the amount of excess air involved in the process: the greater the excess air, the lower the adiabatic flame temperature. If the blades in a turbine can withstand a certain maximum temperature, we can determine the excess air needed so that the maximum allowable blade temperature is not exceeded. We will find that an iterative (trial-and-error) procedure is needed to find the adiabatic flame temperature. A quick approximation to the adiabatic flame temperature is found by assuming the products to be completely N_2. An example will illustrate.

The adiabatic flame temperature is calculated assuming complete combustion, no heat transfer from the combustion chamber, and no dissociation of the products into other chemical species. Each of these effects tends to reduce the adiabatic flame temperature. Consequently, the adiabatic flame temperature that we will calculate represents the maximum possible flame temperature for the specified percentage of theoretical air.

If a significant amount of heat transfer does occur from the combustion chamber, we can account for it by including the following term in the energy equation:

$$Q = UA(T_P - T_E) \tag{9.12}$$

where U = overall heat-transfer coefficient (specified in $\text{kW/m}^2 \cdot \text{K}$)
T_E = temperature of environment
T_P = temperature of products
A = surface area of combustion chamber

EXAMPLE 9.10

Propane is burned with 250% theoretical air; both are at 25°C and 1 atm. Predict the adiabatic flame temperature in the steady-flow combustion chamber.

Solution

The combustion with theoretical air is

$$C_3H_8 + 5(O_2 + 3.76N_2) \Rightarrow 3CO_2 + 4H_2O + 18.8N_2$$

For 250% theoretical air we have

$$C_3H_8 + 12.5(O_2 + 3.76N_2) \Rightarrow 3CO_2 + 4H_2O + 7.5O_2 + 47N_2$$

Since $Q = 0$ for an adiabatic process we demand that $H_R = H_P$. The enthalpy of the reactants, at 25°C, is $H_R = -103\ 850$ kJ.

The temperature of the products is unknown; and we cannot obtain the enthalpies of the components of the products from the tables without knowing the temperatures. This requires a trial-and-error solution. To obtain an initial guess, we assume the products to be composed entirely of nitrogen:

$$H_R = H_P = -103\,850 = 3(-393\,520) + 4(-241\,820) + 61.5(h_P - 8670)$$

where we have noted that the products contain 61.5 mol of gas (at 25°C, $h_{\text{nitrogen}} =$ 8670 kJ/kmol from Table E.2). This gives $h_P = 43\ 400$ kJ/kmol, which suggests a temperature of about 1380 K (take T_P a little less than that predicted by the all-nitrogen assumption). Using this temperature we check using the actual products:

$$-103\,850 = 3(-393\,520 + 64\,120 - 9360) + 4(-241\,820 + 52\,430 - 9900)$$
$$+ 7.5(44\,920 - 8680) + 47(42\,920 - 8670) = 68\,110$$

The temperature is obviously too high. We select a lower value, $T_P = 1300$ K. There results

$$-103\,850 = 3(393\,520 + 59\,520 - 9360) + 4(-241\,820 + 48\,810 - 9900)$$
$$+ 7.5(44\,030 - 8680) + 47(40\,170 - 8670) = -96\,100$$

We use the above two results for 1380 and 1300 K and, assuming a linear relationship, predict that

$$T_P = 1300 - \left[\frac{103\,850 - 96\,100}{68\,110 - (-96\,100)}\right](1380 - 1300) = 1296\text{ K}$$

EXAMPLE 9.11

Propane is burned with theoretical air; both are at 25°C and 1 atm in a steady-flow combustion chamber. Predict the adiabatic flame temperature.

Solution

The combustion with theoretical air is

$$C_3H_8 + 5(O_2 + 3.76N_2) \Rightarrow 3CO_2 + 4H_2O + 18.8N_2$$

For the adiabatic process the first law takes the form $H_R = H_P$ Hence, assuming the products to be composed entirely of nitrogen,

$$-103\,850 = 3(-393\,520) + 4(-241\,820) + 25.8(h_P - 8670)$$

where the products contain 25.8 mol gas. This gives $h_P = 87\,900$ kJ/kmol, which suggests a temperature of about 2600 K. With this temperature we find, using the actual products:

$$-103\,850 = 3(-393\,520 + 137\,400 - 9360) + 4(-241\,820 + 114\,300 - 9900)$$
$$+ 18.8(86\,600 - 8670) = 119\,000$$

At 2400 K there results

$$-103\,850 = 3(-393\,520 + 125\,200 - 9360) + 4(-241\,820 + 103\,500 - 9900)$$
$$+ 18.8(79\,320 - 8670) = -97\,700$$

A straight line extrapolation gives $T_P = 2394$ K.

Quiz No. 1

1. The fuel C_4H_{10} combines with stoichiometric air. Provide the correct values for x, y, z in the reaction $C_4H_{10} + w(O_2 + 3.76N_2) \rightarrow xCO_2 + yH_2O + zN_2$.
   (A) (4, 5, 24.44)
   (B) (5, 6, 30.08)
   (C) (4, 10, 24.44)
   (D) (5, 10, 24.44)
2. Methane (CH_4) is burned with stoichiometric air and the products are cooled to 20°C assuming complete combustion at 100 kPa. Calculate the air-fuel ratio.
   (A) 16.9
   (B) 17.2
   (C) 17.8
   (D) 18.3
3. For the combustion of the methane of Prob. 2, what is the percentage of CO_2 by weight in the products?
   (A) 15.1%
   (B) 17.2%
   (C) 18.1%
   (D) 18.5%
4. For the reaction of the methane of Prob. 2, the dew-point temperature of the products is nearest
   (A) 48°C
   (B) 54°C
   (C) 57°C
   (D) 59°C
5. Ethane (C_2H_6) undergoes complete combustion at 95 kPa with 180% theoretical air. Find the air-fuel ratio.
   (A) 23
   (B) 25
   (C) 27
   (D) 29

6. For the combustion of the ethane of Prob. 5, what is the percentage of CO_2 by weight in the products?
   (A) 5.83%
   (B) 6.35%
   (C) 8.61%
   (D) 11.7%
7. Calculate the mass flow rate of methane (CH_4) required if the inlet air flow rate is 20 m^3/min at 20°C and 100 kPa using stoichiometric air.
   (A) 4.32 kg/min
   (B) 2.87 kg/min
   (C) 1.38 kg/min
   (D) 1.09 kg/min
8. An air-fuel ratio of 25 is used in an engine that burns octane (C_8H_{18}). Find the percentage of excess air required.
   (A) 65.4%
   (B) 69.4%
   (C) 72.4%
   (D) 76.4%
9. Carbon reacts with oxygen to form carbon dioxide in a steady-flow chamber. Calculate the energy involved. Assume the reactants and products are at 25°C and 1 atm.
   (A) −399 300 kJ/kmol
   (B) −393 500 kJ/kmol
   (C) −389 300 kJ/kmol
   (D) −375 200 kJ/kmol
10. Liquid propane (C_3H_8) undergoes combustion with air; both are at 25°C and 1 atm. Calculate the heat transfer if the products from a steady-flow combustor are at 1000 K with 100% theoretical air.
    (A) 1.44×10^6 kJ/kmol
    (B) 1.65×10^6 kJ/kmol
    (C) 1.79×10^6 kJ/kmol
    (D) 1.93×10^6 kJ/kmol

Quiz No. 2

1. The fuel C_5H_{12} combines with stoichiometric air. Provide the correct values for x, y, z in the reaction $C_5H_{12} + w(O_2 + 3.76N_2) \rightarrow xCO_2 + yH_2O + zN_2$.

   (A) (4, 5, 24.44)

   (B) (5, 6, 30.08)

   (C) (5, 12, 30.08)

   (D) (4, 12, 30.08)

2. Ethane (C_2H_6) is burned with stoichiometric air and the products are cooled to 20°C assuming complete combustion at 100 kPa. Calculate the air-fuel ratio.

   (A) 17.9

   (B) 17.2

   (C) 16.8

   (D) 16.1

3. For the combustion of the ethane of Prob. 2, what is the percentage of CO_2 by weight in the products?

   (A) 15.1%

   (B) 17.2%

   (C) 18.1%

   (D) 18.5%

4. For the reaction of the ethane of Prob. 2, the dew-point temperature of the products is nearest

   (A) 48°C

   (B) 54°C

   (C) 56°C

   (C) 59°C

5. Propane (C_3H_8) undergoes complete combustion at 95 kPa with 180% theoretical air. The air-fuel ratio is nearest.

   (A) 28.2

   (B) 26.1

   (C) 24.9

   (D) 22.5

6. For the combustion of the propane of Prob. 5, what is the percentage of CO_2 by weight in the products?
   (A) 5.83%
   (B) 6.35%
   (C) 6.69%
   (C) 7.94%
7. Calculate the mass flow rate of propane (C_3H_8) required if the inlet air flow rate is 20 m^3/min at 20°C and 100 kPa using stoichiometric air.
   (A) 2.70 kg/min
   (B) 1.93 kg/min
   (C) 1.79 kg/min
   (D) 1.52 kg/min
8. Butane (C_4H_{10}) is burned with 50% excess air. If 5% of the carbon in the fuel is converted to CO, calculate the air-fuel ratio.
   (A) 23.2
   (B) 25.8
   (C) 27.1
   (D) 29.9
9. Hydrogen reacts with oxygen to form water vapor in a steady-flow chamber. Calculate the energy involved. Assume the reactants and products are at 25°C and 1 atm.
   (A) 376 700 kJ/kmol
   (B) 483 600 kJ/kmol
   (C) 523 900 kJ/kmol
   (D) 587 800 kJ/kmol
10. Liquid propane (C_3H_8) undergoes complete combustion with air; both are at 25°C and 1 atm. Calculate the heat transfer if the products from a steady-flow combustor are at 1000 K and the percent theoretical air is 200%.
    (A) 1.18×10^6 kJ/kmol
    (B) 1.45×10^6 kJ/kmol
    (C) 1.69×10^6 kJ/kmol
    (D) 1.83×10^6 kJ/kmol

APPENDIX A

Conversion of Units

Length	Force	Mass	Velocity
1 cm = 0.3931 in	1 lbf = 0.4448 × 10^6 dyne	1 oz = 28.35 g	1 mph = 1.467 ft/sec
1 m = 3.281 ft	1 kip = 1000 lbf	1 lbm = 0.4536 kg	1 mph = 0.8684 knot
1 km = 0.6214 mi	1 N = 0.2248 lbf	1 slug = 32.17 lbm	1 ft/sec = 0.3048 m/s
1 in = 2.54 cm	1 N = 10^5 dyn	1 slug = 14.59 kg	1 km/h = 0.2778 m/s
1 ft = 0.3048 m		1 kg = 2.205 lbm	1 knot = 1.688 ft/sec
1 mi = 1.609 km			
1 mi = 5280 ft			
1 mi = 1760 yd			

Power	Work and Heat	Pressure	Volume
1 hp = 550 ft · lbf/sec	1 hp = 550 ft · lbf/sec	1 psi = 2.036 in Hg	1 ft^3 = 7.481 gal (U.S.)
1 hp = 2545 Btu/hr	1 ft · lbf = 1.356 J	1 psi = 27.7 in H_2O	1 ft^3 = 0.02832 m^3
1 hp = 0.7455 kW	1 cal = 3.088 ft · lbf	1 atm = 29.92 in Hg	1 gal (U.S.) = 231 in^3
1 W = 1 J/s	1 cal = 0.003968 Btu	1 atm = 33.93 ft H_2O	1 gal (Brit.) = 1.2 gal
1 W = 3.412 Btu/hr	1 Btu = 1055 J	1 atm =101.3 kPa	1 m^3 = 1000 L
1 kW = 1.341 hp	1 Btu = 0.2930 Wh	1 atm = 1.0133 bar	1 L = 0.03531 ft^3
1 ton = 12 000 Btu/hr	1 Btu = 778 ft·lbf	1 atm = 14.7 psi	1 L = 0.2642 gal
1 ton = 3.517 kW	1 kWh = 3412 Btu	1 in Hg = 0.4912 psi	1 m^3 = 264.2 gal
	1 therm = 10^5 Btu	1 ft H_2O = 0.4331 psi	1 m^3 = 35.31 ft^3
	1 quad = 10^{15} Btu	1 psi = 6.895 kPa	1 ft^3 = 28.32 L
		1 kPa = 0.145 psi	1 in^3 = 16.387 cm^3

APPENDIX B

Material Properties

Table B.1 The U.S. Standard Atmosphere

Altitude, m	Temperature, °C	Pressure, P/P_0	Density, ρ/ρ_0
0	15.2	1.000	1.000
1 000	9.7	0.8870	0.9075
2 000	2.2	0.7846	0.8217
3 000	−4.3	0.6920	0.7423
4 000	−10.8	0.6085	0.6689
5 000	−17.3	0.5334	0.6012
6 000	−23.8	0.4660	0.5389
7 000	−30.3	0.4057	0.4817
8 000	−36.8	0.3519	0.4292
10 000	−49.7	0.2615	0.3376
12 000	−56.3	0.1915	0.2546
14 000	−56.3	0.1399	0.1860
16 000	−56.3	0.1022	0.1359
18 000	−56.3	0.07466	0.09930
20 000	−56.3	0.05457	0.07258
30 000	−46.5	0.01181	0.01503

$P_0 = 101.3$ kPa, $\rho_0 = 1.225$ kg/m^3

Table B.2 Various Ideal Gases

Gas	Chemical Formula	Molar Mass	R, kJ/kg·K	C_p, kJ/kg·K	C_v, kJ/kg·K	k
Air	—	28.97	0.2870	1.003	0.717	1.400
Argon	Ar	39.95	0.2081	0.520	0.312	1.667
Butane	C_4H_{10}	58.12	0.1430	1.716	1.573	1.091
Carbon dioxide	CO_2	44.01	0.1889	0.842	0.653	1.289
Carbon monoxide	CO	28.01	0.2968	1.041	0.744	1.400
Ethane	C_2H_6	30.07	0.2765	1.766	1.490	1.186
Ethylene	C_2H_4	28.05	0.2964	1.548	1.252	1.237
Helium	He	4.00	2.0770	5.198	3.116	1.667
Hydrogen	H_2	2.02	4.1242	14.209	10.085	1.409
Methane	CH_4	16.04	0.5184	2.254	1.735	1.299
Neon	Ne	20.18	0.4120	1.020	0.618	1.667
Nitrogen	N_2	28.01	0.2968	1.042	0.745	1.400
Octane	C_8H_{18}	114.23	0.0728	1.711	1.638	1.044
Oxygen	O_2	32.00	0.2598	0.922	0.662	1.393
Propane	C_3H_8	44.10	0.1886	1.679	1.491	1.126
Steam	H_2O	18.02	0.4615	1.872	1.411	1.327

Table B.3 Critical Point Constants

Substance	Formula	Molar Mass	T_{cr}, K	P_{cr}, MPa	v, m^3/kmol	Z_{cr}
Air		28.97	133	3.77	0.0883	0.30
Ammonia	NH_3	17.03	405.5	11.28	0.0724	0.243
Argon	Ar	39.94	151	4.86	0.0749	0.290
Butane	C_4H_{10}	58.12	425.2	3.80	0.2547	0.274
Carbon dioxide	CO_2	44.01	304.2	7.39	0.0943	0.275
Carbon monoxide	CO	28.01	133	3.50	0.0930	0.294
Ethane	C_2H_6	30.07	305.5	4.88	0.148	0.284
Ethylene	C_2H_4	28.05	282.4	5.12	0.1242	0.271
Helium	He	4.00	5.3	0.23	0.0578	0.302
Hydrogen	H_2	2.02	33.3	1.30	0.0649	0.304
Methane	CH_4	16.04	191.1	4.64	0.0993	0.290
Neon	Ne	20.18	44.5	2.73	0.0417	0.308
Nitrogen	N_2	28.02	126.2	3.39	0.0899	0.291
Oxygen	O_2	32.00	154.8	5.08	0.078	0.308
Propane	C_3H_8	44.09	370.0	4.26	0.1998	0.277
R134a	CF_3CH_2F	102.03	374.3	4.07	0.2478	0.324
Water	H_2O	18.02	647.4	22.1	0.0568	0.233

SOURCE: K. A. Kobe and R. E. Lynn, Jr., *Chem. Rev.*, 52: 117–236 (1953).

Table B.4 Specific Heats of Liquids and Solids

Liquids					
Substance	**State**	C_p	**Substance**	**State**	C_p
Water	1 atm, 25°C	4.177	Glycerin	1 atm, 10°C	2.32
Ammonia	sat., –20°C	4.52	Bismuth	1 atm, 425°C	0.144
	sat., 50°C	5.10	Mercury	1 atm, 10°C	0.138
Freon 12	sat., –20°C	0.908	Sodium	1 atm, 95°C	1.38
	sat., 50°C	1.02	Propane	1 atm, 0°C	2.41
Benzene	1 atm, 15°C	1.80	Ethyl alcohol	1 atm, 25°C	2.43
Solids					
Substance	T, °C	C_p	**Substance**	T, °C	C_p
Ice	–11	2.033	Lead	–100	0.118
	–2.2	2.10		0	0.124
Aluminum	–100	0.699		100	0.134
	0	0.870	Copper	–100	0.328
	100	0.941		0	0.381
Iron	20	0.448		100	0.393
Silver	20	0.233			

C_p kJ/kg·K

Table B.5 Enthalpy of Formation and Enthalpy of Vaporization

Substance	Formula	$\bar{h}_f^{\circ}$, kJ/kmol	$\bar{h}_{fg}$, kJ/kmol
Carbon	$C(s)$	0	
Hydrogen	$H_2(g)$	0	
Nitrogen	$N_2(g)$	0	
Oxygen	$O_2(g)$	0	
Carbon monoxide	$CO(g)$	−110 530	
Carbon dioxide	$CO_2(g)$	−393 520	
Water	$H_2O(g)$	−241 820	
Water	$H_2O(l)$	−285 830	44 010
Hydrogen peroxide	$H_2O_2(g)$	−136 310	61 090
Ammonia	$NH_3(g)$	−46 190	
Oxygen	$O_2(g)$	249 170	
Hydrogen	$H_2(g)$	218 000	
Nitrogen	$N_2(g)$	472 680	
Hydroxyl	$OH(g)$	39 040	
Methane	$CH_4(g)$	−74 850	
Acetylene (Ethyne)	$C_2H_2(g)$	226 730	
Ethylene (Ethene)	$C_2H_4(g)$	52 280	
Ethane	$C_2H_6(g)$	−84 680	
Propylene (Propene)	$C_3H_6(g)$	20 410	
Propane	$C_3H_8(g)$	−103 850	15 060
n-Butane	$C_4H_{10}(g)$	−126 150	21 060
Benzene	$C_6H_6(g)$	82 930	33 830
Methyl alcohol	$CH_3OH(g)$	−200 890	37 900
Ethyl alcohol	$C_2H_5OH(g)$	−235 310	42 340

25°C, 1 atm

SOURCES: JANAF Thermochemical Tables, NSRDS-NBS-37, 1971; *Selected Values of Chemical Thermodynamic Properties,* NBS Technical Note 270-3, 1968; and API Res. Project 44, Carnegie Press, Carnegie Institute of Technology, Pittsburgh, 1953.

Table B.6 Enthalpy of Combustion and Enthalpy of Vaporization

Substance	Formula	– HHV, kJ/kmol	$\bar{h}_{fg}$, kJ/kmol
Hydrogen	$H_2(g)$	–285 840	
Carbon	$C(s)$	–393 520	
Carbon monoxide	$CO(g)$	–282 990	
Methane	$CH_4(g)$	–890 360	
Ethylene	$C_2H_4(g)$	–1 410 970	
Ethane	$C_2H_6(g)$	–1 559 900	
Propane	$C_3H_8(g)$	–2 220 000	15 060
n-Butane	$C_4H_{10}(g)$	–2 877 100	21 060
Benzene	$C_6H_6(g)$	–3 301 500	33 830
Methyl alcohol	$CH_3OH(g)$	–764 540	37 900

25°C, 1 atm

SOURCE: Kenneth Wark, *Thermodynamics*, 3d ed., McGraw-Hill, New York, 1981, Table A-23M.

Table B.7 Constants for van der Waals and Redlich-Kwong Equations

	Van der Waals equation		Redlich-Kwong equation	
Substance	a, kPa·m^6/kg^2	b, m^3/kg	a, kPa·m^6·K$^{1/2}$/kg^2	b, m^3/kg
Air	0.1630	0.00127	1.905	0.000878
Ammonia	1.468	0.00220	30.0	0.00152
Carbon dioxide	0.1883	0.000972	3.33	0.000674
Carbon monoxide	0.1880	0.00141	2.20	0.000978
Freon 12	0.0718	0.000803	1.43	0.000557
Helium	0.214	0.00587	0.495	0.00407
Hydrogen	6.083	0.0132	35.5	0.00916
Methane	0.888	0.00266	12.43	0.00184
Nitrogen	0.1747	0.00138	1.99	0.000957
Oxygen	0.1344	0.000993	1.69	0.000689
Propane	0.481	0.00204	9.37	0.00141
Water	1.703	0.00169	43.9	0.00117

APPENDIX C

Steam Tables

Table C.1 Properties of Saturated H_2O—Temperature Table

T, °C	P, MPa	v, m³/kg		u, kJ/kg		h, kJ/kg			s, kJ/kg·K		
		v_f	v_g	u_f	u_g	h_f	h_{fg}	h_g	s_f	s_{fg}	s_g
0.01	0 0.000611	0.001000	206.1	0.0	2375.3	0.0	2501.3	2501.3	0.0000	9.1571	9.1571
2	0.0007056	0.001000	179.9	8.4	2378.1	8.4	2496.6	2505.0	0.0305	9.0738	9.1043
5	0.0008721	0.001000	147.1	21.0	2382.2	21.0	2489.5	2510.5	0.0761	8.9505	9.0266
10	0.001228	0.001000	106.4	42.0	2389.2	42.0	2477.7	2519.7	0.1510	8.7506	8.9016
20	0.002338	0.001002	57.79	83.9	2402.9	83.9	2454.2	2538.1	0.2965	8.3715	8.6680
30	0.004246	0.001004	32.90	125.8	2416.6	125.8	2430.4	2556.2	0.4367	8.0174	8.4541
40	0.007383	0.001008	19.52	167.5	2430.1	167.5	2406.8	2574.3	0.5723	7.6855	8.2578
50	0.01235	0.001012	12.03	209.3	2443.5	209.3	2382.8	2592.1	0.7036	7.3735	8.0771
60	0.01994	0.001017	7.671	251.1	2456.6	251.1	2358.5	2609.6	0.8310	7.0794	7.9104
70	0.03119	0.001023	5.042	292.9	2469.5	293.0	2333.8	2626.8	0.9549	6.8012	7.7561
80	0.04739	0.001029	3.407	334.8	2482.2	334.9	2308.8	2643.7	1.0754	6.5376	7.6130

Table C.1 Properties of Saturated H_2O—Temperature Table (*Continued*)

T, °C	*P*, MPa	*v*, m³/kg		*u*, kJ/kg		*h*, kJ/kg			*s*, kJ/kg·K		
		v_f	v_g	u_f	u_g	h_f	h_{fg}	h_g	s_f	s_{fg}	s_g
90	0.07013	0.001036	2.361	376.8	2494.5	376.9	2283.2	2660.1	1.1927	6.2872	7.4799
100	0.1013	0.001044	1.673	418.9	2506.5	419.0	2257.0	2676.0	1.3071	6.0486	7.3557
110	0.1433	0.001052	1.210	461.1	2518.1	461.3	2230.2	2691.5	1.4188	5.8207	7.2395
120	0.1985	0.001060	0.8919	503.5	2529.2	503.7	2202.6	2706.3	1.5280	5.6024	7.1304
130	0.2701	0.001070	0.6685	546.0	2539.9	546.3	2174.2	2720.5	1.6348	5.3929	7.0277
140	0.3613	0.001080	0.5089	588.7	2550.0	589.1	2144.8	2733.9	1.7395	5.1912	6.9307
150	0.4758	0.001090	0.3928	631.7	2559.5	632.2	2114.2	2746.4	1.8422	4.9965	6.8387
160	0.6178	0.001102	0.3071	674.9	2568.4	675.5	2082.6	2758.1	1.9431	4.8079	6.7510
170	0.7916	0.001114	0.2428	718.3	2576.5	719.2	2049.5	2768.7	2.0423	4.6249	6.6672
180	1.002	0.001127	0.1941	762.1	2583.7	763.2	2015.0	2778.2	2.1400	4.4466	6.5866
190	1.254	0.001141	0.1565	806.2	2590.0	807.5	1978.8	2786.4	2.2363	4.2724	6.5087
200	1.554	0.001156	0.1274	850.6	2595.3	852.4	1940.8	2793.2	2.3313	4.1018	6.4331
210	1.906	0.001173	0.1044	895.5	2599.4	897.7	1900.8	2798.5	2.4253	3.9340	6.3593
220	2.318	0.001190	0.08620	940.9	2602.4	943.6	1858.5	2802.1	2.5183	3.7686	6.2869
230	2.795	0.001209	0.07159	986.7	2603.9	990.1	1813.9	2804.0	2.6105	3.6050	6.2155
240	3.344	0.001229	0.05977	1033.2	2604.0	1037.3	1766.5	2803.8	2.7021	3.4425	6.1446
250	3.973	0.001251	0.05013	1080.4	2602.4	1085.3	1716.2	2801.5	2.7933	3.2805	6.0738
260	4.688	0.001276	0.04221	1128.4	2599.0	1134.4	1662.5	2796.9	2.8844	3.1184	6.0028
270	5.498	0.001302	0.03565	1177.3	2593.7	1184.5	1605.2	2789.7	2.9757	2.9553	5.9310
280	6.411	0.001332	0.03017	1227.4	2586.1	1236.0	1543.6	2779.6	3.0674	2.7905	5.8579
290	7.436	0.001366	0.02557	1278.9	2576.0	1289.0	1477.2	2766.2	3.1600	2.6230	5.7830
300	8.580	0.001404	0.02168	1332.0	2563.0	1344.0	1405.0	2749.0	3.2540	2.4513	5.7053
310	9.856	0.001447	0.01835	1387.0	2546.4	1401.3	1326.0	2727.3	3.3500	2.2739	5.6239
320	11.27	0.001499	0.01549	1444.6	2525.5	1461.4	1238.7	2700.1	3.4487	2.0883	5.5370
330	12.84	0.001561	0.01300	1505.2	2499.0	1525.3	1140.6	2665.9	3.5514	1.8911	5.4425
340	14.59	0.001638	0.01080	1570.3	2464.6	1594.2	1027.9	2622.1	3.6601	1.6765	5.3366
350	16.51	0.001740	0.008815	1641.8	2418.5	1670.6	893.4	2564.0	3.7784	1.4338	5.2122
360	18.65	0.001892	0.006947	1725.2	2351.6	1760.5	720.7	2481.2	3.9154	1.1382	5.0536
370	21.03	0.002213	0.004931	1844.0	2229.0	1890.5	442.2	2332.7	4.1114	0.6876	4.7990
374.14	22.088	0.003155	0.003155	2029.6	2029.6	2099.3	0.0	2099.3	4.4305	0.0000	4.4305

SOURCES: Keenan, Keyes, Hill, and Moore, *Steam Tables,* Wiley, New York, 1969; G. J. Van Wylen and R. E. Sonntag, *Fundamentals of Classical Thermodynamics*, Wiley, New York, 1973.

Table C.2 Properties of Saturated H_2O—Pressure Table

P, MPa	T, °C	v, m³/kg		u, kJ/kg		h, kJ/kg			s, kJ/kg·K		
		v_f	v_g	u_f	u_g	h_f	h_{fg}	h_g	s_f	s_{fg}	s_g
0.0006	0.01	0.001000	206.1	0.0	2375.3	0.0	2501.3	2501.3	0.0000	9.1571	9.1571
0.0008	3.8	0.001000	159.7	15.8	2380.5	15.8	2492.5	2508.3	0.0575	9.0007	9.0582
0.001	7.0	0.001000	129.2	29.3	2385.0	29.3	2484.9	2514.2	0.1059	8.8706	8.9765
0.0012	9.7	0.001000	108.7	40.6	2388.7	40.6	2478.5	2519.1	0.1460	8.7639	8.9099
0.0014	12.0	0.001001	93.92	50.3	2391.9	50.3	2473.1	2523.4	0.1802	8.6736	8.8538
0.0016	14.0	0.001001	82.76	58.9	2394.7	58.9	2468.2	2527.1	0.2101	8.5952	8.8053
0.002	17.5	0.001001	67.00	73.5	2399.5	73.5	2460.0	2533.5	0.2606	8.4639	8.7245
0.003	24.1	0.001003	45.67	101.0	2408.5	101.0	2444.5	2545.5	0.3544	8.2240	8.5784
0.004	29.0	0.001004	34.80	121.4	2415.2	121.4	2433.0	2554.4	0.4225	8.0529	8.4754
0.006	36.2	0.001006	23.74	151.5	2424.9	151.5	2415.9	2567.4	0.5208	7.8104	8.3312
0.008	41.5	0.001008	18.10	173.9	2432.1	173.9	2403.1	2577.0	0.5924	7.6371	8.2295
0.01	45.8	0.001010	14.67	191.8	2437.9	191.8	2392.8	2584.6	0.6491	7.5019	8.1510
0.012	49.4	0.001012	12.36	206.9	2442.7	206.9	2384.1	2591.0	0.6961	7.3910	8.0871
0.014	52.6	0.001013	10.69	220.0	2446.9	220.0	2376.6	2596.6	0.7365	7.2968	8.0333
0.016	55.3	0.001015	9.433	231.5	2450.5	231.5	2369.9	2601.4	0.7719	7.2149	7.9868
0.018	57.8	0.001016	8.445	241.9	2453.8	241.9	2363.9	2605.8	0.8034	7.1425	7.9459
0.02	60.1	0.001017	7.649	251.4	2456.7	251.4	2358.3	2609.7	0.8319	7.0774	7.9093
0.03	69.1	0.001022	5.229	289.2	2468.4	289.2	2336.1	2625.3	0.9439	6.8256	7.7695
0.04	75.9	0.001026	3.993	317.5	2477.0	317.6	2319.1	2636.7	1.0260	6.6449	7.6709
0.06	85.9	0.001033	2.732	359.8	2489.6	359.8	2293.7	2653.5	1.1455	6.3873	7.5328
0.08	93.5	0.001039	2.087	391.6	2498.8	391.6	2274.1	2665.7	1.2331	6.2023	7.4354
0.1	99.6	0.001043	1.694	417.3	2506.1	417.4	2258.1	2675.5	1.3029	6.0573	7.3602
0.12	104.8	0.001047	1.428	439.2	2512.1	439.3	2244.2	2683.5	1.3611	5.9378	7.2980
0.14	109.3	0.001051	1.237	458.2	2517.3	458.4	2232.0	2690.4	1.4112	5.8360	7.2472
0.16	113.3	0.001054	1.091	475.2	2521.8	475.3	2221.2	2696.5	1.4553	5.7472	7.2025
0.18	116.9	0.001058	0.9775	490.5	2525.9	490.7	2211.1	2701.8	1.4948	5.6683	7.1631
0.2	120.2	0.001061	0.8857	504.5	2529.5	504.7	2201.9	2706.6	1.5305	5.5975	7.1280
0.3	133.5	0.001073	0.6058	561.1	2543.6	561.5	2163.8	2725.3	1.6722	5.3205	6.9927
0.4	143.6	0.001084	0.4625	604.3	2553.6	604.7	2133.8	2738.5	1.7770	5.1197	6.8967
0.6	158.9	0.001101	0.3157	669.9	2567.4	670.6	2086.2	2756.8	1.9316	4.8293	6.7609

Table C.2 Properties of Saturated H_2O—Pressure Table (*Continued*)

P, MPa	*T*, °C	*v*, m³/kg		*u*, kJ/kg		*h*, kJ/kg			*s*, kJ/kg·K		
		v_f	v_g	u_f	u_g	h_f	h_{fg}	h_g	s_f	s_{fg}	s_g
0.8	170.4	0.001115	0.2404	720.2	2576.8	721.1	2048.0	2769.1	2.0466	4.6170	6.6636
1	179.9	0.001127	0.1944	761.7	2583.6	762.8	2015.3	2778.1	2.1391	4.4482	6.5873
2	212.4	0.001177	0.09963	906.4	2600.3	908.8	1890.7	2799.5	2.4478	3.8939	6.3417
4	250.4	0.001252	0.04978	1082.3	2602.3	1087.3	1714.1	2801.4	2.7970	3.2739	6.0709
6	275.6	0.001319	0.03244	1205.4	2589.7	1213.3	1571.0	2784.3	3.0273	2.8627	5.8900
8	295.1	0.001384	0.02352	1305.6	2569.8	1316.6	1441.4	2758.0	3.2075	2.5365	5.7440
10	311.1	0.001452	0.01803	1393.0	2544.4	1407.6	1317.1	2724.7	3.3603	2.2546	5.6149
12	324.8	0.001527	0.01426	1472.9	2513.7	1491.3	1193.6	2684.9	3.4970	1.9963	5.4933
14	336.8	0.001611	0.01149	1548.6	2476.8	1571.1	1066.5	2637.6	3.6240	1.7486	5.3726
16	347.4	0.001711	0.00931	1622.7	2431.8	1650.0	930.7	2580.7	3.7468	1.4996	5.2464
18	357.1	0.001840	0.00749	1698.9	2374.4	1732.0	777.2	2509.2	3.8722	1.2332	5.1054
20	365.8	0.002036	0.00583	1785.6	2293.2	1826.3	583.7	2410.0	4.0146	0.9135	4.9281
22.09	374.14	0.00316	0.00316	2029.6	2029.6	2099.3	0.0	2099.3	4.4305	0.0	4.4305

Table C.3 Superheated Steam

T	*v*	*u*	*h*	*s*	*v*	*u*	*h*	*s*	*v*	*u*	*h*	*s*
	P = 0.010 MPa (45.81°C)				**P = 0.050 MPa (81.33°C)**				**P = 0.10 MPa (99.63°C)**			
Sat.	14.67	2438	2585	8.150	3.240	2484	2646	7.594	1.694	2506	2676	7.359
50	14.87	2444	2593	8.175	—	—	—	—	—	—	—	—
100	17.20	2516	2688	8.448	3.418	2512	2682	7.695	1.696	2507	2676	7.361
150	19.51	2588	2783	8.688	3.889	2586	2780	7.940	1.936	2583	2776	7.613
200	21.82	2661	2880	8.904	4.356	2660	2878	8.158	2.172	2658	2875	7.834
250	24.14	2736	2977	9.100	4.820	2735	2976	8.356	2.406	2734	2974	8.033
300	26.44	2812	3076	9.281	5.284	2811	3076	8.537	2.639	2810	3074	8.216
400	31.06	2969	3280	9.608	6.209	2969	3279	8.864	3.103	2968	3278	8.544
500	35.68	3132	3489	9.898	7.134	3132	3489	9.155	3.565	3132	3488	8.834
600	40.30	3302	3705	10.16	8.057	3302	3705	9.418	4.028	3302	3705	9.098

Table C.3 Superheated Steam (*Continued*)

T	*v*	*u*	*h*	*s*	*v*	*u*	*h*	*s*	*v*	*u*	*h*	*s*
700	44.91	3480	3929	10.40	8.981	3479	3928	9.660	4.490	3479	3928	9.340
800	49.52	3664	4159	10.63	9.904	3664	4159	9.885	4.952	3664	4159	9.565
900	54.14	3855	4396	10.84	10.83	3855	4396	10.10	5.414	3855	4396	9.777
1000	58.76	4053	4641	11.04	11.75	4053	4641	10.30	5.875	4053	4640	9.976
	***P* = 0.20 MPa (120.2°C)**				***P* = 0.30 MPa (133.6°C)**				***P* = 0.40 MPa (143.6°C)**			
Sat	.8857	2530	2707	7.127	.6058	2544	2725	6.992	.4625	2554	2739	6.896
150	.9596	2577	2769	7.280	.6339	2571	2761	7.078	.4708	2564	2753	6.930
200	1.080	2654	2870	7.507	.7163	2651	2866	7.312	.5342	2647	2860	7.171
250	1.199	2731	2971	7.709	.7964	2729	2968	7.517	.5951	2726	2964	7.379
300	1.316	2809	3072	7.893	.8753	2807	3069	7.702	.6548	2805	3067	7.566
400	1.549	2967	3277	8.222	1.032	2966	3276	8.033	.7726	2964	3273	7.898
500	1.781	3131	3487	8.513	1.187	3130	3486	8.325	.8893	3129	3485	8.191
600	2.013	3301	3704	8.777	1.341	3301	3703	8.589	1.006	3300	3702	8.456
700	2.244	3479	3928	9.019	1.496	3478	3927	8.832	1.122	3478	3926	8.699
800	2.475	3663	4158	9.245	1.650	3663	4158	9.058	1.237	3662	4157	8.924
900	2.706	3854	4396	9.457	1.804	3854	4395	9.269	1.353	3854	4395	9.136
1000	2.937	4052	4640	9.656	1.958	4052	4640	9.470	1.468	4052	4639	9.336
	***P* = 0.50 MPa (151.9°C)**				***P* = 0.60 MPa (158.8°C)**				***P* = 0.80 MPa (170.4°C)**			
Sat.	.3749	2561	2749	6.821	.3157	2567	2757	6.760	.2404	2577	2769	6.663
200	.4249	2643	2855	7.059	.3520	2639	2850	6.966	.2608	2631	2839	6.816
250	.4744	2724	2961	7.271	.3938	2721	2957	7.182	.2931	2716	2950	7.038
300	.5226	2803	3064	7.460	.4344	2801	3062	7.372	.3241	2797	3056	7.233
350	.5701	2883	3168	7.632	.4742	2881	3166	7.546	.3544	2878	3162	7.409
400	.6173	2963	3272	7.794	.5137	2962	3270	7.708	.3843	2960	3267	7.572
500	.7109	3128	3484	8.087	.5920	3128	3483	8.002	.4433	3126	3481	7.867
600	.8041	3300	3702	8.352	.6697	3299	3701	8.267	.5018	3298	3699	8.133
700	.8969	3478	3926	8.595	.7472	3477	3925	8.511	.5601	3476	3924	8.377
800	.9896	3662	4157	8.821	.8245	3662	4156	8.737	.6181	3661	4156	8.603
900	1.082	3854	4395	9.033	.9017	3853	4394	8.949	.6761	3853	4394	8.815
1000	1.175	4052	4639	9.233	.9788	4052	4639	9.148	.7340	4051	4638	9.015

Table C.3 Superheated Steam (*Continued*)

T	v	u	h	s	v	u	h	s	v	u	h	s
	P = 1.00 MPa (179.9°C)				**P = 1.20 MPa (188.0°C)**				**P = 1.40 MPa (195.1°C)**			
Sat.	.1944	2584	2778	6.586	.1633	2589	2785	6.5233	.1408	2593	2790	6.469
200	.2060	2622	2828	6.694	.1693	2613	2816	6.5898	.1430	2603	2803	6.498
250	.2327	2710	2943	6.925	.1923	2704	2935	6.8294	.1635	2698	2927	6.747
300	.2579	2793	3051	7.123	.2138	2789	3046	7.0317	.1823	2785	3040	6.953
350	.2825	2875	3158	7.301	.2345	2872	3154	7.2121	.2003	2869	3150	7.136
400	.3066	2957	3264	7.465	.2548	2955	3261	7.3774	.2178	2952	3258	7.303
500	.3541	3124	3478	7.762	.2946	3123	3476	7.6759	.2521	3321	3474	7.603
600	.4011	3297	3698	8.029	.3339	3296	3696	7.9435	.2860	3294	3695	7.871
700	.4478	3475	3923	8.273	.3729	3474	3922	8.1881	.3195	3474	3921	8.116
800	.4943	3660	4155	8.500	.4118	3660	4154	8.4148	.3528	3659	4153	8.843
900	.5407	3852	4393	8.712	.4505	3852	4392	8.6272	.3861	3851	4392	8.556
	P = 1.60 MPa (201.4°C)				**P = 1.80 MPa (207.2°C)**				**P = 2.00 MPa (212.4°C)**			
Sat.	.1238	2596	2794	6.422	.1104	2598	2797	6.3794	.099 6	2600	2800	6.341
250	.1418	2692	2919	6.673	.1250	2686	2911	6.6066	.1114	2680	2902	6.545
300	.1586	2781	3035	6.884	.1402	2777	3029	6.8226	.1255	2773	3024	6.766
350	.1746	2866	3145	7.069	.1546	2863	3141	7.0100	.1386	2860	3137	6.956
400	.1900	2950	3254	7.237	.1685	2948	3251	7.1794	.1512	2945	3248	7.128
500	.2203	3119	3472	7.539	.1955	3118	3470	7.4825	.1757	3116	3468	7.432
600	.2500	3293	3693	7.808	.2220	3292	3692	7.7523	.1996	3291	3690	7.702
700	.2794	3473	3920	8.054	.2482	3473	3918	7.9983	.2232	3471	3917	7.949
800	.3086	3658	4152	8.281	.2742	3658	4151	8.2258	.2467	3657	4150	8.176
900	.3377	3850	4391	8.494	.3001	3850	4390	8.4386	.2700	3849	4389	8.390
	P = 2.50 MPa (234.0°C)				**P = 3.00 MPa (233.9°C)**				**P = 3.50 MPa (242.6°C)**			
Sat	.0800	2603	2803	6.258	.0667	2604	2804	6.1869	.05701	2604	2803	6.125
250	.0870	2663	2880	6.408	.0706	2644	2856	6.2872	.0587	2624	2829	6.175
300	.0989	2762	3009	6.644	.0811	2750	2994	6.5390	.0684	2738	2978	6.446
350	.1098	2852	3126	6.840	.0905	2844	3115	6.7428	.0768	2835	3104	6.658
400	.1201	2939	3239	7.015	.0994	2933	3231	6.9212	.0845	2926	3222	6.840

Table C.3 Superheated Steam (*Continued*)

T	*v*	*u*	*h*	*s*	*v*	*u*	*h*	*s*	*v*	*u*	*h*	*s*
450	.1301	3026	3351	7.175	.1079	3020	3344	7.0834	.0920	3015	3337	7.005
500	.1400	3112	3462	7.323	.1162	3108	3456	7.2338	.0992	3103	3451	7.157
600	.1593	3288	3686	7.596	.1324	3285	3682	7.5085	.1132	3282	3678	7.434
700	.1783	3468	3914	7.844	.1484	3466	3912	7.7571	.1270	3464	3909	7.684
800	.1972	3655	4148	8.072	.1641	3654	4146	7.9862	.1406	3652	4144	7.913
900	.2159	3848	4388	8.285	.1798	3846	4386	8.1999	.1540	3845	4384	8.128
	P = 4.0 MPa (250.4°C)				**P = 4.5 MPa (257.5°C)**				**P = 5.0 MPa (264.0°C)**			
Sat.	.0498	2602	2801	6.070	.0441	2600	2798	6.020	.0394	2597	2794	5.973
300	.0588	2725	2961	6.362	.0514	2712	2943	6.283	.0453	2698	2924	6.208
350	.0664	2827	3092	6.582	.0584	2818	3081	6.513	.0519	2809	3068	6.449
400	.0734	2920	3214	6.769	.0648	2913	3205	6.705	.0578	2907	3196	6.646
450	.0800	3010	3330	6.936	.0707	3005	3323	6.875	.0633	3000	3316	6.819
500	.0864	3100	3445	7.090	.0765	3095	3440	7.030	.0686	3091	3434	6.976
600	.0988	3279	3674	7.369	.0876	3276	3670	7.311	.0787	3273	3666	7.259
700	.1110	3462	3906	7.620	.0985	3460	3903	7.563	.0885	3458	3900	7.512
800	.1229	3650	4142	7.850	.1091	3648	4139	7.794	.0981	3647	4137	7.744
900	.1347	3844	4382	8.065	.1196	3842	4381	8.009	.1076	3841	4379	7.959
	P = 6.0 MPa (275.6°C)				**P = 7.0 MPa (285.9°C)**				**P = 8.0 MPa (295.1°C)**			
Sat.	.0324	2590	2784	5.889	.0274	2580	2772	5.813	.0235	2570	2758	5.743
300	.0362	2667	2884	6.067	.0295	2632	2838	3.930	.0243	2591	2785	5.791
350	.0422	2790	3043	6.334	.0352	2769	3016	6.228	.0300	2748	2987	6.130
400	.0474	2893	3177	6.541	.0399	2879	3158	6.448	.0343	2864	3138	6.363
450	.0521	2989	3302	6.719	.0442	2978	3287	6.633	.0382	2967	3272	6.555
500	.0566	3082	3422	6.880	.0481	3073	3410	6.798	.0418	3064	3398	6.724
550	.0610	3175	3541	7.029	.0520	3167	3531	6.949	.0452	3160	3521	6.872
600	.0652	3267	3658	7.168	.0556	3261	3650	7.089	.0484	3254	3642	7.021
700	.0735	3453	3894	7.423	.0628	3448	3888	7.348	.0548	3444	3882	7.281
800	.0816	3643	4133	7.657	.0698	3640	4128	7.582	.0610	3636	4124	7.517
900	.0896	3838	4375	7.873	.0767	3835	4372	7.799	.0670	3832	4368	7.735

Table C.3 Superheated Steam (*Continued*)

T	*v*	*u*	*h*	*s*	*v*	*u*	*h*	*s*	*v*	*u*	*h*	*s*
	P = 9.0 MPa (303.4°C)				**P = 10.0 MPa (311.1°C)**				**P = 12.5 MPa (327.9°C)**			
Sat.	.0205	2558	2742	5.677	.0180	2544	2725	5.614	.0135	2505	2674	5.462
325	.0233	2647	2856	5.871	.0199	2610	2809	5.757	—	—	—	—
350	.0258	2724	2957	6.036	.0224	2699	2923	5.944	.0161	2625	2826	5:712
400	.0299	2848	3118	6.285	.0264	2832	3096	6.212	.0200	2789	3039	6.042
450	.0335	2955	3257	6.484	.0298	2943	3241	6.419	.0230	2912	3200	6.272
500	.0368	3055	3336	6.658	.0328	3046	3374	6.597	.0256	3022	3342	6.462
550	.0399	3152	3511	6.814	.0356	3145	3501	6.756	.0280	3125	3475	6.629
600	.0428	3248	3634	6.959	.0384	3242	3625	6.903	.0303	3225	3604	6.781
650	.0457	3344	3755	7.094	.0410	3338	3748	7.040	.0325	3324	3730	6.922
700	.0486	3439	3876	7.222	.0436	3435	3870	7.169	.0346	3423	3855	7.054
800	.0541	3632	4119	7.460	.0486	3629	4115	7.408	.0387	3620	4104	7.296
900	.0595	3829	4364	7.678	.0535	3826	4361	7.627	.0427	3819	4352	7.518
	P = 15.0 MPa (342.2°C)				**P = 17.5 MPa (354.8°C)**				**P = 20.0 MPa (365.8°C)**			
Sat.	.0103	2456	2610	5.310	.0079	2390	2529	5.142	.00583	2293	2410	4.927
400	.0156	2741	2976	5.881	.0124	2685	2903	5.721	.00994	2619	2818	5.554
450	.0184	2880	3156	6.140	.0152	2844	3110	6.018	.01270	2806	3060	5.902
500	.0208	2997	3309	6.344	.0174	2970	3274	6.238	.01477	2943	3238	6.140
550	.0229	3105	3449	6.520	.0193	3084	3421	6.423	.01656	3062	3394	6.335
600	.0249	3209	3582	6.678	.0211	3192	3560	6.587	.01818	3174	3538	6.505
650	.0268	3310	3712	6.822	.0223	3296	3694	6.736	.01969	3281	3675	6.658
700	.0286	3411	3840	6.957	.0243	3399	3825	6.874	.02113	3386	3809	6.799
800	.0321	3611	4092	7.204	.0274	3602	4081	7.124	.02385	3593	4070	7.054
900	.0355	3812	4344	7.428	.0303	3805	4335	7.351	.02645	3798	4326	7.283

T in °C, *v* in m^3/kg, *u* and *h* in kJ/kg, *s* in kJ/kg · K

Table C.4 Compressed Liquid

T	v	u	h	s	v	u	h	s	v	u	h	s
	P = 5 MPa (264.0°C)				P = 10 MPa (311.1°C)				P = 15 MPa (342.4°C)			
0	0.000998	0.04	5.04	0.0001	0.000995	0.09	10.04	0.0002	0.000993	0.15	15.05	0.0004
20	0.001000	83.65	88.65	0.296	0.000997	83.36	93.33	0.2945	0.000995	83.06	97.99	0.2934
40	0.001006	167.0	172.0	0.570	0.001003	166.4	176.4	0.5686	0.001001	165.8	180.78	0.5666
60	0.001015	250.2	255.3	0.828	0.001013	249.4	259.5	0.8258	0.001010	248.5	263.67	0.8232
80	0.001027	333.7	338.8	1.072	0.001024	332.6	342.8	1.0688	0.001022	331.5	346.81	1.0656
100	0.001041	417.5	422.7	1.303	0.001038	416.1	426.5	1.2992	0.001036	414.7	430.28	1.2955
120	0.001058	501.8	507.1	1.523	0.001055	500.1	510.6	1.5189	0.001052	498.4	514.19	1.5145
140	0.001077	586.8	592.2	1.734	0.001074	584.7	595.4	1.7292	0.001071	582.7	598.72	1.7242
160	0.001099	672.6	678.1	1.938	0.001095	670.1	681.1	1.9317	0.001092	667.7	684.09	1.9260
180	0.001124	759.6	765.2	2.134	0.001120	756.6	767.8	2.1275	0.001116	753.8	770.50	2.1210
200	0.001153	848.1	853.9	2.326	0.001148	844.5	856.0	2.3178	0.001143	841.0	858.2	2.3104
	P = 20 MPa (365.8°C)				P = 30 MPa				P = 50 MPa			
0	0.000990	0.19	20.01	0.0004	0.000986	0.25	29.82	0.0001	0.000977	0.20	49.03	0.0014
20	0.000993	82.77	102.6	0.2923	0.000989	82.17	111.8	0.2899	0.000980	81.00	130.02	0.2848
40	0.000999	165.2	185.2	0.5646	0.000995	164.0	193.9	0.5607	0.000987	161.9	211.21	0.5527
60	0.001008	247.7	267.8	0.8206	0.001004	246.1	276.2	0.8154	0.000996	243.0	292.79	0.8052
80	0.001020	330.4	350.8	1.0624	0.001016	328.3	358.8	1.0561	0.001007	324.3	374.70	1.0440
100	0.001034	413.4	434.1	1.2917	0.001029	410.8	441.7	1.2844	0.001020	405.9	456.89	1.2703
120	0.001050	496.8	517.8	1.5102	0.001044	493.6	524.9	1.5018	0.001035	487.6	539.39	1.4857
140	0.001068	580.7	602.0	1.7193	0.001062	576.9	608.8	1.7098	0.001052	569.8	622.35	1.6915
160	0.001088	665.4	687.1	1.9204	0.001082	660.8	693.3	1.9096	0.001070	652.4	705.92	1.8891
180	0.001112	751.0	773.2	2.1147	0.001105	745.6	778.7	2.1024	0.001091	735.7	790.25	2.0794
200	0.001139	837.7	860.5	2.3031	0.001130	831.4	865.3	2.2893	0.001115	819.7	875.5	2.2634

T in °C, v in m^3/kg, u and h in kJ/kg, s in kJ/kg · K

Table C.5 Saturated Solid—Vapor

		v, m^3/kg		u, kJ/kg			h, kJ/kg			s, kJ/kg·K		
T, °C	P, kPa	Sat. Solid $v_i \times 10^3$	Sat. Vapor v_g	Sat. Solid u_i	Subl. u_{ig}	Sat. Vapor u_g	Sat. Solid h_i	Subl. h_{ig}	Sat. Vapor h_g	Sat. Solid s_i	Subl. s_{ig}	Sat. Vapor s_g
0.01		1.091	206.1	−333.4	2709	2375	−333.4	2835	2501	−1.221	10.378	9.156
0	0.611	1.091	206.3	−333.4	2709	2375	−333.4	2835	2501	−1.221	10.378	9.157
−2	0.518	1.090	241.7s	−337.6	2710	2373	−337.6	2835	2498	−1.237	10.456	9.219
−4	0.438	1.090	283.8	−341.8	2712	2370	−341.8	2836	2494	−1.253	10.536	9.283
−6	0.369	1.090	334.2	−345.9	2713	2367	−345.9	2836	2490	−1.268	10.616	9.348
−8	0.611	1.089	394.4	−350.0	2714	2364	−350.0	2837	2487	−1.284	10.698	9.414
−10	0.260	1.089	466.7	−354.1	2716	2361	−354.1	3837	2483	−1.299	10.781	9.481
−12	0.218	1.089	553.7	−358.1	2717	2359	−358.1	2837	2479	−1.315	10.865	9.550
−14	0.182	1.088	658.8	−362.2	2718	2356	−362.2	2838	2476	−1.331	10.950	9.619
−16	0.151	1.088	786.0	−366.1	2719	2353	−366.1	2838	2472	−1.346	11.036	9.690
−20	0.104	1.087	1129	−374.0	2722	2348	−374.0	2838	2464	−1.377	11.212	9.835
−24	0.070	1.087	1640	−381.8	2724	2342	−381.8	2839	2457	−1.408	11.394	9.985
−28	0.047	1.086	2414	−389.4	2726	2336	−389.4	2839	2450	−1.439	11.580	10.141

SOURCES: Keenan, Keyes, Hill, and Moore, *Steam Tables,* Wiley, New York, 1969; G. J. Van Wylen and R. E. Sonntag, *Fundamentals of Classical Thermodynamics,* Wiley, New York, 1973.

APPENDIX D

R134a

Table D.1 Properties of Saturated R134a—Temperature Table

T, °C	P, kPa	v, m³/kg		u, kJ/kg		h, kJ/kg			s, kJ/kg · K	
		v_f	v_g	u_f	u_g	h_f	h_{fg}	h_g	s_f	s_g
−40	51.64	0.7055	0.3569	−0.04	204.45	0.00	222.88	222.88	0.0000	0.9560
−36	63.32	0.7113	0.2947	4.68	206.73	4.73	220.67	225.40	0.0201	0.9506
−32	77.04	0.7172	0.2451	9.47	209.01	9.52	218.37	227.90	0.0401	0.9456
−28	93.05	0.7233	0.2052	14.31	211.29	14.37	216.01	230.38	0.0600	0.9411
−24	111.60	0.7296	0.1728	19.21	213.57	19.29	213.57	232.85	0.0798	0.9370
−20	132.99	0.7361	0.1464	24.17	215.84	24.26	211.05	235.31	0.0996	0.9332
−16	157.48	0.7428	0.1247	29.18	218.10	29.30	208.45	237.74	0.1192	0.9298
−12	185.40	0.7498	0.1068	34.25	220.36	34.39	205.77	240.15	0.1388	0.9267
−8	217.04	0.7569	0.0919	39.38	222.60	39.54	203.00	242.54	0.1583	0.9239
−4	252.74	0.7644	0.0794	44.56	224.84	44.75	200.15	244.90	0.1777	0.9213

Table D.1 Properties of Saturated R134a—Temperature Table (*Continued*)

T, °C	P, kPa	v, m³/kg		u, kJ/kg		h, kJ/kg			s, kJ/kg·K	
		v_f	v_g	u_f	u_g	h_f	h_{fg}	h_g	s_f	s_g
0	292.82	0.7721	0.0689	49.79	227.06	50.02	197.21	247.23	0.1970	0.9190
4	337.65	0.7801	0.0600	55.08	229.27	55.35	194.19	249.53	0.2162	0.9169
8	387.56	0.7884	0.0525	60.43	231.46	60.73	191.07	251.80	0.2354	0.9150
12	442.94	0.7971	0.0460	65.83	233.63	66.18	187.85	254.03	0.2545	0.9132
16	504.16	0.8062	0.0405	71.29	235.78	71.69	184.52	256.22	0.2735	0.9116
20	571.60	0.8157	0.0358	76.80	237.91	77.26	181.09	258.36	0.2924	0.9102
24	645.66	0.8257	0.0317	82.37	240.01	82.90	177.55	260.45	0.3113	0.9089
28	726.75	0.8362	0.0281	88.00	242.08	88.61	173.89	262.50	0.3302	0.9076
32	815.28	0.8473	0.0250	93.70	244.12	94.39	170.09	264.48	0.3490	0.9064
36	911.68	0.8590	0.0223	99.47	246.11	100.25	166.15	266.40	0.3678	0.9053
40	1016.4	0.8714	0.0199	105.30	248.06	106.19	162.05	268.24	0.3866	0.9041
44	1129.9	0.8847	0.0177	111.22	249.96	112.22	157.79	270.01	0.4054	0.9030
48	1252.6	0.8989	0.0159	117.22	251.79	118.35	153.33	271.68	0.4243	0.9017
52	1385.1	0.9142	0.0142	123.31	253.55	124.58	148.66	273.24	0.4432	0.9004

T in °C, v in m³/kg, u and h in kJ/kg, s in kJ/kg · K

SOURCE: Tables D.1 through D.3 are based on equations from D. P. Wilson and R. S. Basu, "Properties of a New Stratospherically Safe Working Fluid—Refrigerant 134a," *ASHRAE Trans.*, Vol. 94, Pt. 2, 1988, pp. 2095–2118.

Table D.2 Properties of Saturated R134a—Pressure Table

P	T	v_f	v_g	u_f	u_g	h_f	h_{fg}	h_g	s_f	s_g
60	−37.07	0.7097	0.3100	3.14	206.12	3.46	221.27	224.72	0.0147	0.9520
80	−31.21	0.7184	0.2366	10.41	209.46	10.47	217.92	228.39	0.0440	0.9447
100	−26.43	0.7258	0.1917	16.22	212.18	16.29	215.06	231.35	0.0678	0.9395
120	−22.36	0.7323	0.1614	21.23	214.50	21.32	212.54	233.86	0.0879	0.9354
140	−18.80	0.7381	0.1395	25.66	216.52	25.77	210.27	236.04	0.1055	0.9322
160	−15.62	0.7435	0.1229	29.66	218.32	29.78	208.19	237.97	0.1211	0.9295
180	−12.73	0.7485	0.1098	33.31	219.94	33.45	206.26	239.71	0.1352	0.9273
200	−10.09	0.7532	0.0993	36.69	221.43	36.84	204.46	241.30	0.1481	0.9253
240	−5.37	0.7618	0.0834	42.77	224.07	42.95	201.14	244.09	0.1710	0.9222
280	−1.23	0.7697	0.0719	48.18	226.38	48.39	198.13	246.52	0.1911	0.9197

Table D.2 Properties of Saturated R134a—Pressure Table (*Continued*)

P	T	v_f	v_g	u_f	u_g	h_f	h_{fg}	h_g	s_f	s_g
320	2.48	0.7770	0.0632	53.06	228.43	53.31	195.35	248.66	0.2089	0.9177
360	5.84	0.7839	0.0564	57.54	230.28	57.82	192.76	250.58	0.2251	0.9160
400	8.93	0.7904	0.0509	61.69	231.97	62.00	190.32	252.32	0.2399	0.9145
500	15.74	0.8056	0.0409	70.93	235.64	71.33	184.74	256.07	0.2723	0.9117
600	21.58	0.8196	0.0341	78.99	238.74	79.48	179.71	259.19	0.2999	0.9097
800	31.33	0.8454	0.0255	92.75	243.78	93.42	170.73	264.15	0.3459	0.9066
1000	39.39	0.8695	0.0202	104.42	247.77	105.29	162.68	267.97	0.3838	0.9043

Table D.3 Superheated R134a

T	v	u	h	s	v	u	h	s
	P = 0.10 MPa (−26.43°C)				**P = 0.14 MPa (−18.80°C)**			
Sat.	0.19170	212.18	231.35	0.9395	0.13945	216.52	236.04	0.9322
−20	0.19770	216.77	236.54	0.9602	—	—	—	—
−10	0.20686	224.01	244.70	0.9918	0.14519	223.03	243.40	0.9606
0	0.21587	231.41	252.99	1.0227	0.15219	230.55	251.86	0.9922
10	0.22473	238.96	261.43	1.0531	0.15875	238.21	260.43	1.0230
20	0.23349	246.67	270.02	1.0829	0.16520	246.01	269.13	1.0532
30	0.24216	254.54	278.76	1.1122	0.17155	253.96	277.97	1.0828
40	0.25076	262.58	287.66	1.1411	0.17783	262.06	286.96	1.1120
50	0.25930	270.79	296.72	1.1696	0.18404	270.32	296.09	1.1407
60	0.26779	279.16	305.94	1.1977	0.19020	278.74	305.37	1.1690
70	0.27623	287.70	315.32	1.2254	0.19633	287.32	314.80	1.1969
80	0.28464	296.40	324.87	1.2528	0.20241	296.06	324.39	1.2244

Table D.3 Superheated R134a (*Continued*)

T	v	u	h	s	v	u	h	s
	P = 0.20 MPa (−10.09°C)				**P = 0.24 MPa (−5.37°C)**			
Sat.	0.09933	221.43	241.30	0.9253	0.08343	224.07	244.09	0.9222
0	0.10438	229.23	250.10	0.9582	0.08574	228.31	248.89	0.9399
10	0.10922	237.05	258.89	0.9898	0.08993	236.26	257.84	0.9721
20	0.11394	244.99	267.78	1.0206	0.09399	244.30	266.85	1.0034
30	0.11856	253.06	276.77	1.0508	0.09794	252.45	275.95	1.0339
40	0.12311	261.26	285.88	1.0804	0.10181	260.72	285.16	1.0637
50	0.12758	269.61	295.12	1.1094	0.10562	269.12	294.47	1.0930
60	0.13201	278.10	304.50	1.1380	0.10937	277.67	303.91	1.1218
70	0.13639	286.74	314.02	1.1661	0.11307	286.35	313.49	1.1501
80	0.14073	295.53	323.68	1.1939	0.11674	295.18	323.19	1.1780
	P = 0.28 MPa (−1.23°C)				**P = 0.32 MPa (2.48°C)**			
Sat.	0.07193	226.38	246.52	0.9197	0.06322	228.43	248.66	0.917
0	0.07240	227.37	247.64	0.9238	—	—	—	—
10	0.07613	235.44	256.76	0.9566	0.06576	234.61	255.65	0.942
20	0.07972	243.59	265.91	0.9883	0.06901	242.87	264.95	0.974
30	0.08320	251.83	275.12	1.0192	0.07214	251.19	274.28	1.006
40	0.08660	260.17	284.42	1.0494	0.07518	259.61	283.67	1.036
50	0.08992	268.64	293.81	1.0789	0.07815	268.14	293.15	1.066
60	0.09319	277.23	303.32	1.1079	0.08106	276.79	302.72	1.095
70	0.09641	285.96	312.95	1.1364	0.08392	285.56	312.41	1.124
80	0.09960	294.82	322.71	1.1644	0.08674	294.46	322.22	1.152
	P = 0.40 MPa (8.93°C)				**P = 0.60 MPa (21.58°C)**			
Sat.	0.05089	231.97	252.32	0.9145	0.03408	238.74	259.19 C	0.9097
10	0.05119	232.87	253.35	0.9182	—	—	—	—
20	0.05397	241.37	262.96	0.9515	—	—	—	—
30	0.05662	249.89	272.54	0.9837	0.03581	246.41	267.89	0.9388
40	0.05917	258.47	282.14	1.0148	0.03774	255.45	278.09	0.9719

Table D.3 Superheated R134a (*Continued*)

T	*v*	*u*	*h*	*s*	*v*	*u*	*h*	*s*
50	0.06164	267.13	291.79	1.0452	0.03958	264.48	288.23	1.0037
60	0.06405	275.89	301.51	1.0748	0.04134	273.54	298.35	1.0346
70	0.06641	284.75	311.32	1.1038	0.04304	282.66	308.48	1.0645
80	0.06873	293.73	321.23	1.1322	0.04469	291.86	318.67	1.0938
90	0.07102	302.84	331.25	1.1602	0.04631	301.14	328.93	1.1225
100	0.07327	312.07	341.38	1.1878	0.04790	310.53	339.27	1.1505
	P = 0.80 MPa (31.33°C)				**P = 1.00 MPa (39.39°C)**			
Sat.	0.02547	243.78	264.15	0.9066	0.02020	247.77	267.97	0.9043
40	0.02691	252.13	273.66	0.9374	0.02029	248.39	268.68	0.9066
50	0.02846	261.62	284.39	0.9711	0.02171	258.48	280.19	0.9428
60	0.02992	271.04	294.98	1.0034	0.02301	268.35	291.36	0.9768
70	0.03131	280.45	305.50	1.0345	0.02423	278.11	302.34	1.0093
80	0.03264	289.89	316.00	1.0647	0.02538	287.82	313.20	1.0405
90	0.03393	299.37	326.52	1.0940	0.02649	297.53	324.01	1.0707
100	0.03519	308.93	337.08	1.1227	0.02755	307.27	334.82	1.1000
110	0.03642	318.57	347.71	1.1508	0.02858	317.06	345.65	1.1286
120	0.03762	328.31	358.40	1.1784	0.02959	326.93	356.52	1.1567
	P = 1.2 MPa (46.32°C)				**P = 1.4 MPa (52.43°C)**			
Sat.	0.01663	251.03	270.99	0.9023	0.01405	253.74	273.40	0.9003
50	0.01712	254.98	275.52	0.9164	—	—	—	—
60	0.01835	265.42	287.44	0.9527	0.01495	262.17	283.10	0.9297
70	0.01947	275.59	298.96	0.9868	0.01603	272.87	295.31	0.9658
80	0.02051	285.62	310.24	1.0192	0.01701	283.29	307.10	0.9997
90	0.02150	295.59	321.39	1.0503	0.01792	293.55	318.63	1.0319
100	0.02244	305.54	332.47	1.0804	0.01878	303.73	330.02	1.0628
110	0.02335	315.50	343.52	1.1096	0.01960	313.88	341.32	1.0927
120	0.02423	325.51	354.58	1.1381	0.02039	324.05	352.59	1.1218

APPENDIX E

Ideal-Gas Tables

Table E.1 Properties of Air

T, K	h, kJ/kg	P_r	u, kJ/kg	v_r	$s°$, kJ/kg·K
220	219.97	0.4690	156.82	1346	1.3910
240	240.02	0.6355	171.13	1084	1.4782
260	260.09	0.8405	185.45	887.8	1.5585
280	280.13	1.0889	199.75	738.0	1.6328
300	300.19	1.3860	214.07	621.2	1.7020
340	340.42	2.149	242.82	454.1	1.8279
380	380.77	3.176	271.69	343.4	1.9401
420	421.26	4.522	300.69	266.6	2.0414
460	462.02	6.245	329.97	211.4	2.1341
500	503.02	8.411	359.49	170.6	2.2195
540	544.35	11.10	389.34	139.7	2.2991
580	586.04	14.38	419.55	115.7	2.3735

Table E.1 Properties of Air (*Continued*)

T, K	h, kJ/kg	P_r	u, kJ/kg	v_r	$s°$, kJ/kg·K
620	628.07	18.36	450.09	96.92	2.4436
660	670.47	23.13	481.01	81.89	2.5098
700	713.27	28.80	512.33	69.76	2.5728
740	756.44	35.50	544.02	59.82	2.6328
780	800.03	43.35	576.12	51.64	2.6901
820	843.98	52.49	608.59	44.84	2.7450
860	888.27	63.09	641.40	39.12	2.7978
900	932.93	75.29	674.58	34.31	2.8486
940	977.92	89.28	708.08	30.22	2.8975
980	1023.2	105.2	741.98	26.73	2.9447
1020	1068.9	123.4	776.10	23.72	2.9903
1060	1114.9	143.9	810.62	21.14	3.0345
1100	1161.1	167.1	845.33	18.896	3.0773
1140	1207.6	193.1	880.35	16.946	3.1188
1180	1254.3	222.2	915.57	15.241	3.1592
1220	1301.3	254.7	951.09	13.747	3.1983
1260	1348.6	290.8	986.90	12.435	3.2364
1300	1396.0	330.9	1022.8	11.275	3.2734
1340	1443.6	375.3	1058.9	10.247	3.3096
1380	1491.4	424.2	1095.3	9.337	3.3447
1420	1539.4	478.0	1131.8	8.526	3.3790
1460	1587.6	537.1	1168.5	7.801	3.4125
1500	1636.0	601.9	1205.4	7.152	3.4452
1540	1684.5	672.8	1242.4	6.569	3.4771
1580	1733.2	750.0	1279.6	6.046	3.5083
1620	1782.0	834.1	1317.0	5.574	3.5388
1700	1880.1	1025	1392.7	4.761	3.5979
1800	2003.3	1310	1487.2	3.944	3.6684
1900	2127.4	1655	1582.6	3.295	3.7354
2000	2252.1	2068	1678.7	2.776	3.7994

Table E.2 Properties of Nitrogen, N_2

T, K	$\bar{h}$, kJ/kmol	$\bar{u}$, kJ/kmol	$\bar{s}^\circ$, kJ/kmol·K
220	6 391	4 562	182.639
240	6 975	4 979	185.180
260	7 558	5 396	187.514
280	8 141	5 813	189.673
298	8 669	6 190	191.502
300	8 723	6 229	191.682
320	9 306	6 645	193.562
340	9 888	7 061	195.328
360	10 471	7 478	196.995
380	11 055	7 895	198.572
400	11 640	8 314	200.071
440	12 811	9 153	202.863
480	13 988	9 997	205.424
520	15 172	10 848	207.792
560	16 363	11 707	209.999
600	17 563	12 574	212.066
640	18 772	13 450	214.018
680	19 991	14 337	215.866
720	21 220	15 234	217.624
760	22 460	16 141	219.301
800	23 714	17 061	220.907
840	24 974	17 990	222.447
860	25 610	18 459	223.194
900	26 890	19 407	224.647
920	27 532	19 883	225.353
940	28 178	20 362	226.047
960	28 826	20 844	226.728
980	29 476	21 328	227.398
1020	30 784	22 304	228.706
1060	32 101	23 288	229.973
1100	33 426	24 280	231.199
1200	36 777	26 799	234.115
1300	40 170	29 361	236.831
1400	43 605	31 964	239.375
1600	50 571	37 268	244.028
1800	57 651	42 685	248.195
2000	64 810	48 181	251.969
2200	72 040	53 749	255.412
2400	79 320	59 366	258.580

Table E.3 Properties of Oxygen, O_2

T, K	$\bar{h}$, kJ/kmol	$\bar{u}$, kJ/kmol	$\bar{s}°$, kJ/kmol·K	T, K	$\bar{h}$, kJ/kmol	$\bar{u}$, kJ/kmol	$\bar{s}°$, kJ/kmol·K
220	6 404	4 575	196.171	720	21 845	15 859	232.291
240	6 984	4 989	198.696	760	23 178	16 859	234.091
260	7 566	5 405	201.027	800	24 523	17 872	235.810
280	8 150	5 822	203.191	840	25 877	18 893	237.462
298	8 682	6 203	205.033	880	27 242	19 925	239.051
300	8 736	6 242	205.213	920	28 616	20 967	240.580
320	9 325	6 664	207.112	960	29 999	22 017	242.052
340	9 916	7 090	208.904	1000	31 389	23 075	243.471
360	10 511	7 518	210.604	1040	32 789	24 142	244.844
380	11 109	7 949	212.222	1080	34 194	25 214	246.171
400	11 711	8 384	213.765	1120	35 606	26 294	247.454
420	12 314	8 822	215.241	1160	37 023	27 379	248.698
440	12 923	9 264	216.656	1200	38 447	28 469	249.906
460	13 535	9 710	218.016	1240	39 877	29 568	251.079
480	14 151	10 160	219.326	1280	41 312	30 670	252.219
500	14 770	10 614	220.589	1320	42 753	31 778	253.325
520	15 395	11 071	221.812	1360	44 198	32 891	254.404
540	16 022	11 533	222.997	1400	45 648	34 008	255.454
560	16 654	11 998	224.146	1600	52 961	39 658	260.333
580	17 290	12 467	225.262	1800	60 371	45 405	264.701
600	17 929	12 940	226.346	2000	67 881	51 253	268.655
640	19 219	13 898	228.429	2200	75 484	57 192	272.278
680	20 524	14 871	230.405	2400	83 174	63 219	275.625

Table E.4 Properties of Carbon Dioxide, CO_2

T, K	$\bar{h}$, kJ/kmol	$\bar{u}$, kJ/kmol	$\bar{s}°$, kJ/kmol·K
220	6 601	4 772	202.966
240	7 280	5 285	205.920
260	7 979	5 817	208.717
280	8 697	6 369	211.376
298	9 364	6 885	213.685
300	9 431	6 939	213.915
320	10 186	7 526	216.351
340	10 959	8 131	218.694
360	11 748	8 752	220.948
380	12 552	9 392	223.122
400	13 372	10 046	225.225
420	14 206	10 714	227.258
440	15 054	11 393	229.230
460	15 916	12 091	231.144
480	16 791	12 800	233.004
500	17 678	13 521	234.814
520	18 576	14 253	236.575
540	19 485	14 996	238.292
560	20 407	15 751	239.962
580	21 337	16 515	241.602
600	22 280	17 291	243.199
640	24 190	18 869	246.282
680	26 138	20 484	249.233

T, K	$\bar{h}$, kJ/kmol	$\bar{u}$, kJ/kmol	$\bar{s}°$, kJ/kmol·K
720	28 121	22 134	252.065
760	30 135	23 817	254.787
800	32 179	25 527	257.408
840	34 251	27 267	259.934
880	36 347	29 031	262.371
920	38 467	30 818	264.728
960	40 607	32 625	267.007
1000	42 769	34 455	269.215
1040	44 953	36 306	271.354
1080	47 153	38 174	273.430
1120	49 369	40 057	275.444
1160	51 602	41 957	277.403
1200	53 848	43 871	279.307
1240	56 108	45 799	281.158
1280	58 381	47 739	282.962
1320	60 666	49 691	284.722
1360	62 963	51 656	286.439
1400	65 271	53 631	288.106
1600	76 944	63 741	295.901
1800	88 806	73 840	302.884
2000	100 804	84 185	309.210
2200	112 939	94 648	314.988
2400	125 152	105 197	320.302

Table E.5 Properties of Carbon Monoxide, CO

T, K	$\bar{h}$, kJ/kmol	$\bar{u}$, kJ/kmol	$\bar{s}^\circ$, kJ/kmol·K
220	6 391	4 562	188.683
240	6 975	4 979	191.221
260	7 558	5 396	193.554
280	8 140	5 812	195.713
300	8 723	6 229	197.723
320	9 306	6 645	199.603
340	9 889	7 062	201.371
360	10 473	7 480	203.040
380	11 058	7 899	204.622
400	11 644	8 319	206.125
420	12 232	8 740	207.549
440	12 821	9 163	208.929
460	13 412	9 587	210.243
480	14 005	10 014	211.504
520	15 197	10 874	213.890
560	16 399	11 743	216.115
600	17 611	12 622	218.204
640	18 833	13 512	220.179
680	20 068	14 414	222.052
720	21 315	15 328	223.833
760	22 573	16 255	225.533
780	23 208	16 723	226.357
800	23 844	17 193	227.162
840	25 124	18 140	228.724
880	26 415	19 099	230.227
920	27 719	20 070	231.674
960	29 033	21 051	233.072
1000	30 355	22 041	234.421
1040	31 688	23 041	235.728
1080	33 029	24 049	236.992
1120	34 377	25 865	238.217
1160	35 733	26 088	239.407
1200	37 095	27 118	240.663
1240	38 466	28 426	241.686
1280	39 844′	29 201	242.780
1320	41 226	30 251	243.844
1360	42 613	31 306	244.880
1400	44 007	32 367	245.889
1600	51 053	37 750	250.592
1800	58 191	43 225	254.797
2000	65 408	48 780	258.600
2200	72 688	54 396	262.065
2400	80 015	60 060	265.253

SOURCE: JANAF Thermochemical Tables, NSRDS-NBS-37, 1971.

Table E.6 Properties of Water, H_2O

T, K	$\bar{h}$, kJ/kmol	$\bar{u}$, kJ/kmol	$\bar{s}°$, kJ/kmol·K	T, K	$\bar{h}$, kJ/kmol	$\bar{u}$, kJ/kmol	$\bar{s}°$, kJ/kmol·K
220	7 295	5 466	178.576	760	26 358	20 039	221.720
240	7 961	5 965	181.471	780	27 125	20 639	222.717
260	8 627	6 466	184.139	800	27 896	21 245	223.693
280	9 296	6 968	186.616	820	28 672	21 855	224.651
298	9 904	7 425	188.720	840	29 454	22 470	225.592
300	9 966	7 472	188.928	860	30 240	23 090	226.517
320	10 639	7 978	191.098	880	31 032	23 715	227.426
340	11 314	8 487	193.144	900	31 828	24 345	228.321
360	11 992	8 998	195.081	920	32 629	24 980	229.202
380	12 672	9 513	196.920	940	33 436	25 621	230.070
400	13 356	10 030	198.673	960	34 247	26 265	230.924
420	14 043	10 551	200.350	980	35 061	26 913	231.767
440	14 734	11 075	201.955	1000	35 882	27 568	232.597
460	15 428	11 603	203.497	1020	36 709	28 228	233.415
480	16 126	12 135	204.982	1040	37 542	28 895	234.223
500	16 828	12 671	206.413	1060	38 380	29 567	235.020
520	17 534	13 211	207.799	1080	39 223	30 243	235.806
540	18 245	13 755	209.139	1100	40 071	30 925	236.584
560	18 959	14 303	210.440	1120	40 923	31 611	237.352
580	19 678	14 856	211.702	1140	41 780	32 301	238.110
600	20 402	15 413	212.920	1160	42 642	32 997	238.859
620	21 130	15 975	214.122	1180	43 509	33 698	239.600
640	21 862	16 541	215.285	1200	44 380	34 403	240.333
660	22 600	17 112	216.419	1220	45 256	35 112	241.057
680	23 342	17 688	217.527	1240	46 137	35 827	241.773
700	24 088	18 268	218.610	1260	47 022	36 546	242.482
720	24 840	18 854	219.668	1280	47 912	37 270	243.183
740	25 597	19 444	220.707	1300	48 807	38 000	243.877

Table E.6 Properties of Water, H_2O (*Continued*)

T, K	$\bar{h}$, kJ/kmol	$\bar{u}$, kJ/kmol	$\bar{s}°$, kJ/kmol·K
1320	49 707	38 732	244.564
1340	50 612	39 470	245.243
1360	51 521	40 213	245.915
1400	53 351	41 711	247.241
1440	55 198	43 226	248.543
1480	57 062	44 756	249.820
1520	58 942	46 304	251.074
1560	60 838	47 868	252.305
1600	62 748	49 445	253.513
1700	67 589	53 455	256.450
1800	72 513	57 547	259.262
1900	77 517	61 720	261.969
2000	82 593	65 965	264.571
2100	87 735	70 275	267.081
2200	92 940	74 649	269.500
2300	98 199	79 076	271.839
2400	103 508	83 553	274.098
2500	108 868	88 082	276.286
2600	114 273	92 656	278.407
2700	119 717	97 269	280.462
2800	125 198	101 917	282.453
2900	130 717	106 605	284.390
3000	136 264	111 321	286.273
3100	141 846	116 072	288.102
3150	144 648	118 458	288.9
3200	147 457	120 851	289.884
3250	150 250	123 250	290.7

APPENDIX F

Psychrometric Chart

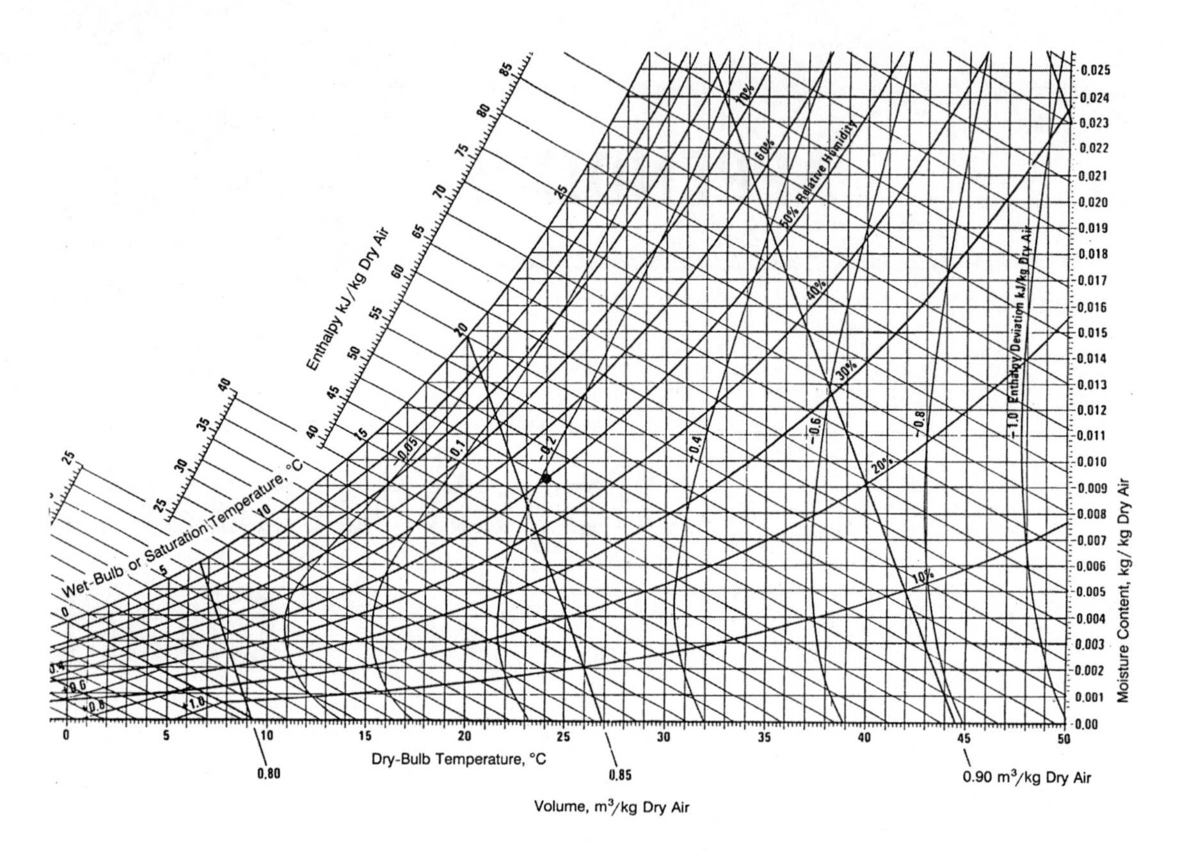

Psychrometric chart. $P = 101$ kPa. (Carrier Corporation.)

APPENDIX G

Compressibility Chart

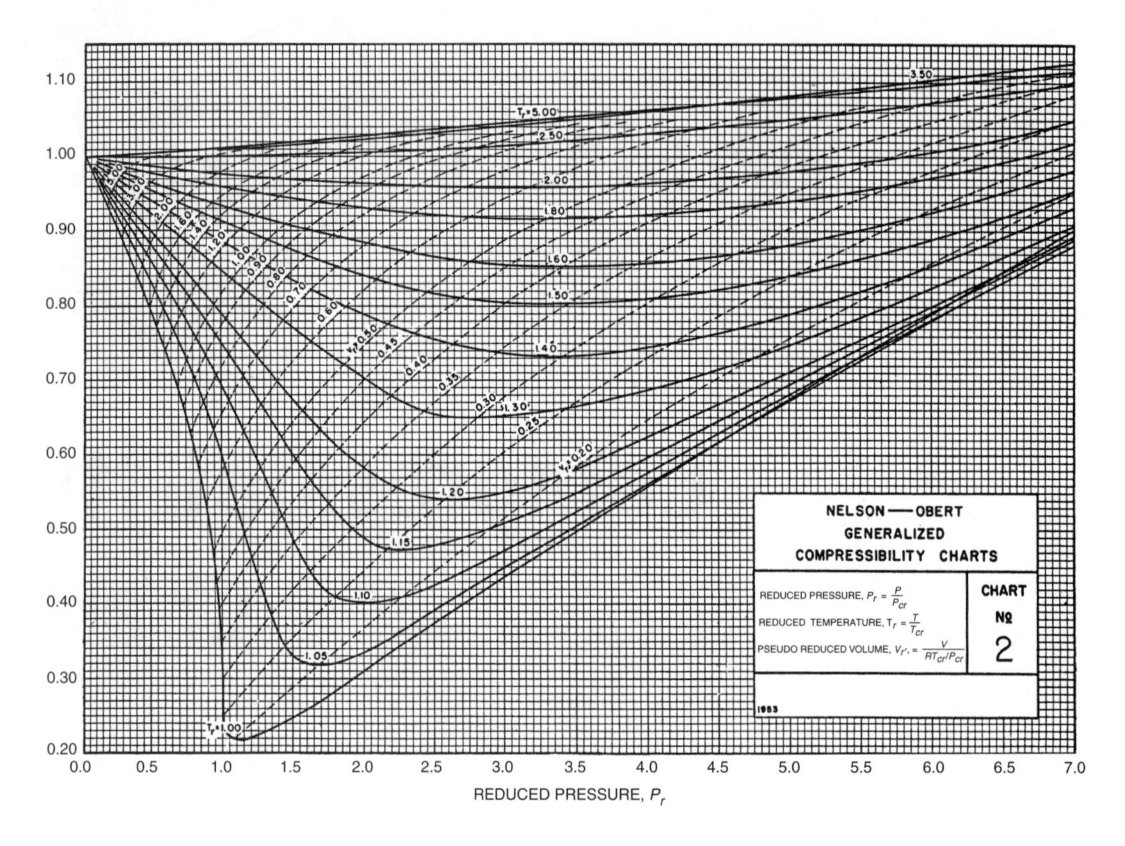

Compressibility chart. (V. M. Faires: *Problems on Thermodynamics*, 4th ed. NY: Macmillan, 1962.)

Final Exams

Final Exam* No. 1

1. Thermodynamics does not include the study of energy
   - (A) Storage
   - (B) Utilization
   - (C) Transfer
   - (D) Transformation
2. A joule can be converted to which of the following?
   - (A) $\text{Pa}\cdot\text{m}^2$
   - (B) $\text{N}\cdot\text{kg}$
   - (C) Pa/m^2
   - (D) $\text{Pa}\cdot\text{m}^3$

---

*I propose the following grades: D, 11–14; C, 15–20; B, 21–25; A, 26–40.

3. A gage pressure of 400 kPa acting on a 4-cm-diameter piston is resisted by an 800-N/m spring. How much is the spring compressed?

   (A) 0.628 m

   (B) 0.951 m

   (C) 1.32 m

   (D) 1.98 m

4. Estimate the enthalpy of steam at 288°C and 2 MPa.

   (A) 2931 kJ/kg

   (B) 2957 kJ/kg

   (C) 2972 kJ/kg

   (D) 2994 kJ/kg

5. Saturated water vapor is heated in a rigid container from 200 to 600°C. The final pressure is nearest

   (A) 3.30 MPa

   (B) 3.25 MPa

   (C) 3.20 MPa

   (D) 3.15 MPa

6. A 0.2-m^3 tire is filled with 25°C air until $P = 280$ kPa. The mass of air in the tire is nearest

   (A) 7.8 kg

   (B) 0.888 kg

   (C) 0.732 kg

   (D) 0.655 kg

7. Find the work produced if air expands from 0.2 to 0.8 m^3 if $P = 0.2 + 0.4V$ kPa.

   (A) 0.48 kJ

   (B) 0.42 kJ

   (C) 0.36 kJ

   (D) 0.24 kJ

8. Ten kilograms of saturated liquid water expands until $T_2 = 200°C$ while $P = \text{const} = 500$ kPa. The work is nearest

   (A) 1920 kJ

   (B) 2120 kJ

   (C) 2340 kJ

   (D) 2650 kJ

9. If $P_1 = 400$ kPa and $T_1 = 400°C$, what is T_2 when the 0.2-m² piston hits the stops?

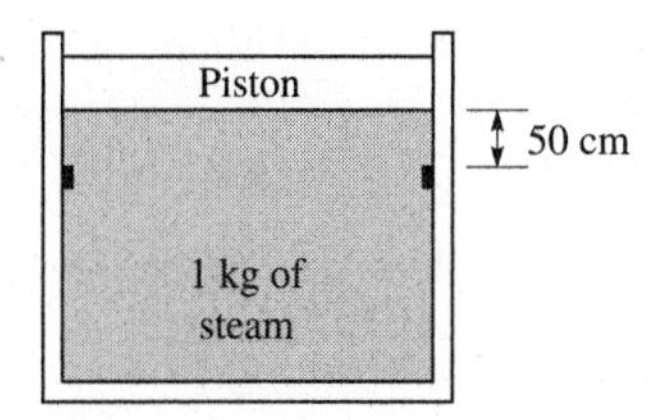

   (A) 314°C

   (B) 315°C

   (C) 316°C

   (D) 317°C

10. Calculate the heat released for the steam of Prob. 9.

   (A) 190 kJ

   (B) 185 kJ

   (C) 180 kJ

   (D) 175 kJ

11. After the piston hits the stops in Prob. 9, how much additional heat is released before $P = 100$ kPa?

   (A) 1580 kJ

   (B) 1260 kJ

   (C) 930 kJ

   (D) 730 kJ

12. One kilogram of air is compressed at $T = \text{const} = 100°\text{C}$ until the volume is halved. The rejected heat is
    (A) 42 kJ
    (B) 53 kJ
    (C) 67 kJ
    (D) 74 kJ
13. Five kilograms of copper at 200°C is submerged in 20 kilogram of water at 10°C in an insulated container. After time passes, the temperature will be nearest
    (A) 14°C
    (B) 16°C
    (C) 18°C
    (D) 20°C
14. Which of the following is not a control volume?
    (A) Insulated tank
    (B) Car radiator
    (C) Compressor
    (D) Turbine
15. A nozzle accelerates steam at 4 MPa and 500°C to 1 MPa and 300°C. If $V_1 = 20$ m/s, the exiting velocity is nearest
    (A) 575 m/s
    (B) 750 m/s
    (C) 825 m/s
    (D) 890 m/s
16. Calculate the mass flow rate at the nozzle exit of Prob. 15.
    (A) 2.9 kg/s
    (B) 2.3 kg/s
    (C) 1.8 kg/s
    (D) 1.1 kg/s
17. Calculate the power needed to raise the pressure of liquid water 4 MPa assuming the temperature remains constant and the mass flux is 5 kg/s.
    (A) 20 kW
    (B) 32 kW
    (C) 48 kW
    (D) 62 kW

18. A Carnot engine operates between reservoirs at 20 and 200°C. If 10 kW of power is produced, the rejected heat is nearest
    (A) 26.3 kJ/s
    (B) 20.2 kJ/s
    (C) 16.3 kJ/s
    (D) 12.0 kJ/s
19. It is proposed that a thermal engine can operate between ocean layers at 10 and 27°C and produce 10 kW of power while discharging 9900 kJ/min of heat. Such an engine is
    (A) Impossible
    (B) Reversible
    (C) Possible
    (D) Probable
20. Select the Kelvin-Planck statement of the second law.
    (A) An engine cannot produce more heat than the heat it receives.
    (B) A refrigerator cannot transfer heat from a low-temperature reservoir to a high-temperature reservoir without work.
    (C) An engine cannot produce work without discharging heat.
    (D) An engine discharges heat if the work is less than the heat it receives.
21. A Carnot refrigerator requires 10 kW to remove 20 kJ/s from a 20°C reservoir. What is T_H?
    (A) 440 K
    (B) 400 K
    (C) 360 K
    (D) 340 K
22. Find ΔS_{net} of a 10-kg copper mass at 100°C if submerged in a 20°C lake.
    (A) 1.05 kJ/K
    (B) 0.73 kJ/K
    (C) 0.53 kJ/K
    (D) 0.122 kJ/K

23. Find the maximum work output from a steam turbine if $P_1 = 10$ MPa, $T_1 = 600°C$, and $P_2 = 40$ kPa.
    (A) 1625 kJ/kg
    (B) 1545 kJ/kg
    (C) 1410 kJ/kg
    (D) 1225 kJ/kg
24. Find the minimum work input required by an air compressor if $P_1 =$ 100 kPa, $T_1 = 27°C$, and $P_2 = 800$ kPa.
    (A) 276 kJ/kg
    (B) 243 kJ/kg
    (C) 208 kJ/kg
    (D) 187 kJ/kg
25. Find the maximum outlet velocity for an air nozzle that operates between 400 and 100 kPa if $T_1 = 27°C$.
    (A) 570 m/s
    (B) 440 m/s
    (C) 380 m/s
    (D) 320 m/s
26. The efficiency of an adiabatic steam turbine operating between 10 MPa, 600°C, and 40 kPa is 80%. What is u_2?
    (A) 2560 kJ/kg
    (B) 2484 kJ/kg
    (C) 2392 kJ/kg
    (D) 2304 kJ/kg
27. Power cycles are idealized. Which of the following is not an idealization?
    (A) Friction is neglected.
    (B) Heat transfer does not occur across a finite temperature difference.
    (C) Pipes connecting components are insulated.
    (D) Pressure drops in pipes are neglected.

28. In the analysis of the Otto cycle, which do we not assume?
    (A) The combustions process is a constant-pressure process.
    (B) All processes are quasiequilibrium processes.
    (C) The exhaust process is replaced by a heat-rejection process.
    (D) The specific heats are assumed constant.
29. The four temperatures in an Otto cycle are $T_1 = 300$ K, $T_2 = 754$ K, $T_3 = 1600$ K, and $T_4 = 637$ K. If $P_1 = 100$ kPa, find P_2. Assume constant specific heats.
    (A) 2970 kPa
    (B) 2710 kPa
    (C) 2520 kPa
    (D) 2360 kPa
30. The heat added in the Otto cycle of Prob. 29 is nearest
    (A) 930 kJ/kg
    (B) 850 kJ/kg
    (C) 760 kJ/kg
    (D) 610 kJ/kg
31. The compression ratio of the Otto cycle of Prob. 29 is nearest
    (A) 11
    (B) 10
    (C) 9
    (D) 8
32. The four temperatures in an Brayton cycle are $T_1 = 300$ K, $T_2 = 500$ K, $T_3 = 1000$ K, and $T_4 = 600$ K. Find the heat added. Assume constant specific heats.
    (A) 200 kJ/kg
    (B) 300 kJ/kg
    (C) 400 kJ/kg
    (D) 500 kJ/kg

33. The pressure ratio of the Brayton cycle of Prob. 32 is nearest
    (A) 2
    (B) 4
    (C) 6
    (D) 8
34. The back work ratio w_C/w_T of the Brayton cycle of Prob. 32 is nearest
    (A) 0.8
    (B) 0.7
    (C) 0.6
    (D) 0.5
35. A Rankine cycle operates between 10 MPa and 20 kPa. If the high temperature of the steam is 600°C, the heat input is nearest
    (A) 3375 kJ/kg
    (B) 3065 kJ/kg
    (C) 2875 kJ/kg
    (D) 2650 kJ/kg
36. The turbine output of the Rankine cycle of Prob. 35 is nearest
    (A) 1350 kJ/kg
    (B) 1400 kJ/kg
    (C) 1450 kJ/kg
    (D) 1500 kJ/kg
37. The steam in the Rankine cycle of Prob. 35 is reheated at 4 MPa to 600°C. The total heat input is nearest
    (A) 3850 kJ/kg
    (B) 3745 kJ/kg
    (C) 3695 kJ/kg
    (D) 3625 kJ/kg
38. If the mass flux is 2400 kg/min, the pump power requirement in the Rankine cycle of Prob. 35 is nearest
    (A) 1580 kW
    (B) 1260 kW
    (C) 930 kW
    (D) 730 kW

39. Given $T = 25°C$ and $\omega = 0.015$, find ϕ. Assume $P_{atm} = 100$ kPa. (Use equations.)
    (A) 72.3%
    (B) 72.9%
    (C) 73.6%
    (D) 74.3%
40. Air at 5°C and 90% humidity is heated to 25°C. Estimate the final humidity.
    (A) 20%
    (B) 25%
    (C) 30%
    (D) 35%

Final Exam No. 2

1. Select the correct statement for a liquid.
    (A) The enthalpy depends on pressure only.
    (B) The enthalpy depends on temperature only.
    (C) The specific volume depends on pressure only.
    (D) The internal energy depends on temperature only.
2. In a quasiequilibrium process:
    (A) The pressure remains constant.
    (B) The pressure varies with temperature.
    (C) The pressure is everywhere constant at an instant.
    (D) The pressure force results in a positive work input.
3. Which of the following first-law statements is wrong?
    (A) The internal energy change equals the work of a system for an adiabatic process.
    (B) The heat transfer equals the enthalpy change for an adiabatic process.
    (C) The heat transfer equals the quasiequilibrium work of a system for a constant-volume process in which the internal energy remains constant.
    (D) The net heat transfer equals the work output for an engine operating on a cycle.

4. Which of the following statements about work for a quasiequilibrium processs is wrong?
   (A) Work is energy crossing a boundary.
   (B) The differential of work is inexact.
   (C) Work is the area under a curve on the *P*-*T* diagram.
   (D) Work is a path function.
5. The gage pressure at a depth of 20 m in water is nearest
   (A) 100 kPa
   (B) 200 kPa
   (C) 300 kPa
   (D) 400 kPa
6. One kilogram of steam in a cylinder requires 320 kJ of heat while $P = \text{const} = 1$ MPa. Find T_2 if $T_1 = 300°C$.
   (A) 480°C
   (B) 450°C
   (C) 420°C
   (D) 380°C
7. The work required in Prob. 6 is nearest
   (A) 85 kJ
   (B) 72 kJ
   (C) 66 kJ
   (D) 59 kJ
8. Two hundred kilograms of a solid at 0°C is added to 42 kg of water at 80°C. If $C_{p,s} = 0.8$ kJ/kg·K, the final temperature is nearest
   (A) 58°C
   (B) 52°C
   (C) 48°C
   (D) 42°C
9. The temperature of steam at 1600 kPa and $u = 3000$ kJ/kg is nearest
   (A) 470°C
   (B) 450°C
   (C) 430°C
   (D) 420°C

10. The enthalpy of steam at 400 kPa and $v = 0.7\ \text{m}^3/\text{kg}$ is nearest

    (A) 3125 kJ/kg

    (B) 3135 kJ/kg

    (C) 3150 kJ/kg

    (D) 3165 kJ/kg

11. The heat needed to raise the temperature of 0.1 m^3 of saturated steam from 120 to 500°C in a rigid tank is nearest

    (A) 306 kJ

    (B) 221 kJ

    (C) 152 kJ

    (D) 67 kJ

12. Estimate C_p for steam at 4 MPa and 350°C.

    (A) 2.62 kJ/kg·K

    (B) 2.53 kJ/kg·K

    (C) 2.44 kJ/kg·K

    (D) 2.32 kJ/kg·K

13. A cycle involving a system has 3 processes. Find $W_{1\text{-}2}$.

    (A) 50 kJ

    (B) 40 kJ

    (C) 30 kJ

    (D) 20 kJ

U_{before}	Q	W	U_{after}
1→2	40		30
2→3		−20	
3→1	10		
10			

14. Methane gas is heated at $P = \text{const} = 200$ kPa from 0 to 300°C. The required heat is

    (A) 766 kJ/kg

    (B) 731 kJ/kg

    (C) 692 kJ/kg

    (D) 676 kJ/kg

15. $Q_{2\text{-}3}$ for the cycle shown is nearest
    (A) −500 kJ
    (B) −358 kJ
    (C) −306 kJ
    (D) −278 kJ

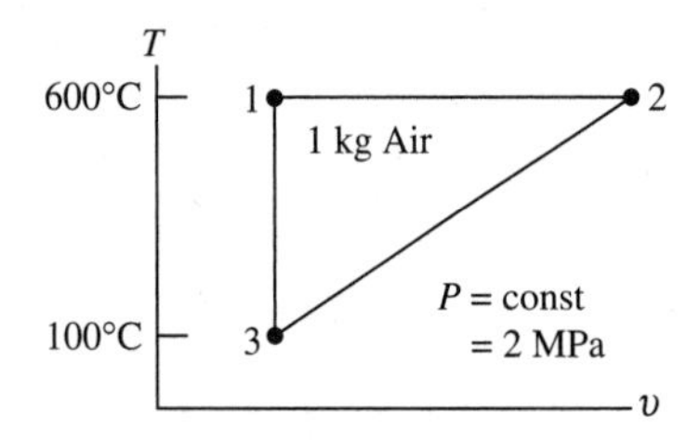

16. $W_{1\text{-}2}$ for the cycle shown in Prob. 15 is nearest
    (A) 500 kJ
    (B) 358 kJ
    (C) 213 kJ
    (D) 178 kJ
17. $Q_{1\text{-}2}$ for the cycle shown in Prob. 15 is nearest
    (A) 500 kJ
    (B) 358 kJ
    (C) 213 kJ
    (D) 178 kJ
18. R134a enters an expansion valve as saturated liquid at 16°C and exits at −20°C. Its quality at the exit is nearest
    (A) 0.3
    (B) 0.26
    (C) 0.22
    (D) 0.20
19. The work needed to adiabatically compress 1 kg/s of air from 20 to 200°C in a steady-flow process is nearest
    (A) 180 kW
    (B) 130 kW
    (C) 90 kW
    (D) 70 kW

20. If 200 kJ/kg of heat is added to air in a steady-flow heat exchanger, the temperature rise will be
    (A) 200°C
    (B) 220°C
    (C) 240°C
    (D) 260°C
21. A nozzle accelerates air from 10 m/s to 400 m/s. The temperature drop is
    (A) 20°C
    (B) 40°C
    (C) 60°C
    (D) 80°C
22. The net entropy change in the universe during any real process is
    (A) Zero
    (B) ≥ 0
    (C) ≤ 0
    (D) Any of these
23. An isentropic process is
    (A) Adiabatic and reversible.
    (B) Reversible but may not be adiabatic.
    (C) Adiabatic but may not be reversible.
    (D) Always reversible.
24. Which of the following second law statements is incorrect?
    (A) Heat must be rejected from a heat engine.
    (B) The entropy of an isolated process must remain constant or rise.
    (C) The entropy of a hot block decreases as it cools.
    (D) Work must be input if energy is transferred from a cold body to a hot body.
25. An inventor claims to extract 100 kJ of energy from 30°C sea water and generate 7 kJ of heat while rejecting 93 kJ to a 10°C stratum. The scheme fails because
    (A) The temperature of the sea water is too low.
    (B) It is a perpetual-motion machine.
    (C) It violates the first law.
    (D) It violates the second law.

26. Given the entropy relationship $\Delta s = C_p \ln T_2/T_1$. For which of the following is it incorrect?

    (A) Air, $P = \text{const}$

    (B) Water

    (C) A reservoir

    (D) Copper

27. Calculate the maximum power output of an engine operating between 600 and 20°C if the heat input is 100 kJ/s.

    (A) 83.3 kW

    (B) 77.3 kW

    (C) 66.4 kW

    (D) 57.3 kW

28. Ten kilograms of iron at 300°C is chilled in a large volume of ice and water. If $C_{\text{iron}} = 0.45$ kJ/kg·K, the net entropy change is nearest

    (A) 0.41 kJ/K

    (B) 0.76 kJ/K

    (C) 1.20 kJ/K

    (D) 1.61 kJ/K

29. A farmer has a pressurized air tank at 2000 kPa and 20°C. Estimate the exiting temperature from a valve. Assume an isentropic process.

    (A) −150°C

    (B) −120°C

    (C) −90°C

    (D) −40°C

30. Three temperatures in a diesel cycle are $T_{\text{inlet}} = T_1 = 300$ K, $T_2 = 950$ K, and $T_3 = 2000$ K. Estimate the heat input assuming constant specific heats.

    (A) 1150 kJ/kg

    (B) 1050 kJ/kg

    (C) 950 kJ/kg

    (D) 850 kJ/kg

31. The compression ratio for the diesel cycle of Prob. 30 is nearest
    (A) 19.8
    (B) 18.4
    (C) 17.8
    (D) 17.2
32. The temperature T_4 for the diesel cycle of Prob. 30 is nearest
    (A) 950 K
    (B) 900 K
    (C) 850 K
    (D) 800 K
33. Find q_B in kJ/kg.
    (A) 3400
    (B) 3200
    (C) 2900
    (D) 2600

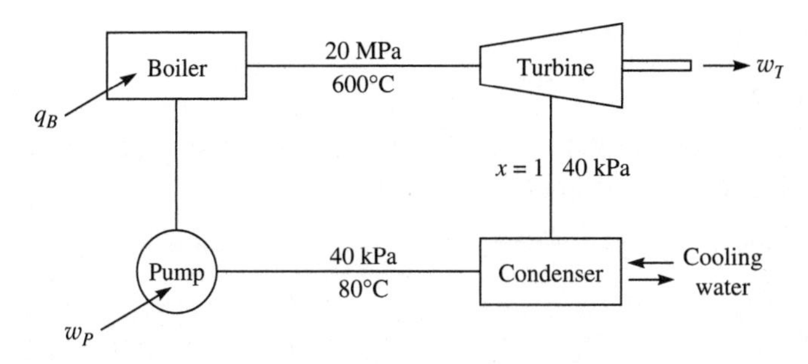

34. Find w_P in kJ/kg. Refer to the figure in Prob. 33.
    (A) 20
    (B) 30
    (C) 40
    (D) 50
35. Find w_T in kJ/kg. Refer to the figure in Prob. 33.
    (A) 1000
    (B) 950
    (C) 925
    (D) 900

36. Calculate the mass flux of the condenser cooling water if $\Delta T = 30°C$ and $\dot{m}_{steam} = 50$ kg/s. Refer to the figure in Prob. 33.

    (A) 1990 kg/s

    (B) 1920 kg/s

    (C) 1840 kg/s

    (D) 1770 kg/s

37. The outside temperature is 35°C. A thermometer reads 28°C when air blows over a wet cloth attached to its bulb. Estimate the humidity.

    (A) 50%

    (B) 55%

    (C) 60%

    (D) 65%

38. Outside air at 10°C and 100% humidity is heated to 30°C. How much heat is needed if $\dot{m}_{air} = 10$ kg/s?

    (A) 185 kJ/s

    (B) 190 kJ/s

    (C) 195 kJ/s

    (D) 205 kJ/s

39. Ethane (C_2H_6) is burned with dry air that contains 5 mol O_2 for each mole of fuel. Calculate the percent of excess air.

    (A) 52.3%

    (B) 46.2%

    (C) 42.9%

    (D) 38.7%

40. The air-fuel ratio (in kg air/kg fuel) of the combustion process in Prob. 39 is nearest

    (A) 17

    (B) 19

    (C) 21

    (D) 23

Solutions to Quizzes and Final Exams

Chapter 1

QUIZ NO. 1

1. B
2. B
3. C
4. D $h = \frac{P}{\gamma} = \frac{101300\ \text{N/m}^2}{6660\ \text{N/m}^3} = 15.21\ \text{m}$

5. A $PA = Kx \qquad 400\,000 \times \pi \times 0.02^2 = 800x \qquad \therefore x = 0.628 \text{ m}$

6. A The processes in an internal combustion engine are all thermodynamically slow except combustion and can be assumed to be quasiequilibrium processes. Heat transfer across a finite temperature difference is always a nonequilibrium process. A temperature difference in a room is also nonequilibrium.

7. D Its mass at sea level is $m = \frac{W_1}{g} = \frac{40}{9.81} = 4.077 \text{ kg}$

$\therefore W_2 = 4.077 \times 9.77 = 39.84 \text{ N}$

8. C $\rho = \frac{1}{v} = \frac{1}{20} = 0.05 \text{ kg/m}^3 \qquad \gamma = \rho g = 0.05 \times 9.81 = 0.4905 \text{ N/m}^3$

9. B $P_{\text{gage}} = \gamma_{\text{Hg}} h = (13.6 \times 9810) \times 0.3 = 40\,000 \text{ Pa}$

$\therefore P = 40 + 101 = 141 \text{ kPa}$

$$\Delta U = \Delta KE = \frac{1}{2} m_1 V_1^2 + \frac{1}{2} m_2 V_2^2 = \frac{1}{2} \times 1700 \times 22.78^2 + \frac{1}{2} \times 1400 \times 25^2$$
$$= 878\,000 \text{ J}$$

10. A If gage B reads zero, the pressure in the two compartments would be equal. So, $P_A > P_C$.

$P_C = P_A - P_B = 400 - 180 = 220 \text{ kPa gage} \quad \therefore P_C = 220 + 100 = 320 \text{ kPa}$

11. D The energy equation provides

$$\frac{1}{2} m \cancel{V_1^2} + mgh_1 = \frac{1}{2} m V_2^2 + mg \cancel{h_2} \qquad \therefore V_2 = \sqrt{2gh_1} = \sqrt{2 \times 9.81 \times 5}$$
$$= 9.90 \text{ m/s}$$

12. A $\Delta PE_{\text{spring}} = \int_0^x Kx\,dx = \frac{1}{2} Kx^2 = \Delta KE = \frac{1}{2} mV^2$

$$2 \times \frac{1}{2} K \times 0.1^2 = \frac{1}{2} \times 1800 \times 16^2$$

$\therefore K = 23 \times 10^6 \text{ N/m} \qquad \text{or} \qquad 23 \text{ MN/m}$

QUIZ NO. 2

1. C
2. C
3. C $\mathrm{J} = \mathrm{N} \cdot \mathrm{m} = \dfrac{\mathrm{N}}{\mathrm{m}^2} \cdot \mathrm{m}^2 \cdot \mathrm{m} = \mathrm{Pa} \cdot \mathrm{m}^3$ since $\mathrm{Pa} = \mathrm{N/m}^2$
4. A Since atmospheric pressure isn't given, it's assumed to be 100 kPa. Then,

   $P = \gamma_{\mathrm{Hg}} h_{\mathrm{Hg}} \quad 178\,000 + 100\,000 = (13.6 \times 9810) \times h_{\mathrm{Hg}} \quad \therefore h_{\mathrm{Hg}} = 2.084 \text{ m}$
5. B $PA = Kx + W \quad P \times \pi \times 0.12^2 = 2000 \times 0.6 + 40 \times 9.81 \quad \therefore P = 35\,200 \,\mathrm{Pa}$

   This pressure is gage pressure so the absolute pressure is

   $P = 35.2 + 100 = 135.2 \text{ kPa}$
6. D $m = \dfrac{9800}{9.79} = 1001 \text{ kg} \qquad W_2 = mg_2 = 1001 \times 9.83 = 9840 \text{ N}$
7. D $F - W = ma \qquad F = 900 \times 30 + 900 \times 9.81 = 35830 \text{ N}$
8. D $W = mg = \dfrac{V}{v} g = \dfrac{200}{10} \times 9.81 = 196.2 \text{ N}$
9. B $P = P_{\mathrm{atm}} + \gamma h = 100\,000 + 9810 \times 0.7 = 106870 \text{ Pa}$ or 107 kPa
10. B $F = P_{\mathrm{gage}} A = \gamma_{\mathrm{Hg}} h_{\mathrm{Hg}} A = (13.6 \times 9810) \times 0.710 \times \pi \times 0.1^2 = 2976 \text{ N}$
11. A The kinetic energy is converted to internal energy:

    $$\Delta U = \Delta KE = \frac{1}{2} mV^2 = \frac{1}{2} \times \frac{4}{9.81} \times 60^2 = 734 \text{ J}$$
12. C $V_1 = \dfrac{82 \times 1000}{3600} = 22.78 \text{ m/s}$ and $V_2 = \dfrac{90 \times 1000}{3600} = 25 \text{ m/s}$

    $$\Delta U = \Delta KE = \left(\frac{1}{2} 1700 \times 22.78^2 + \frac{1}{2} 1400 \times 25^2 \right) = 879\,000 \text{ J}$$

Chapter 2

QUIZ NO. 1

1. B

2. C $V = mv = 4 \times [0.001156 + 0.8(0.1274 - 0.0012)] = 0.409 \text{ m}^3$

3. D $v = v_f + x(v_g - v_f) \qquad 0.2 = 0.0011 + x(0.3044 - 0.0011)$

   $\therefore x = 0.656$

   We found v_g by interpolation ($v_f = 0.0011$ for 0.6 MPa and 0.8 MPa):

   $$v_g = \frac{0.8 - 0.63}{0.8 - 0.6}(0.3157 - 0.2404) + 0.2404 = 0.3044$$

4. A Table C.2: $m = \dfrac{V}{v_f} = \dfrac{1.2}{0.0011} = 1091 \text{ kg}$. When completely vaporized use

   v_g: $V = mv = 1091 \times 0.3157 = 344 \text{ m}^3$

5. A $v_2 = v_1 = 5.042 \text{ m}^3\text{/kg}$. Locate the state where $T_2 = 800°\text{C}$ and

   $v_2 \cong 5.0 \text{ m}^3\text{/kg}$. It is superheat at $P_2 \cong 0.10$ MPa ($v_2 = 4.95 \cong 5.0 \text{ m}^3\text{/kg}$)

6. B Interpolate to find $v_g = 0.69 \text{ m}^3\text{/kg}$. Then, with $v_f = 0.0011$,

   $v = 0.0011 + 0.6(0.69 - 0.0011) = 0.414 \text{ m}^3\text{/kg} \quad V = 3 \times 0.414 = 1.24 \text{ m}^3$

7. B $m_f = \dfrac{V_f}{v_f} = \dfrac{0.01 \times A}{0.001061} = 9.425A \qquad m_g = \dfrac{V_g}{v_g} = \dfrac{1 \times A}{0.8857} = 1.129A$

   $x = \dfrac{m_g}{m_g + m_f} = \dfrac{1.129}{1.129 + 9.425} = 0.107$ (Note: the A's divide out.)

8. A $v_1 = 0.0011 + 0.6(0.4625 - 0.0011) = 0.2779 \text{ m}^3\text{/kg}$

   $\therefore m = \dfrac{V}{v_1} = \dfrac{10}{0.2779} = 36.0 \text{ kg}$

   $v_2 = v_1 = 0.2779 = 0.0011 + x_2(0.6058 - 0.0011) \qquad \therefore x_2 = 0.4577$

   $m_g = x_2 \times m = 0.4577 \times 36.0 = 16.5 \text{ kg}$

9. B Use absolute values: $m=\frac{PV}{RT}=\frac{(280+100)\times 0.2}{0.287\times(25+273)}=0.8886\text{ kg}$

10. A $\rho_{\text{outside}}=\frac{P}{RT}=\frac{100}{0.287\times 253}=1.377\text{ kg/m}^3$

$$\rho_{\text{inside}}=\frac{100}{0.287\times 293}=1.189\text{ kg/m}^3 \quad \therefore \Delta\rho=1.377-1.189=0.188\text{ kg/m}^3$$

11. C $P=\frac{RT}{v-b}-\frac{a}{v^2}=\frac{0.297\times 220}{0.04-0.00138}-\frac{0.1747}{0.04^2}=1580\text{ kPa}$

12. C $P=\frac{mRT}{V}=\frac{10\times 0.462\times 673}{0.734}=4240\text{ kPa}$

Table C.3 with $T=400°\text{C}$ and $v=0.0734\text{ m}^3/\text{kg}$ gives $P=4.0\text{ MPa}$

$$\therefore \text{error}=\frac{4.24-4}{4}\times 100=6\%$$

QUIZ NO. 2

1. A

2. C The temperature is less than 212.4°C so it is compressed liquid. Therefore, the v_f at 200°C (ignore the pressure) must be used:

$$V=mv_f=10\times 0.001156=0.01156\text{ m}^3 \quad \text{or} \quad 11.6\text{ L}$$

3. C $v=v_f+x(v_g-v_f)=0.001156+0.8(0.1274-0.001156)=0.1022\text{ m}^3/\text{kg}$

4. A $v_g=\frac{1}{10}(0.07159-0.08620)+0.08620=0.08474 \qquad v_f=0.0012$

$$v=0.0012+0.85(0.08474-0.0012)=0.0722\text{ m}^3/\text{kg}$$

5. D $v_g=\frac{3}{5}(2.361-3.407)+3.407=2.78 \qquad v_f=0.001$

$$v=\frac{10}{5}=2=0.001+x(2.78-0.001) \qquad \therefore x=0.72$$

$$P=\frac{3}{5}(70.1-47.4)+47.4=61\text{ kPa}$$

6. B $v = v_f + x(v_g - v_f) = 0.001043 + 0.5(1.694 - 0.001043) = 0.848$

With $v_2 = 0.848\ \text{m}^3/\text{kg}$ and $T_2 = 200°\text{C}$, we find state 2 in the superheat region near $P_2 = 0.26$ MPa.

7. A $m_f = \dfrac{V_f}{v_f} = \dfrac{0.002}{0.00106} = 1.887\ \text{kg} \qquad m_g = \dfrac{V_g}{v_g} = \dfrac{1.2}{0.8919} = 1.345\ \text{kg}$

$$x = \frac{m_g}{m_g + m_f} = \frac{1.345}{1.345 + 1.887} = 0.416$$

8. B $P = P_{\text{atm}} + \dfrac{W}{A} = 100\,000 + \dfrac{48 \times 9.81}{\pi \times 0.01^2} = 1.6 \times 10^6\ \text{Pa} \quad \text{or} \quad 1.6\ \text{MPa}$

Table C.3: $v = 0.1452\ \text{m}^3/\text{kg} \qquad \therefore V = mv = 2 \times 0.1452 = 0.2904\ \text{m}^3$

9. C Since the temperature could be low and the pressure is quite high, we should check the Z factor. But first we estimate the temperature, assuming $Z \cong 0.9$ (just a guess):

$$T = \frac{PV}{mZR} = \frac{2000 \times 0.040}{2 \times 0.9 \times 0.287} = 155\ \text{K}$$

Use Table B.3 and App. G:

$$P_R = \frac{P}{P_c} = \frac{2}{3.77} = 0.53 \qquad T_R = \frac{T}{T_c} = \frac{155}{133} = 1.17 \qquad \therefore Z = 0.9$$

The guess was quite good so $T = 155$ K. Usually, a second iteration is needed.

10. B Assume the mass and the volume do not change. Then,

$$\frac{P_2}{P_1} = \frac{mRT_2V_1}{mRT_1V_2} \cong \frac{T_2}{T_1} \qquad \therefore P_2 = (240 + 100) \times \frac{65 + 273}{-30 + 273}$$

$$= 473 \quad \text{or} \quad 373\ \text{kPa}$$

11. D $T_R = \dfrac{220}{126.2} = 1.74 \qquad P_R = \dfrac{1590}{3390} = 0.469 \qquad \therefore Z = 0.98$

$$P = \frac{ZRT}{v} = \frac{0.98 \times 0.297 \times 220}{0.04} = 1600\ \text{kPa}$$

12. A $P = \dfrac{mRT}{V} = \dfrac{10 \times 0.462 \times 373}{17} = 101.4$ kPa. Table C.3 with $T = 100°$C

and $v = 1.7$ m³/kg gives $P = 0.10$ MPa. $\therefore$ error $= \dfrac{101-100}{100} \times 100 = 1\%$

Steam can be treated as an ideal gas when the state is near the saturation curve at low pressures.

Chapter 3

QUIZ NO. 1

1. C

2. A Use Table C.3: $W = Pm(v_2 - v_1) = 800 \times 10 \times (0.3843 - 0.2404) = 1151$ kJ

3. D $PA = P_{atm}A + mg \qquad \therefore P = P_{atm} + \dfrac{mg}{A} = 100\,000 + \dfrac{64 \times 9.81}{\pi \times 0.08^2} = 131\,200$ Pa

$$W_{1\text{-}2} = PA\Delta h = 131.2 \times \pi \times 0.08^2 \times 0.02 = 0.0528 \text{ kJ}$$

4. B

5. C $P_1A = W + P_{atm}A \qquad P_1 \times \pi \times 0.05^2 = 16 \times 9.81 + 100000 \times \pi \times 0.05^2$

$\therefore P_1 = 120000$ Pa or 120 kPa

$$v_1 = 0.001 + 0.2(1.428 - 0.001) = 0.286 \text{ m}^3/\text{kg}$$

$$m = \frac{V}{v} = \frac{\pi \times 0.05^2 \times 0.05}{0.286} = 0.00137 \text{ kg}$$

6. A At 120 kPa: $v_2 = \dfrac{V_2}{m} = \dfrac{\pi \times 0.05^2 \times 0.08}{0.001373} = 0.458 \text{ m}^3/\text{kg}$

At 120°C: $0.458 = 0.001 + x_2(1.428 - 0.001) \qquad \therefore x_2 = 0.320$ or 32%

7. D $v_3 = v_2 = 0.458$ and $T_2 = 400°$C. Interpolate in Table C.3:

$$P_3 = \frac{0.5137 - 0.458}{0.5137 - 0.3843}(0.8 - 0.6) + 0.6 = 0.69 \text{ MPa}$$

8. A There is no work after the stops are hit:

$$W = P(v_2 - v_1) = 120(0.458 - 0.286) = 20.6 \text{ kJ}$$

9. A $W = P(V_2 - V_1) = 200(0.02 - 0.2) = -36$ kJ

10. C For an isothermal process, $P_1V_1 = P_2V_2$.

$$W = mRT_1 \ln\frac{V_2}{V_1} = P_1V_1 \ln\frac{P_1}{P_2} = 100 \times 1000 \times 10^{-6} \ln\frac{100}{2000} = 0.30 \text{ kJ}$$

11. D Plot the P-V diagram. Six vertical rectangles will provide the work:

$$W = \Sigma P\Delta V \cong (34\,000 + 27\,500 + 24\,000 + 22\,500 + 21\,000 + 2000) \times 10^{-6}$$
$$= 0.13 \text{ kJ}$$

12. B $W = \int_0^{0.6} 10x^2 dx = 10 \times \frac{1}{3} \times 0.6^3 = 0.72 \text{ N} \cdot \text{m}$ or 0.72 J

13. C The rotational speed must be in rad/s:

$$\dot{W} = T\omega + V_i = 20 \times \frac{400 \times 2\pi}{60} + 20 \times 10 = 1038 \text{ J/s}$$

14. D $\dot{W} = V_i = 200 \times 12 = 2400$ W or $\frac{2400}{746} = 3.22$ hp

QUIZ NO. 2

1. D

2. C $W = P(V_2 - V_1) = P\left(\frac{mRT_2}{P} - \frac{mRT_1}{P}\right) = mR(T_2 - T_1)$

$$= 10 \times 0.287(400 - 170) = 660 \text{ kJ}$$

3. A $W = mP(v_2 - v_1) = 10 \times 400(0.5342 - 0.001) = 2130$ kJ

4. B Use Table C.2 at $P = 0.2$ MPa:

$$v_2 = v_1 = v_f + x(v_g - v_f) = 0.0011 + 0.1(0.8857 - 0.0011) = 0.08956$$

At $T_2 = 300°\text{C}$ and $v_2 = 0.090 \text{ m}^3/\text{kg}$, Table C.3 provides $P_2 \cong 2.75$ MPa.

5. D $P_1 = \frac{W}{A} + P_{\text{atm}} = \frac{64 \times 9.8}{\pi \times 0.1^2} + 100\,000 = 120\,000$ Pa

$$m_f = \frac{V}{v} = \frac{\pi \times 0.1^2 \times 0.0004}{0.001047} = 0.0120\,\text{kg}$$

$$m_g = \frac{\pi \times 0.1^2 \times 0.036}{1.428} = 0.000792\,\text{kg}$$

$$\therefore x_1 = \frac{0.000792}{0.012 + 0.000792} = 0.0619$$

6. C $v_2 = \frac{V_2}{m} = \frac{\pi \times 0.1^2 \times 0.0564}{0.012 + 0.000792} = 0.1385 = 0.00105 + x_2(1.428 - 0.00105)$

$$\therefore x_2 = 0.96$$

7. B $v_3 = v_2 = 0.1385$ and $T_3 = 220°\text{C}$. Go to Table C.3:

$$\left.\begin{matrix} T = 220°\text{C}, P = 1.5, v = 0.1403 \\ T = 220°\text{C}, P = 2.0, v = 0.1209 \end{matrix}\right\} \quad \therefore P_3 = \frac{0.1403 - 0.1385}{0.1403 - 0.1209} \times 0.5 + 1.5$$

$$= 1.52 \text{ MPa}$$

8. A $W = W_{1\text{-}2} + \cancel{W_{2\text{-}3}} = P(V_2 - V_1) = 120 \times \pi \times 0.1^2 \times 0.02 = 0.0754 \text{ kJ}$

9. D $W = \int P dV = \int \frac{mRT}{V} dV = mRT \int \frac{dV}{V}$

$$= P_1 V_1 \ln \frac{V_2}{V_1} = 200 \times 0.2 \times \ln \frac{0.02}{0.2} = -92 \text{ kJ}$$

10. C $W = W_{1\text{-}2} + W_{2\text{-}3} = P(V_2 - V_1) + P_2 V_2 \ln \frac{V_3}{V_2}$

$$= 200(0.3 - 0.1) + 200 \times 0.3 \ln \frac{0.5}{0.3} = 70.6 \text{ kJ}$$

11. A A sketch of the cycle will help visualize the processes.

$$W_{1\text{-}2} = pm(v_2 - v_1) = 200 \times 2 \times \left(\frac{0.287 \times 873}{200} - \frac{0.287 \times 373}{200} \right) = 287 \text{ kJ}$$

$$W_{3\text{-}1} = mRT \ln \frac{v_2}{v_1} = 2 \times 0.287 \times 373 \ln \frac{0.287 \times 873/200}{0.287 \times 373/200} = -182 \text{ kJ}$$

$$W_{2\text{-}3} = 0 \qquad \therefore W_{\text{net}} = 287 + (-182) + 0 = 105 \text{ kJ}$$

12. A $W = \int_{V_1}^{V_2} P\,dV = \int_{0.1}^{V_2} \frac{C}{V^2} dV \qquad C = P_1 V_1^2 = 100 \times 0.1^2 = 1$

$$V_2 = \sqrt{C/P_2} = \sqrt{1/800} = 0.03536 \text{ m}^3$$

$$\therefore W = -C\left(V_2^{-1} - 0.1^{-1}\right) = -1\left(\frac{1}{0.03536} - \frac{1}{0.1}\right) = -18.3 \text{ kJ}$$

13. B $W = Vi\Delta t = 120 \times 10 \times (10 \times 60) = 720\,000 \text{ J}$

14. D $W = \dot{W}\Delta t = Vi\Delta t = 110 \times 12 \times 600 = 792\,000 \text{ J}$

Chapter 4

QUIZ NO. 1

1. B
2. A For a rigid container the boundary work is zero:

$$q - \cancel{w} = \Delta u = u_2 - u_1 = 2960 - 2553.6 = 406 \text{ kJ/kg}$$

where we used $v_2 = v_1$ and interpolated for u_2 in Table C.3 as follows:

$$\left.\begin{aligned} v_2 &= v_1 = 0.4625 \\ T_2 &= 400°\text{C} \end{aligned}\right\} \qquad \therefore u_2 = 2960 \text{ kJ/kg}$$

3. C $Q - \cancel{W} = m\Delta u = \frac{V}{v}(u_2 - u_1) = \frac{0.3}{0.206}(3655.3 - 2621.9) = 1505 \text{ kJ}$

4. D $v_1 = 0.0011 + 0.95(0.2404 - 0.0011) = 0.2284$

$$\therefore m = \frac{V}{v} = \frac{1200 \times 10^{-6}}{0.2284} = 0.00525 \text{ kg}$$

$$\begin{aligned} Q &= m(u_2 - u_1) + W \\ &= 0.00525\{2960 - [720 + 0.95(1857)]\} + 800(0.3843 - 0.2284) \\ &= 127 \text{ kJ} \end{aligned}$$

5. C $Q = m(h_2 - h_1) \quad 170 = 1 \times (h_2 - 3108) \quad \therefore h_2 = 3278$ and $T_2 = 402°\text{C}$

We interpolated at 320°C in Table C.3 to find h_1 and then to find T_2.

6. D $W = mP(v_2 - v_1) = 1 \times 400 \times (0.7749 - 0.6784) = 38.6 \text{ kJ}$

We interpolated in Table C.3 to find v_1 and v_2 using $T_2 = 402°\text{C}$.

7. D In Table C.3 we observe that $u = 2949$ kJ/kg is between 1.6 MPa and 1.8 MPa but closer to 1.6 MPa so the closest answer is 1.7 MPa.

8. A The mass of the cubes is $m_i = \dfrac{V}{v} = \dfrac{16 \times 8 \times 10^{-6}}{0.00109} = 0.1174 \text{ kg}$ where we used Table C.5 for v. The heat gained by the ice is lost by the water:

$$m_i\left[(C_p)_i(\Delta T)_i + h_{if} + (C_p)_w(\Delta T)_{iw}\right] = m_w(C_p)_w(\Delta T)_w$$

$$0.1174\left[2.1 \times 10 + 330 + 4.18(T_2 - 0)\right] = (1000 \times 10^{-3}) \times 4.18(20 - T_2)$$

$$\therefore T_2 = 9.07°\text{C}$$

9. B $Q = m\,(u_2 - u_1) + mRT\ln\dfrac{P_1}{P_2} = 2 \times 0.287 \times 303\ln\dfrac{210}{2000} = -392 \text{ kJ}$

10. C $T_2 = T_1\left(\dfrac{P_2}{P_1}\right)^{(k-1)/k} = 293 \times 6^{0.2857} = 488.9 \text{ K}$

$$-W = \Delta U = mC_v(T_2 - T_1) = 2 \times 0.717 \times (489 - 293) = 281 \text{ kJ}$$

11. A $W = -\Delta U = -mC_v(T_2 - T_1) = -5 \times 0.717 \times 100 = -358 \text{ kJ}$

12. C $q = h_2 - h_1 = C_p(T_2 - T_1) = 2.254 \times (300 - 0) = 676 \text{ kJ/kg}$

13. D $T_2 = T_1\left(\dfrac{P_2}{P_1}\right)^{(k-1)/k} = 373\left(\dfrac{10\,000}{100}\right)^{0.4/1.4} = 1390 \text{ K}$

$$q - w = C_v(T_2 - T_1) = 0.717(1390 - 373) = 729 \text{ kJ/kg}$$

14. C $v_1 = 0.0011 + 0.5(0.4625 - 0.0011) = 0.2318 \text{ m}^3\text{/kg}$

$$u_1 = 604.3 + 0.5(2553.6 - 604.3) = 1579 \text{ kJ/kg}$$

$$Q - W = m(u_2 - u_1) \qquad 800 = \frac{0.15}{0.2318}(u_2 - 1579) \qquad \therefore u_2 = 2815 \text{ kJ/kg}$$

Search Table C.3 for the location of state 2, near 300°C and 1 MPa.

15. B

16. B $h_2 = h_1 = 3501$. Interpolate in Table C.3 at 0.8 MPa: $T_2 \cong 510°\text{C}$

17. B $0 = C_p\Delta T + \dfrac{V_2^2 - V_1^2}{2} \qquad 1.00\Delta T = \dfrac{20^2 - 200^2}{2 \times 1000} \qquad \therefore \Delta T = 19.8°\text{C}$

18. C For an incompressible liquid: $\dot{W}_P = \dot{m}\dfrac{\Delta P}{\rho} = 4 \times \dfrac{6000 - 100}{1000} = 23.6$ kW

19. A $V_2 = \dfrac{\rho_1 A_1 V_1}{\rho_2 A_2} = \dfrac{\dfrac{4000}{0.287 \times 573} \times (10 \times 10^{-4}) \times 150}{\dfrac{400}{0.287 \times 373} \times (50 \times 10^{-4})} = 195$ m/s

20. D $V_1 = \dfrac{20}{1000 \times \pi \times 0.1^2} = 0.637$ m/s $\qquad V_2 = \dfrac{20}{1000 \times \pi \times 0.06^2} = 1.768$ m/s

$$\dot{W}_P = \dot{m}\left(\frac{P_2 - P_1}{\rho} + \frac{V_2^2 - V_1^2}{2}\right) = 20\left(\frac{4000}{1000} + \frac{1.768^2 - 0.637^2}{2}\right) = 107 \text{ kW}$$

21. D $\dot{W}_T = -\rho \times A_1 V_1 \left(\cancel{\dfrac{V_2^2 - V_1^2}{2}} + \dfrac{P_2 - P_1}{\rho}\right) = -1000 \times 20 \times \dfrac{0 - 300}{1000} = 6000$ kW

22. A $T_2 = 298\left(\dfrac{500}{80}\right)^{0.4/1.4} = 503$ K

$$\dot{m} = \rho_2 A_2 V_2 = \frac{500}{0.287 \times 503} \times \pi \times 0.05^2 \times 100 = 2.72 \text{ kg/s}$$

$$\cancel{\dot{Q}} - (-\dot{W}_C) = \dot{m}C_p(T_2 - T_1) \qquad \therefore \dot{W}_C = 2.72 \times 1.00(503 - 293) = 558 \text{ kW}$$

23. C $\dot{m} = \rho_1 A_1 V_1 = \dfrac{800}{0.287 \times 473} \times \pi \times 0.05^2 \times 20 = 0.9257$ kg/s

$$\dot{Q} = \dot{m}C_p(T_2 - T_1) + \dot{W}_S = 0.9257 \times 1.0(200 - 20) + (-400) = -233 \text{ kJ/s}$$

24. A $0 = \dfrac{V_2^2 - V_1^2}{2} + C_p(T_2 - T_1)$

$$0 = \frac{15^2 - 200^2}{2 \times 1000} + 1.042(T_2 - 20) \qquad \therefore T_2 = -0.91°\text{C}$$

QUIZ NO. 2

1. A
2. A Use $Q - W = \Delta E$ and $\Sigma Q = \Sigma W$ and $\Sigma \Delta E = 0$.
3. D $Q - \cancel{W} = m\Delta u = \dfrac{V}{v}(u_2 - u_1)$

$$1000 = \frac{0.2}{0.0011 + 0.8(0.3157 - 0.0011)}\{u_2 - [669.9 + 0.8(2567.4 - 669.9)]\}$$

$\therefore u_2 = 3452$ kJ/kg and $v_2 = 0.2528$ m³/kg. Check Table C.3:

$$\left.\begin{array}{l} P_2 = 1.5 \text{ MPa}, v_2 = 0.2528, u_2 = 3216 \quad \therefore T_2 = 556°\text{C} \\ P_2 = 2.0 \text{ MPa}, v_2 = 0.2528, u_2 = 3706 \quad \therefore T_2 = 826°\text{C} \end{array}\right\} \quad \therefore T_2 = 686°\text{C}$$

4. D $v_1 = v_2 = 0.00103 + 0.5(3.407 - 0.001) = 1.704$ m³/kg

$$Q = m(u_2 - u_1) \qquad 800 = \frac{2}{1.704}\{u_2 - [334.8 + 0.5(2482.2 - 334.8)]\}$$

$\therefore u_2 = 2090$ kJ/kg and with $v_2 = 1.704$ m³/kg, $T_2 = 80°$C since $u_2 < u_g$

5. B $Q = m(h_2 - h_1) \qquad 20 = \dfrac{0.004}{0.2039}(h_2 - 3256) \qquad \therefore h_2 = 4276$ kJ/kg

With $P_2 = 1.5$ MPa and $h_2 = 4280$ kJ/kg, we find $T_2 = 824°$C.

6. C $q = h_2 - h_1 = 3273 - 604.7 = 2668$ kJ/kg

7. C $C_p = \dfrac{\Delta h}{\Delta T} = \dfrac{3213.6 - 2960.7}{400 - 300} = 2.53$ kJ/kg·°C using central differences.

8. B $m_c C_{p,c}(T_2 - 0) = m_w C_{p,w}(30 - T_2) \qquad 20 \times 0.38 \times T_2 = 10 \times 4.18 \times (30 - T_2)$

$\therefore T_2 = 25.4°$C. The energy lost by the water is gained by the copper.

9. D $Q = W = mRT \ln\dfrac{V_2}{V_1} = 1 \times 0.287 \times 373 \ln 0.5 = -74.2$ kJ

10. D $\not{Q} - W = m(u_2 - u_1) \quad W_{\text{paddle wheel}} = mC_v\Delta T = 5 \times 0.717 \times 100 = 358$ kJ

11. A $m = \dfrac{P\not{V}}{RT} = \dfrac{200 \times 2}{2.077 \times 323} = 0.596$ kg $\qquad T_2 = T_1\dfrac{P_2}{P_1} = 323\dfrac{800}{200} = 1292$ K

$$Q - \not{W} = mC_v\Delta T = 0.596 \times 3.116(1292 - 323) = 1800 \text{ kJ}$$

12. D $T_2 = T_1\left(\dfrac{P_2}{P_1}\right)^{(k-1)/k} = 293 \times 8^{0.2857} = 530.7$ K or 258°C

13. A $Q - \not{W} = mC_v(T_2 - T_1)$ $\quad m = \frac{P_1V_1}{RT_1} = \frac{800 \times 8000 \times 10^{-6}}{0.287 \times 373} = 0.05978 \text{ kg}$

$Q = 0.05978 \times 0.717(93.2 - 373) = -12.0 \text{ kJ}$ where

$T_2 = 373\frac{200}{800} = 93.2 \text{ K}$

14. C $T_2 = T_1\left(\frac{P_2}{P_1}\right)^{(n-1)/n} = 373\left(\frac{100}{600}\right)^{0.2/1.2} = 276.7 \text{ K}$

$w = \frac{P_2v_2 - P_1v_1}{1-n} = \frac{R(T_2 - T_1)}{1-n} = \frac{0.287(276.7 - 373)}{1-1.2} = 138.2 \text{ kJ/kg}$

$q = C_v\Delta T + w = 0.745(276.7 - 373) + 138.2 = 66.5 \text{ kJ/kg}$

15. D See Eqs. (4.51) and (4.49).

16. D $T_2 = T_1\left(\frac{P_2}{P_1}\right)^{(k-1)/k} = 293 \times 8^{0.2857} = 530.7 \text{ K}$ or $258°\text{C}$

17. B $\dot{m} = \rho_1A_1V_1 = \frac{P_1}{RT_1}A_1V_1$ $\quad 2 = \frac{100}{0.287 \times 293} \times \pi \times 0.1^2 \times V_1$ $\quad \therefore V_1 = 53.5 \text{ m/s}$

18. A $\dot{m}_s\Delta h_s = \dot{m}_wC_p\Delta T_w$ $\quad 10 \times 2373 = 400 \times 4.18 \times \Delta T$ $\quad \therefore \Delta T = 14.2°\text{C}$

19. A $\left.\begin{array}{l} P_2 = 0.6 \text{ MPa} \\ h_2 = 3634 \text{ kJ/kg} \end{array}\right\}$ $\quad \therefore T_2 = 570°\text{C}$

20. C $\not{\dot{Q}} - \dot{W}_S = \dot{m}C_p(T_2 - T_1)$ $\quad 5 = \dot{m} \times 1.00(530.7 - 293)$ $\quad \therefore \dot{m} = 0.021 \text{ kg/s}$

21. B $h_1 = 3445.2 \text{ kJ/kg}$ $\quad h_2 = h_f + x_2h_{fg} = 251.4 + 0.9(2358.3) = 2374 \text{ kJ/kg}$

$\dot{W}_T = -\dot{m}(h_2 - h_1) = -6(2374 - 3445) = 6430 \text{ kW}$

22. B $-w_T = h_2 - h_1 = C_p(T_2 - T_1) = 1.0 \times (-20 - 100)$ $\quad \therefore w_T = 120 \text{ kJ/kg}$

23. C $T_2 = T_1\left(\frac{P_2}{P_1}\right)^{(k-1)/k} = 468\left(\frac{585}{85}\right)^{0.4/1.4} = 812 \text{ K}$ or $539°\text{C}$

$$0=\left(\frac{V_2^2-V_1^2}{2}+C_p(T_2-T_1)\right)$$

$$=\left(\frac{V_2^2-100^2}{2\times1000}+1.0(812-468)\right)\qquad \therefore V_2=835\text{ m/s}$$

24. A $\dot{m}_a C_{pa}(T_{2a}-T_{1a})=\dot{m}_w C_{pw}(T_{2w}-T_{1w})$

$$5\times1.0\times200=\dot{m}_w\times4.18\times10\qquad \therefore \dot{m}_w=23.9\text{ kg/s}$$

Chapter 5

QUIZ NO. 1

1. A $\eta_{\text{max}}=1-\dfrac{T_L}{T_H}=1-\dfrac{283}{300}=0.05667\qquad \dot{Q}_{\text{in}}=\dot{W}+\dot{Q}_{\text{out}}$

$$\eta_{\text{proposed}}=\frac{\dot{W}}{\dot{Q}_{\text{in}}}=\frac{10}{10+9900/60}=0.0571\qquad \therefore \text{impossible}$$

2. C $\text{COP}=\dfrac{T_H}{T_H-T_L}=\dfrac{293}{293-253}=7.325$

$$\text{COP}=7.325=\frac{Q_H}{W}=\frac{2000}{W}\qquad \therefore W=273\text{ kJ/h}$$

3. D

4. B $\eta=\dfrac{\text{output}}{\text{input}}=\dfrac{500}{100\times6000/3600}=0.3$

5. B Let T be the unknown intermediate temperature. Then

$$\eta_1=1-\frac{T}{600}\qquad\text{and}\qquad \eta_2=1-\frac{300}{T}$$

It is given that $\eta_1=\eta_2+0.2\eta_2$. Substitute and find

$$1-\frac{T}{600}=1.2\left(1-\frac{300}{T}\right)\qquad \therefore T^2+120T-216\,300=0$$

$$\therefore T=\frac{-120\pm\sqrt{120^2-4(-216\,300)}}{2}=409\text{ K}\qquad\text{or}\qquad 136°\text{C}$$

6. C $\text{COP} = \frac{T_H}{T_H - T_L} = \frac{293}{293-248} = 6.511 = \frac{\dot{Q}_H}{\dot{W}} = \frac{1800/60}{\dot{W}}$

$\therefore \dot{W} = 4.61 \text{ kW} \quad \text{or} \quad 6.18 \text{ hp}$

7. C

8. A $T_1 = \frac{P_1V_1}{mR} = \frac{200\times 0.8}{2\times 0.287} = 278.7 \text{ K}$

$$\therefore \Delta S = m\left(C_p \ln\frac{T_2}{T_1} - R\ln\frac{P_2}{P_1}\right) = 2\times 1.0\times \ln\frac{773}{278.7} = 2.04 \text{ kJ/K}$$

9. D $\Delta S = m\left(C_v \ln\frac{T_2}{T_1} + R\ln\frac{v_2}{v_1}\right) \qquad T_2 = T_1\frac{P_2}{P_1}$ since $V = \text{const}$

$$\Delta S = 2\times 0.717\ln\frac{1500}{300} = 2.31 \text{ kJ/K} \quad \text{since} \quad T_2 = 300\times\frac{600}{120} = 1500 \text{ K}$$

10. B $-W = mC_v\Delta T \quad 200 = \frac{400\times 0.2}{0.287\times 313}\times 0.717(T_2 - 313) \quad \therefore T_2 = 626.2 \text{ K}$

$$\Delta S = m\left(C_v \ln\frac{T_2}{T_1} + R\ln\frac{v_2}{v_1}\right) = \frac{400\times 0.2}{0.287\times 313}\times 0.717\ln\frac{626.2}{313} = 0.443 \text{ kJ/K}$$

11. A $V_1 = \frac{mRT_1}{P_1} = \frac{0.2\times 0.287\times 313}{150} = 0.1198 \text{ m}^3$

$$\text{and} \quad V_2 = V_1\left(\frac{P_1}{P_2}\right)^{1/k} = 0.1198\left(\frac{150}{600}\right)^{1/1.4} = 0.0445 \text{ m}^3$$

12. C $m = \frac{P_1V_1}{RT_1} = \frac{100\times 2}{0.287\times 293} = 2.38 \text{ kg}$

$$\therefore \Delta S = -2.38\times 0.287\ln\frac{50}{100} = 0.473 \text{ kJ/K}$$

13. D $-W = mC_v(T_2 - T_1) = 10\times 0.717\times(356.6 - 773) = -2986 \text{ kJ}$

$$\text{where} \quad T_2 = T_1\left(\frac{p_2}{p_1}\right)^{(k-1)/k} = 773\left(\frac{400}{6000}\right)^{0.2857} = 356.6 \text{ K}$$

14. D First find T_2: $T_2 = T_1\left(\dfrac{P_2}{P_1}\right)^{(k-1)/k} = 673\left(\dfrac{2000}{400}\right)^{0.2857} = 1065.9 \text{ K}$

Since $Q = 0$: $-W = mC_v(T_2 - T_1) = 2\times 0.717\times(1065.9-673) = 563 \text{ kJ}$

15. D The heat gained by the ice is lost by the water: $\Delta h_{\text{melt}} \times m_i = m_w C_{p,w}\Delta T_w$.

(We assume not all the ice melts.) $m_i = \dfrac{20\times 4.18\times 20}{340} = 4.918 \text{ kg}$

The entropy change is:

$$\Delta S_i + \Delta S_w = \frac{Q}{T_i} + m_w C_{p,w}\ln\frac{T_2}{293}$$

$$= \frac{4.918\times 340}{273} + 20\times 4.18\times \ln\frac{273}{293} = 0.214 \text{ kJ/K}$$

16. A $m_c C_{p,c}\Delta T_c = m_w C_{p,w}\Delta T_w \qquad 5\times 0.385(100 - T_2) = 10\times 4.18(T_2 - 10)$

$\therefore T_2 = 14.0°\text{C}$ and $\Delta S_c = m_c C_{p,c}$

$\ln\dfrac{T_2}{(T_1)_c} = 5\times 0.385\ln\dfrac{287}{373} = -0.505 \text{ kJ/K}$

$$\Delta S_w = m_w C_{p,w}\ln\frac{T_2}{(T_1)_w} = 10\times 4.18\ln\frac{287}{283} = 0.587 \text{ kJ/K}$$

$\Delta S_{\text{universe}} = \Delta S_c + \Delta S_w = -0.505 + 0.587 = 0.082 \text{ kJ/K}$

17. B Use enthalpies from Tables C.2 and C.3:

$w_T = h_1 - h_2 = 3658 - 2637 = 1021 \text{ kJ/kg}$

18. C $s_{2'} = s_1 = 7.1678 = 1.0260 + x_{2'}\times 6.6449 \qquad \therefore x_{2'} = 0.924$

$h_{2'} = 317.6 + 0.924\times 2319 = 2460 \text{ kJ/kg}$

$w_{T,\max} = 3660 - 2460 = 1200 \text{ kJ/kg} \qquad \therefore \eta = \dfrac{w_{T,\text{actual}}}{w_{T,\text{isentropic}}} = \dfrac{1020}{1200} = 0.85$

19. B $V_2^2 - V_1^2 = 2(h_1 - h_2) = 2\times(2874 - h_2)\times 1000$ (must change kJ to J)

$s_{2'} = s_1 = 7.759 \text{ kJ/kg}\cdot\text{K} \quad P_2 = 100 \text{ kPa}$ therefore, $h_2 = 2840 \text{ kJ/kg}$

$\eta = \dfrac{(V_2^2 - V_1^2)_a}{(V_2^2 - V_1^2)_s} \qquad 0.85 = \dfrac{V_2^2 - 20^2}{2\times(2874 - 2840)\times 1000} \qquad \therefore V_2 \cong 230 \text{ m/s}$

20. A $s_{2'} = s_1 = 7.808 = 0.6491 + x_{2'}(7.502) \qquad \therefore x_{2'} = 0.954$

$h_{2'} = 192 + 0.954(2393) = 2476 \text{ kJ/kg}$

$\therefore \eta = \dfrac{4000}{4(3693 - 2476)} = 0.82 \quad \text{or} \quad 82\%$

QUIZ NO. 2

1. A $\eta = 1 - \dfrac{T_L}{T_H} = 1 - \dfrac{293}{373} = 0.214$

2. D

3. C $\eta = 1 - \dfrac{T_L}{T_H} = 1 - \dfrac{293}{473} = 0.3805 = \dfrac{\dot{W}}{\dot{Q}_L + \dot{W}} = \dfrac{10}{\dot{Q}_L + 10} \qquad \therefore \dot{Q}_L = 16.3 \text{ kW}$

4. B $V = 100 \times 1000 / 3600 = 27.78 \text{ m/s}$. Be careful with the units.

$$\eta = \frac{\text{Drag} \times \text{velocity}}{\text{Input power}} = \frac{\frac{1}{2}\rho V^2 A C_D \times V \text{ (J/s)}}{\dot{W}}$$

$$= \frac{\frac{1}{2} \times 1.23 \times 27.78^2 \times 2 \times 0.28 \times 27.78}{\frac{1}{13} \times 27.78 \times 10^{-3} \times 740 \times 9000} = 0.519$$

5. A The maximum possible efficiency is $\eta = 1 - \dfrac{T_L}{T_H} = 1 - \dfrac{293}{1173} = 0.75$

$\eta_{\text{actual}} = \dfrac{\text{output}}{\text{input}} = \dfrac{43 \times 0.746}{2500/60} = 0.77 \qquad \therefore \text{impossible}$

6. D $\text{COP}_{\text{ref}} = \dfrac{Q_L}{W} = \dfrac{T_L}{T_H - T_L} \quad \dfrac{10^{-5}}{W} = \dfrac{2 \times 10^{-6}}{293 - 2 \times 10^{-6}} \qquad \therefore W = 1465 \text{ kJ}$

7. C $\Delta S = mC_v \ln \dfrac{T_2}{T_1} = 1 \times 0.717 \times \ln \dfrac{573}{293} = 0.481 \text{ kJ/K}$

8. C

9. B $\Delta S = mC_p \ln\frac{T_2}{T_1} - \frac{P_1V_1}{T_1}\ln\frac{P_2}{P_1} = -\frac{6000\times 500\times 10^{-6}}{1073}\ln\frac{200}{6000} = 9.51 \text{ kJ/K}$

10. A $m = \frac{PV}{RT} = \frac{80\times 4}{0.297\times 300} = 3.59 \text{ kg}$

$$5.2 = 3.59\left(0.745\ln\frac{400}{300} + 0.297\ln\frac{V_2}{4}\right) \qquad \therefore V_2 = 255 \text{ m}^3$$

11. D $\Delta s = -0.287\ln\frac{4000}{50} = -1.26 \text{ kJ/kg}\cdot\text{K}$

12. B $m = \frac{P_1V_1}{RT_1} = \frac{200\times 2}{0.287\times 320} = 4.36 \text{ kg} \qquad Q - W = m\Delta u$

$$500 - \frac{-40\times 40}{1000}\times 10\times 60 = 4.36\times 0.717(T_2 - 320) \qquad \therefore T_2 = 787 \text{ K}$$

$$\Delta S = mC_v \ln\frac{T_2}{T_1} = 4.36\times 0.717\times \ln\frac{787}{320} = 2.81 \text{ kJ/K}$$

13. A $T_2 = T_1\left(\frac{P_2}{P_1}\right)^{(k-1)/k} = 500\left(\frac{500}{20}\right)^{0.2857} = 199 \text{ K}$

$$W = -4\times 0.717(199 - 500) = 863 \text{ kJ}$$

14. D $s_1 = 2.047 + 0.85(4.617) = 5.971 \quad h_1 = 721 + 0.85(2048) = 2462$

$$q = h_2 - h_1 \quad 2000 = h_2 - 2462 \quad \therefore h_2 = 4462 \quad \left.\begin{matrix} P_2 = 800 \\ h_2 = 4462 \end{matrix}\right\} \quad \therefore s_2 = 8.877$$

$$\therefore \Delta s = 8.877 - 5.971 = 2.91 \text{ kJ/kg}\cdot\text{K}$$

15. B $u_1 = 2507 \text{ kJ/kg} \quad s_1 = s_2 = 7.356 = 0.832 + x_2(7.077) \quad \therefore x_2 = 0.922$

$$u_2 = 251 + 0.922\times 2205 = 2284 \text{ kJ/kg}$$

$$\therefore W = m(u_1 - u_2) = 2(2507 - 2284) = 447 \text{ kJ}$$

16. A The final temperature is 0°C. The heat lost by the iron is

$$Q = m_{\text{Fe}}C_p(T_1 - T_2) = 10\times 0.448\times 300 = 1344 \text{ kJ}$$

$$\Delta S = m_{\text{Fe}}C_p \ln\frac{T_2}{T_1} + \frac{Q}{T_{\text{ice}}} = 10\times 0.448\times \ln\frac{273}{673} + \frac{1344}{273} = 0.881 \text{ kJ/K}$$

17. A Assume that all the ice does not melt:

$$20\times1.9\times5+330\times m=10\times4.18(20)\quad\therefore m=1.96\text{ kg}\quad\text{and}\quad T_2=0°\text{C}$$

$$\Delta S=m_iC_{p,i}\ln\frac{T_{2i}}{T_{1i}}+\frac{Q_i}{T_i}+m_wC_{p,w}\ln\frac{T_{2w}}{T_{1w}}$$

$$=5\times1.9\times\ln\frac{273}{253}+\frac{1.96}{273}\times330+10\times4.18\times\ln\frac{273}{293}=0.064\text{ kJ/K}$$

18. C $P_1=\dfrac{2\times0.287\times573}{2.0}=164.5\text{ kPa}\quad T_2=573\times\dfrac{120}{164.5}=418\text{ K}$

$$\Delta S_{\text{air}}=2\times0.717\ln\frac{418}{573}=-0.452\text{ kJ/K}$$

19. B $Q=2\times0.717(573-418)=222\text{ kJ}$

$$\therefore\Delta S_{\text{universe}}=-0.452+\frac{222}{300}=0.289\text{ kJ/K}$$

20. C $h_2=2666\qquad x_2=1.0\qquad \left.\begin{matrix}T_1=600°\text{C}\\ s_2=s_1=7.435\end{matrix}\right\}\quad\therefore h_1=3677$

$$\dot{W}_T=\dot{m}(h_1-h_2)\qquad 200=\dot{m}(3678-2666)\qquad\therefore\dot{m}=0.198\text{ kg/s}$$

Chapter 6

QUIZ NO. 1

1. B
2. A Refer to Tables C.2 and C.3:

   For the boiler $q_B=h_3-h_2=3658.4-251.4=3407\text{ kJ/kg}$

3. C Refer to Table C.3: $w_T=h_4-h_3=3658.4-2609.7=1048\text{ kJ/kg}$
4. B $w_P=\dfrac{P_2-P_1}{\rho}=\dfrac{6000-20}{1000}=5.98\text{ kJ/kg}$

5. D $\eta = \dfrac{w_{\text{actual}}}{w_{\text{isentropic}}} = \dfrac{h_3 - h_4}{h_3 - h_{4'}} = \dfrac{3658.4 - 2609.7}{3658.4 - 2363} = 0.809$

where we used $s_3 = s_{4'} = 7.1677 = 0.8320 + x_{4'} \times 7.0766$

$\therefore x_{4'} = 0.8953$ and $h_{4'} = 251.4 + 0.8953 \times 2358.3 = 2363 \text{ kJ/kg}$

6. D $\dot{m}_s(h_4 - h_1) = \dot{m}_w C_p \Delta T_w$

$2(2609.7 - 251.4) = \dot{m}_w \times 4.18 \times 10 \qquad \therefore \dot{m}_w = 112.8 \text{ kg/s}$

7. C Refer to Fig. 6.4:

$h_1 = 251 \quad h_3 = 3658 \quad s_3 = 7.168 = s_4 \quad \therefore h_4 = 3023 \quad h_5 = 3267$

$s_5 = 7.572 = s_6 = 0.8319 + 7.0774 x_6 \quad \therefore x_6 = 0.952 \quad \text{and} \quad h_6 = 2497$

$$\eta = \frac{h_3 - h_4 + h_5 - h_6}{h_3 - h_1 + h_5 - h_4} = \frac{3658 - 3023 + 3267 - 2497}{3658 - 251 + 3267 - 3203} = 0.405$$

8. B $\eta = \dfrac{w_{\text{actual}}}{w_{\text{isentropic}}} = \dfrac{h_3 - h_4}{h_3 - h_{4'}} = \dfrac{3658 - h_4}{3658 - 2363} = 0.8 \quad \therefore h_4 = 2622 \quad P_4 = 20 \text{ kPa}$

$2622 > h_g = 2610 \text{ @} 20 \text{ kPa} \qquad \therefore$ superheat (slightly). So, $T_4 \cong 65°\text{C}$

Note: $\Delta h \cong 100$ for $\Delta T \cong 50°\text{C}$ (see Table C.3). So, 10% of 50 = 5°C

9. A Refer to Fig. 6.5: $h_1 = 251 \quad h_3 = 3658 \quad s_3 = 7.168 = s_5 \quad \therefore h_5 = 3023$

$$m_5 = \frac{h_6 - h_2}{h_5 - h_2} = \frac{721 - 251}{3023 - 251} = 0.170 \text{ kg}$$

10. B $\dot{m}_5 = \dfrac{h_6 - h_2}{h_5 - h_2} \dot{m}_6 = \dfrac{604.7 - 251.4}{2860.5 - 251.4} \times 20 = 2.708 \text{ kg/s}$

11. C $h_1 = 233.9 \quad h_4 = h_3 = 105.3 \quad \left. \begin{array}{l} P_2 = 1.0 \text{ MPa} \\ s_2 = s_1 = 0.9354 \end{array} \right\} \quad \therefore h_2 = 277.8 \text{ kJ/kg}$

$\dot{W}_C = \dot{m}(h_2 - h_1) \quad 10 \times 0.746 = \dot{m}(277.8 - 233.9) \quad \therefore \dot{m} = 0.1699 \text{ kg/s}$

$\dot{Q}_E = \dot{m}(h_1 - h_4) = 0.1699 \times (233.9 - 105.3) = 21.8 \text{ kJ/s}$

12. C $\text{COP} = \dfrac{h_1 - h_4}{h_2 - h_1} = \dfrac{233.9 - 105.3}{277.8 - 233.9} = 2.93$

13. A $h_1 = 241.3 \quad h_4 = h_3 = 105.3 \quad \left.\begin{array}{l} P_2 = 1.0 \text{ MPa} \\ s_2 = s_1 = 0.9253 \end{array}\right\} \quad \therefore h_2 = 274.8 \text{ kJ/kg}$

$$\dot{Q}_{\text{Cond}} = \dot{m}(h_2 - h_3). \quad \frac{80\,000}{3600} = \dot{m}(274.8 - 105.3) \quad \therefore \dot{m} = 0.1311 \text{ kg/s}$$

$$\dot{W}_{\text{Comp}} = \dot{m}(h_2 - h_1) = 0.1311 \times (274.8 - 241.3) = 4.39 \text{ kW} \quad \text{or} \quad 3.28 \text{ hp}$$

14. D $Q_f = AV = \dfrac{\dot{m}}{\rho_1} = \dot{m}v_1 = 0.1311 \times 0.0993 = 0.013 \text{ m}^3\text{/s}$

QUIZ NO. 2

1. B

2. A $h_3 = 3468 \text{ kJ/kg} \quad \left.\begin{array}{l} P_4 = 100 \\ s_4 = s_3 = 7.432 \end{array}\right\} \quad \therefore h_4 = 2704 \text{ kJ/kg}$

$$w_T = h_3 - h_4 = 3468 - 2704 = 764 \text{ kJ/kg}$$

3. D $q_B = h_3 - h_2 = 3468 - 417 = 3050 \text{ kJ/kg}$

4. C $\eta = \dfrac{w_T}{q_B} = \dfrac{764}{3050} = 0.25$

5. B $w_P = v\Delta P = \dfrac{\Delta P}{\rho} = \dfrac{2000 - 100}{1000} = 1.9 \text{ kJ/kg}$

$$\therefore \dot{W}_P = 2 \times 1.9 / 0.746 = 5.09 \text{ hp}$$

6. A $h_1 = 417 \quad h_3 = 3468 \quad \left.\begin{array}{l} P_4 = 0.8 \text{ MPa} \\ s_4 = s_3 = 7.432 \end{array}\right\} \quad \therefore h_4 = 3175 \quad h_5 = 3481$

$$\left.\begin{array}{l} P_6 = 0.1 \text{ MPa} \\ s_6 = s_5 = 7.867 \end{array}\right\} \quad \therefore h_6 = 2895 \text{ kJ/kg} \quad \text{(Refer to Fig. 6.4.)}$$

$$\eta = \frac{w_T}{q_B} = \frac{3468 - 3175 + 3481 - 2895}{3468 - 417 + 3481 - 3175} = 0.262$$

7. D $h_1 = 417 \quad h_3 = 3468 \quad h_4 = 2704 \quad \left.\begin{matrix} P_5 = 0.8 \text{ MPa} \\ s_5 = s_3 = 7.432 \end{matrix}\right\} \quad \therefore h_5 = 3175$

$$m_5 = \frac{h_6 - h_2}{h_5 - h_2} m_6 = \frac{721 - 417}{3175 - 417} \times 1 = 0.110 \text{ kg}$$

8. B $\eta = \dfrac{w_T}{q_B} = \dfrac{3468 - 3175 + 0.89(3175 - 2704)}{3468 - 721} = 0.259$

9. A $\eta = \dfrac{w_a}{w_s} \qquad 0.9 = \dfrac{3468 - h_4}{3468 - 2704} \qquad \therefore h_4 = 2780$ and $T_4 = 152°\text{C}$

10. D

11. C $h_1 = 230.4 \quad h_4 = h_3 = 106.2 \quad \left.\begin{matrix} P_2 = 1.016 \text{ MPa} \\ s_2 = s_1 = 0.9411 \end{matrix}\right\} \quad \therefore h_2 = 279.7 \text{ kJ/kg}$

In the R134a table, we have used P_2 = 1.0 MPa, close enough to 1.016 MPa.

$$\dot{Q}_E = \dot{m}(h_1 - h_4) = 0.6 \times (230.4 - 106.2) = 74.5 \text{ kJ/s}$$

12. D $\text{COP} = \dfrac{h_1 - h_4}{h_2 - h_1} = \dfrac{230.4 - 106.2}{279.7 - 230.4} = 2.52$

13. A $h_1 = 241.3 \qquad h_4 = h_3 = 106.2$

$$\dot{Q}_E = \dot{m}(h_1 - h_4) \qquad 10 \times 3.52 = \dot{m}(241.3 - 106.2) \qquad \therefore \dot{m} = 0.261 \text{ kg/s}$$

14. C $\left.\begin{matrix} P_2 = 1.2 \text{ MPa} \\ s_2 = s_1 = 0.9253 \end{matrix}\right\} \quad \therefore h_2 = 279 \text{ kJ/kg}$

$$\therefore \dot{W}_C = 0.261 \times (279 - 241) = 9.9 \text{ kW} \quad \text{or} \quad 13.3 \text{ hp}$$

15. B $h_1 = 231.4 \qquad h_4 = h_3 = 105.3 \qquad \left.\begin{matrix} P_2 = 1.0 \text{ MPa} \\ s_2 = s_1 = 0.9395 \end{matrix}\right\} \quad \therefore h_2 = 279.2 \text{ kJ/kg}$

The maximum temperature change is 12°C since water freezes at 0°C. So

$$\dot{Q}_{\text{Cond}} = \dot{m}_{\text{gw}} C_p (12 - 0) \qquad \frac{60\,000}{3600} = \dot{m}_{\text{gw}} \times 4.18 \times 12 \quad \therefore \dot{m}_{\text{gw}} = 0.332 \text{ kg/s}$$

Chapter 7

QUIZ NO. 1

1. B
2. B $V = Ah = \pi \times 0.1^2 \times 0.2 = 0.00628 \text{ m}^3$

$$\frac{V_c}{V} = 0.05 \qquad \frac{V_c}{0.00628} = 0.05 \qquad \therefore V_c = 3.14 \times 10^{-4} \text{ m}^3$$

and $r = \dfrac{V_1}{V_2} = \dfrac{0.00628 + 3.14 \times 10^{-4}}{3.14 \times 10^{-4}} = 21$

3. D $P_1 = 100 \text{ kPa} \quad P_4 = 100\left(\dfrac{873}{293}\right)^{3.5} = 4566 \text{ kPa}$

$$v_1 = \frac{0.287 \times 293}{100} = 0.8409 \text{ m}^3/\text{kg}$$

$$w_{\text{net}} = \Delta T \Delta s \qquad 200 = (873 - 293) \times 0.287 \ln \frac{P_2}{100} \qquad \therefore P_2 = 332.5 \text{ kPa}$$

$$P_3 = 332.5\left(\frac{873}{293}\right)^{3.5} = 15180 \text{ kPa} \qquad \text{and}$$

$$v_3 = \frac{0.287 \times 873}{15180} = 0.01651 \text{ m}^3/\text{kg}$$

$$\text{MEP} = \frac{w_{\text{net}}}{v_1 - v_3} = \frac{200}{0.8409 - 0.0165} = 243 \text{ kPa}$$

4. C
5. A Refer to Fig. 7.2.

$$v_4 = v_1 = \frac{0.287 \times 295}{85} = 0.9961 \qquad v_3 = v_2 = \frac{0.9961}{8} = 0.1245 \text{ m}^3/\text{kg}$$

$$T_3 = \frac{0.1245 \times 8000}{0.287} = 3470 \text{ K} \qquad T_2 = 295 \times 8^{0.4} = 677.7 \text{ K}$$

$$q_{\text{in}} = C_v(T_3 - T_2) = 0.717 \times (3470 - 677.7) = 2002 \text{ kJ/kg}$$

6. C $T_4 = 3470 \times 8^{-0.4} = 1510 \text{ K} \qquad \therefore q_{\text{out}} = 0.717 \times (1510 - 295) = 871 \text{ kJ/kg}$

$$\text{MEP} = \frac{w_{\text{net}}}{v_1 - v_2} = \frac{q_{\text{in}} - q_{\text{out}}}{v_1 - v_2} = \frac{2002 - 871}{0.9961 - 0.1245} = 1298 \text{ kPa}$$

7\. A $v_4 = v_1 = \dfrac{0.287 \times 303}{85} = 1.023 \quad v_2 = \dfrac{1.023}{16} = 0.06394$ (Refer to Fig. 7.3.)

$$\therefore T_2 = 303 \times 16^{0.4} = 918.5 \text{ K} \qquad P_3 = P_2 = \frac{0.287 \times 918.5}{0.06394} = 4123 \text{ kPa}$$

$$v_3 = \frac{0.287 \times 2273}{4123} = 0.1582 \qquad \therefore r_c = \frac{v_3}{v_2} = \frac{0.1582}{0.06394} = 2.47$$

8\. B $q_{\text{in}} = 1.0(2273 - 918.5) = 1354 \qquad P_4 = 4123\left(\dfrac{0.1582}{1.023}\right)^{1.4} = 302.2 \text{ kPa}$

$$\therefore T_4 = 303 \times \frac{302.2}{85} = 1077 \text{ K} \quad \text{and} \quad q_{\text{out}} = 0.717(1077 - 303) = 555 \text{ kJ/kg}$$

$$\therefore \dot{W}_{\text{net}} = \dot{m}(q_{\text{in}} - q_{\text{out}}) \quad 500 \times 0.746 = \dot{m}(1354 - 555) \quad \therefore \dot{m} = 0.467 \text{ kg/s}$$

9\. D $T_2 = 273 \times 6^{0.2857} = 455.5 \text{ K} \quad \therefore \eta = 1 - \dfrac{273}{455.5} = 0.401$ (Refer to Fig. 7.5.)

10\. C $T_4 = 1273 \times 6^{-0.2857} = 763 \text{ K}$

$$\text{BWR} = \frac{\dot{W}_C}{\dot{W}_T} = \frac{w_C}{w_T} = \frac{T_2 - T_1}{T_3 - T_4} = \frac{455.5 - 273}{1273 - 763} = 0.358$$

11\. B $\eta = 1 - \dfrac{293}{1073} \times 5^{0.2857} = 0.568 \quad T_5 = 1073 \times 5^{0.2857} = 677 \text{ K}$

$$T_2 = 293 \times 5^{0.2857} = 464 \text{ K} \quad q_{\text{in}} = 1.0 \times (1073 - 677) = 396 \text{ kJ/kg}$$

$\dot{W}_{\text{out}} = 4 \times 0.568 \times 396 = 899 \text{ kW}$ (Refer to Fig. 7.6.)

12\. A $\dot{W}_C = 4 \times 1.0 \times (464 - 293) = 684 \text{ kW} \qquad \text{BWR} = \dfrac{684}{899 + 684} = 0.432$

13\. D $w_C = h_3 - h_2 = C_p(T_3 - T_2) = 250 - 10 = 240 \text{ kJ/kg}$ (Refer to Fig. 7.9.)

14\. C $\dfrac{P_3}{P_2} = \left(\dfrac{T_3}{T_2}\right)^{k/(k-1)} = \left(\dfrac{523}{283}\right)^{3.5} = 8.58$

15. A $T_4 = T_1\left(\frac{P_4}{P_1}\right)^{(k-1)/k} = 223 \times 8.58^{0.2857} = 412 \text{ K}$ or $139°\text{C}$

$$\text{COP} = \frac{q_{\text{in}}}{w_C - w_T} = \frac{h_2 - h_1}{h_3 - h_2 - (h_4 - h_1)} = \frac{T_2 - T_1}{T_3 - T_2 - T_4 + T_1}$$

$$= \frac{10 - (-50)}{250 - 10 - 139 + (-50)} = 1.18$$

QUIZ NO. 2

1. A

2. D In Fig. 7.1, process 2-3 is at P = const. not v = const.

$$T_2 = T_1\left(\frac{P_2}{P_1}\right)^{(k-1)/k} = 288\left(\frac{2000}{100}\right)^{0.2857} = 677.8 \text{ K}$$

$$q_{2\text{-}3} = C_p(T_3 - T_2) = 1.0(1673 - 677.8) = 522 \text{ kJ/kg}$$

3. B $v_1 = \frac{0.287 \times 293}{100} = 0.8409 \qquad v_3 = \frac{0.8409}{15} = 0.05606 \text{ m}^3\text{/kg}$

$$\therefore T_3 = \frac{3000 \times 0.05606}{0.287} = 586 \text{ K} \qquad \therefore \eta = 1 - \frac{293}{586} = 0.5$$

4. C $\eta = 1 - \frac{1}{r^{k-1}} = 1 - \frac{1}{9^{0.4}} = 0.585$

5. C $\eta = \frac{\dot{W}_{net}}{\dot{Q}_{in}} \quad 0.585 = \frac{500}{\dot{Q}_{in}} \quad \therefore \dot{Q}_{in} = 855 \text{ kJ/s} \quad \frac{v_1}{v_2} = 9$ and $T_1 = 303 \text{ K}$

$$\therefore T_2 = 303 \times 9^{0.4} = 729.7 \text{ K} \quad \text{Given: } T_3 = 1273 \text{ K}$$

$$\dot{Q}_{\text{in}} = \dot{m}C_v(T_3 - T_2) \quad 855 = \dot{m} \times 0.717(1273 - 729.7) \quad \therefore \dot{m} = 2.19 \text{ kg/s}$$

6. B $v_1 = \frac{0.287(293)}{110} = 0.7645 \quad v_2 = \frac{0.7645}{16} = 0.04778 \quad T_2 = 293 \times 16^{.4} = 888.2 \text{ K}$

$$P_3 = P_2 = \frac{0.287 \times 888.2}{0.04778} = 5335 \quad 1800 = 1 \times (T_3 - 888.2) \quad \therefore T_3 = 2688 \text{ K}$$

$$v_3 = \frac{0.287 \times 2688}{5335} = 0.1446 \text{ m}^3/\text{kg} \qquad \therefore r_c = \frac{v_3}{v_2} = \frac{0.1446}{0.04778} = 3.03$$

7. A $q_{\text{in}} = 1.0(2688 - 888) = 1800 \quad \therefore w_{\text{net}} = 1022$ (Refer to Fig. 7.3.)

$$\text{and MEP} = \frac{1022}{0.7645 - 0.04778} = 1430 \text{ kPa}$$

8. D $T_2 = 303\left(\frac{500}{80}\right)^{0.2857} = 511.5 \text{ K}$ (Refer to Fig. 7.5.)

$$1800 = 1.0 \times (T_3 - 511.5) \qquad \therefore T_3 = 2311.5 \text{ K}$$

$$w_C = 1.0 \times (511.5 - 303) = 208 \text{ kJ/kg}$$

9. D $\eta = 1 - \frac{303}{511.5} = 0.408 \qquad \therefore w_{\text{net}} = \eta q_{\text{in}} = 0.408 \times 1800 = 734 \text{ kJ/kg}$

10. A $\text{BWR} = \frac{w_C}{w_T} = \frac{208}{208 + 734} = 0.221$

11. B Refer to Fig. 7.6.

$T_2 = 511.5$ K $\quad 1800 = 1.0 \times (T_4 - T_3)$ but $T_3 = T_5$ so that $1800 = T_4 - T_5$

$$w_C = 511.5 - 303 = 208.5 \text{ kJ/kg} \qquad \therefore \eta = \frac{1800 - 208.5}{1800} = 0.884$$

12. D

13. C $T_2 = 573 \times 5^{-0.2857} = 362 \text{ K} \qquad T_4 = 253 \times 5^{0.2857} = 401 \text{ K}$

$\dot{W}_C = 2 \times 1.0 \times (573 - 362) = 422 \text{ kW}$ (Refer to Fig. 7.9.)

$$\dot{W}_T = 2 \times 1.0 \times (401 - 253) = 296 \text{ kW} \qquad \dot{W}_{\text{in}} = 422 - 296 = 126 \text{ kW}$$

14. D $\text{COP} = \frac{\text{desired effect}}{\text{energy purchased}} = \frac{2 \times 1.0 \times (362 - 253)}{126} = 1.73$

Chapter 8

QUIZ NO. 1

1. B $\phi = 0.4 = \frac{P_v}{P_g} = \frac{P_v}{4.246} \quad \therefore P_v = 1.698 \text{ kPa}$

   $\omega = 0.622 \times \frac{1.698}{100 - 1.698} = 0.1074$

2. C $P_g = 2.338 \text{ kPa} \quad 0.5 = \frac{P_v}{2.338} \quad \therefore P_v = 1.17 \text{ kPa} \quad \text{and} \quad P_a = 98.83 \text{ kPa}$

   $\omega = 0.622 \times \frac{1.17}{98.83} = 0.00736$

3. A $m_a = \frac{PV}{RT} = \frac{98.83 \times 12 \times 15 \times 3}{0.287 \times 293} = 635 \text{ kg}$

   $m_v = \omega \times m_a = 0.00736 \times 635 = 4.67 \text{ kg}$

4. D $H = m_a(C_p T + \omega h_v) = 635 \times (1.0 \times 20 + 0.00736 \times 2538) = 24\,600 \text{ kJ}$

5. A Since the answers are fairly separated, the psychrometric chart can be used. If they are separated by small amounts, the equations are needed.

6. B The humidity is quite low in Arizona so no dew should appear.

7. C The psychrometric chart is used.

8. D The air must be heated for comfort. The humidity will decrease so moisture must be added.

9. D Use the psychrometric chart and follow a horizontal line from the state of 5°C and 80% humidity.

10. C $m_{w,1} - m_{w,2} = \omega_1 \rho_{\text{air},1} V - \omega_2 \rho_{\text{air},2} V$

    $= \left(0.025 \times \frac{98}{0.287 \times 308} - 0.008 \times \frac{98}{0.287 \times 298}\right) \times 3 \times 10 \times 20$

    $= 11.1 \text{ kg water}$

    Note: We assumed the partial pressures to be 98 kPa, a good estimate.

11. A The pressure is not atmospheric so equations must be used.

$$P_{v1} = 0.2 \times 617.8 = 123.6 \text{ kPa} \qquad v_{v1} = \frac{RT_1}{P_{v1}} = \frac{0.462 \times 433}{123.6} = 1.62 \text{ m}^3\text{/kg}$$

The volume remains constant. At the above specific volume, condensation is observed at $T = 101°\text{C}$.

12. C $P_{a1} = 400 - 123.6 = 276.4 \text{ kPa} \qquad m_a = \dfrac{276.4 \times 2}{0.287 \times 433} = 4.45 \text{ kg}$

$$\omega_1 = 0.622 \times \frac{123.6}{276.4} = 0.278 \qquad \phi_2 = 1.0 \qquad \therefore p_{v2} = 2.338 \text{ kPa}$$

$$P_{a2} = P_{a1}\frac{T_2}{T_1} = 276.4 \times \frac{293}{433} = 187 \text{ kPa} \qquad \therefore \omega_2 = 0.622 \times \frac{2.338}{187} = 0.0078$$

$$\therefore m_{\text{condensed}} = m_a(\omega_1 - \omega_2) = 4.45 \times (0.278 - 0.0078) = 11.20 \text{ kg}$$

13. B $Q = m_a(u_{a2} - u_{a1}) + m_{v2}u_{v2} - m_{v1}u_{v1} + \Delta m_w (h_{fg})_{\text{avg}}$

$$= m_a[C_v(T_2 - T_1) + \omega_2 u_{v2} - \omega_1 u_{v1} + (\omega_2 - \omega_1)(h_{fg})_{\text{avg}}]$$

$$= 4.45 \times [0.717(20 - 160) + 0.0078 \times 2402.9 - 0.278 \times 2568.4$$
$$+ (0.0078 - 0.278) \times 2365] = -6290 \text{ kJ}$$

14. C From the psychrometric chart, $T = 20.7°\text{C} \qquad \phi = 49\%$

15. A $\dot{m}_3 = \dot{m}_2 + \dot{m}_1 = 50 \times \dfrac{100 - 0.4 \times 3.9}{0.287 \times 301} + 30 \times \dfrac{100 - 0.6 \times 1.2}{0.287 \times 283} = 93.6 \text{ kg/min}$

$$\omega_1 = 0.0045 \qquad \omega_2 = 0.0095 \qquad \dot{m}_{w1} = 50 \times 0.0045$$
$$= 0.26 \qquad \dot{m}_{w2} = 30 \times 0.0095 = 0.35 \qquad \therefore \dot{m}_{w,\text{total}} = 0.26 + 0.35$$
$$= 0.61 \text{ kg/min} \qquad \text{and} \qquad \dot{m}_{\text{total}} = 93.6 + 0.61 = 94.2 \text{ kg/min}$$

QUIZ NO. 2

1. D At 30°C we find $P_g = 4.246$ kPa. At 20°C the partial pressure of water vapor is $P_v = 2.338 \text{ kPa} \qquad \therefore \phi = \dfrac{2.338}{4.246} = 0.551$

2. B $P_v = 2.338 \text{ kPa} \qquad \therefore \omega_2 = 0.622 \times \dfrac{2.338}{100 - 2.338} = 0.01489$

$$\omega_1 = \frac{1.0 \times (20 - 30) + 0.01489 \times 2454}{2556 - 83.9} = 0.01074$$

3\. A $0.01074 = 0.622 \times \dfrac{P_v}{100 - P_v}$ $\quad \therefore P_v = 1.708 \text{ kPa} \qquad P_g = 4.246 \text{ kPa}$

$$\therefore \phi = \frac{1.708}{4.246} = 0.402$$

4\. C $h = 1.0 \times 30 + 0.01074 \times 2556 = 57.5 \text{ kJ/kg}$

5\. B $\phi_2 = 20\%$

6\. D $Q = \dot{m}(h_2 - h_1) = 50 \times (38 - 21.2) = 840 \text{ kJ/min} \quad \text{or} \quad 14 \text{ kJ/s}$

7\. C

8\. B From the psychrometric chart, $\omega = 0.0103$. Assume $P_{air} = 98$ kPa.

Then $3 \times 10 \times 20 \times \dfrac{98}{0.287 \times 293} \times 0.0103 = 7.20$ kg water or 7.2 L water

9\. C Use the psychrometric chart. Locate the state of 40°C and 20% humidity. Follow a constant h-line to 80% humidity. Go straight down to $T = 24.5$°C.

10\. D Refer to Fig. 8.4, process *F-G-H-I*.

$$q_{\text{heat}} = h_H - h_I = 35 - 20 = 15 \text{ kJ/kg dry air}$$

11\. A Review Example 8.9.

$$\frac{d_{2\text{-}3}}{d_{3\text{-}1}} = \frac{\dot{m}_1}{\dot{m}_2} \qquad \dot{m}_1 = \rho_1 Q_{f,1} = \frac{98}{0.287 \times 288} \times 40 = 47.4 \text{ kg/min} \qquad \text{and}$$

$$\dot{m}_2 = \rho_2 Q_{f,2} = \frac{98}{0.287 \times 305} \times 20 = 22.4 \text{ kg/min}$$

$$\therefore \frac{d_{2\text{-}3}}{d_{3\text{-}1}} = \frac{47.4}{22.4} = 2.12 \approx 2 \qquad \therefore \phi_3 = 64\%$$

Note: We assumed the partial pressures to be 98 kPa, a good estimate.

12\. D Refer to Prob. 11: $T_3 = 20.5$°C.

13\. B Refer to the psychrometric chart:

$$h_1 = 115, \quad h_2 = 30, \quad h_3 = 45 \qquad \rho = \frac{100 - 0.8 \times 5.98}{0.287 \times 309} = 1.074 \text{ kg/m}^3$$

$$\dot{Q}_{\text{cool}} = 100 \times 1.074 \times (115 - 30) = 9130 \text{ kJ/min}$$

14\. A Using the h's from Prob. 13,

$$\dot{Q}_{\text{heat}} = 100 \times 1.074 \times (45 - 30) = 1610 \text{ kJ/min}$$

Chapter 9

QUIZ NO. 1

1. A $(4, 5, 24.44)$

2. B $CH_4 + 2(O_2 + 3.76N_2) \rightarrow CO_2 + 2H_2O + 7.52N_2$

$$AF = \frac{2 \times 4.76 \times 29}{1 \times 16} = 17.2$$

3. A $$\%\ CO_2 = \frac{1 \times 44}{44 + 36 + 7.52 \times 28} \times 100 = 15.1\%$$

4. D $P_v = \dfrac{2}{10.52} \times 100 = 19\text{ kPa}$ or 0.019 MPa $\therefore T_{dp} = 59°C$

5. D $C_2H_6 + 6.3(O_2 + 3.76N_2) \rightarrow 2CO_2 + 3H_2O + 2.8O_2 + 23.69N_2$

$$AF = \frac{6.3 \times 4.76 \times 29}{1 \times 30} = 29.0$$

6. B $$\%\ CO_2 = \frac{2}{2 + 3 + 2.8 + 23.69} \times 100 = 6.35\%$$

7. C $CH_4 + 2(O_2 + 3.76N_2) \rightarrow CO_2 + 2H_2O + 7.52N_2$

$$AF = \frac{2 \times 4.76 \times 29}{1 \times 16} = 17.26 \qquad \dot{m}_{\text{fuel}} = \frac{20\rho_{\text{air}}}{17.26} = \frac{20}{17.26} \times \frac{100}{0.287 \times 298}$$
$$= 1.38\text{ kg/min}$$

8. A $C_8H_{18} + x(O_2 + 3.76N_2) \rightarrow 8CO_2 + 9H_2O + aN_2 + bO_2$

$$AF = \frac{4.76(29x)}{1 \times 114} = 25 \qquad \therefore x = 20.67 \qquad a = 77.72 \qquad b = 8.17$$

$$\therefore \%\text{ excess air} = \frac{20.67}{12.5} \times 100 - 100 = 65.4\%$$

9. B $C + O_2 \rightarrow CO_2$

$$Q = H_P - H_R = \sum_{\text{prod}} N_i(\bar{h}_f^o)_i - \sum_{\text{react}} N_i(\bar{h}_f^o)_i$$
$$= 1(-393\,520) - 0 - 0 = -393\,520\text{ kJ/kmol}$$

10. A $C_3H_8 + 5(O_2 + 3.76N_2) \rightarrow 3CO_2 + 18.8N_2 + 4H_2O$

$$q = 3(-393520 + 42769 - 9364) + 4(-241820 + 35882 - 9904) + 18.8(30129 - 8669) - (-103850) = -1\,436\,000 \text{ kJ/kmol}$$

QUIZ NO. 2

1. B $(5, 6, 30.08)$
2. D $C_2H_6 + 3.5(O_2 + 3.76N_2) \rightarrow 2CO_2 + 3H_2O + 13.16N_2$

$$AF = \frac{3.5 \times 4.76 \times 29}{1 \times 30} = 16.1$$

3. B $\% \, CO_2 = \dfrac{2 \times 44}{88 + 54 + 13.16 \times 28} \times 100 = 17.24\%$
4. C $P_v = \dfrac{3}{18.16} \times 100 = 16.5 \text{ kPa}$ or 0.0165 MPa $\therefore T_{dp} = 55.9°C$
5. A $C_3H_8 + 9(O_2 + 3.76N_2) \rightarrow 3CO_2 + 4H_2O + 4O_2 + 33.84N_2$

$$AF = \frac{9 \times 4.76 \times 29}{1 \times 44} = 28.2$$

6. C $\% \, CO_2 = \dfrac{3}{3 + 4 + 4 + 33.84} \times 100 = 6.69\%$
7. D $C_3H_8 + 5(O_2 + 3.76N_2) \rightarrow 3CO_2 + 4H_2O + 18.8N_2$

$$AF = \frac{5 \times 4.76 \times 29}{1 \times 44} = 15.7 \qquad \dot{m}_{fuel} = \frac{20\rho_{air}}{15.7} = \frac{20}{15.7} \times \frac{100}{0.287 \times 298} = 1.52 \text{ kg/min}$$

8. A $C_4H_{10} + 9.75(O_2 + 3.76N_2) \rightarrow 3.8CO_2 + 0.2CO + 5H_2O + 36.66N_2 + 3.35O_2$

$$AF = \frac{9.75 \times 4.76 \times 29}{58} = 23.2$$

9. B $2H_2 + O_2 \rightarrow 2H_2O$

$$Q = H_P - H_R = \sum_{\text{prod}} N_i(\bar{h}_f^o)_i - \sum_{\text{react}} N_i(\bar{h}_f^o)_i$$
$$= 2(-241820) - 0 - 0 = -483\,640 \text{ kJ/kmol}$$

10. A $C_3H_8 + 7.5(O_2 + 3.76N_2) \rightarrow 3CO_2 + 28.2N_2 + 4H_2O + 2.5O_2$

$$q = 3(-393\,520 + 42\,769 - 9364) + 4(-241\,820 + 35\,882 - 9904)$$
$$+ 28.2(30\,129 - 8669) + 2.5(31\,389 - 8682) + 103\,850$$
$$= -1\,178\,000 \text{ kJ/kmol}$$

Final Exam No. 1

1. B

2. D $J = N \cdot m = \dfrac{N}{m^2} \cdot m^3 = Pa \cdot m^3$

3. A $PA = Kx \qquad 400\,000\pi \times 0.02^2 = 800x \qquad \therefore x = 0.628 \text{ m}$

4. D $h = \dfrac{38}{50}(3023 - 2902) = 2994 \text{ kJ/kg}$

5. D $v_1 = v_g = 0.1274 = v_2 \qquad T_2 = 600°C$

$$P_2 = \frac{0.1324 - 0.1274}{0.1324 - 0.09885}(1) + 3 = 3.15 \text{ MPa}$$

6. B $m = PV/RT = 380 \times 0.2/(0.287 \times 298) = 0.888 \text{ kg}$

7. D $W = \int PdV = \int_{0.2}^{0.8}(0.2 + 0.4V)dV = 0.2 \times 0.6 + 0.2(0.8^2 - 0.2^2) = 0.24 \text{ kJ}$

8. B $W = mP\Delta v = 10 \times 500(0.443 - 0.0012) = 2120 \text{ kJ}$

9. B $v_1 = 0.7726 \qquad v_2 = 0.7726 - 0.5(0.2)$
$= 0.6826 \quad P_2 = 400 \text{ kPa} \qquad \therefore T_2 = 315°C$

10. D $Q = m\Delta h = 1 \times (3098 - 3273) = -175 \text{ kJ}$

11. A $Q = m\Delta u = 1 \times (2829 - 1246) = 1580$ kJ. We used:

$$v_3 = v_2 = 0.6726 = 0.001 + x_3(1.694 - 0.001) \qquad \therefore x_3 = 0.397, u_3 = 1246$$

12. D $Q = mRT \ln(v_2/v_1) = 1 \times 0.287 \times 373 \ln(1/2) = -74$ kJ

13. A $m_{cu} C_{cu} \Delta T_{cu} = m_w C_w \Delta T_w$

$$5 \times 0.39(200 - T_2) = 20 \times 4.18(T_2 - 10) \qquad \therefore T_2 = 14°\text{C}$$

14. A

15. D $0 = h_2 - h_1 + \dfrac{V_2^2 - V_1^2}{2} = 3051 - 3445 + \dfrac{V_2^2 - 20^2}{2 \times 1000} \qquad \therefore V_2 = 888$ m/s

16. C $\dot{m} = A_1 V_1 / v_1 = \pi \times 0.05^2 \times 20 / 0.08643 = 1.817$ kg/s

17. A $-\dot{W}_S = \dot{m}\Delta h = \dot{m}(u + v\Delta P) = 5(0 + 0.001 \times 4000) = 20$ kW

18. C $\eta = \dfrac{\dot{W}}{\dot{Q}_{in}} = 1 - \dfrac{T_L}{T_H} = 1 - \dfrac{293}{473} = \dfrac{10}{\dot{Q}_{out} + 10} \qquad \therefore \dot{Q}_{out} = 16.3$ kJ/s

19. A $\eta_{max} = 1 - \dfrac{T_L}{T_H} = 1 - \dfrac{283}{300} = 0.0567 \qquad \eta_{engine} = \dfrac{10}{10 + 165}$

$$= 0.0571 \qquad \text{Impossible}$$

20. C

21. A $\text{COP} = \dfrac{20}{10} = \dfrac{T_L}{T_H - T_L} = \dfrac{293}{T_H - 293} \qquad \therefore T_H = 439.5°\text{C}$

22. D $\Delta S_{net} = \Delta S_{cu} + \Delta S_{lake} = 0.39 \times 10 \ln \dfrac{293}{373} + \dfrac{0.39 \times 10 \times (100 - 20)}{293}$

$$= 0.122 \text{ kJ/K}$$

23. C $h_1 - h_2 = 3870 - 2462 = 1408$ kJ/kg where $s_2 = s_1 = 7.170$

$$= 1.026 + 6.645 x_2 \text{ so that } x_2 = 0.9246 \text{ and } h_2 = 2462 \text{ kJ/kg}$$

24. B $T_2 = 300\left(\dfrac{800}{100}\right)^{0.2857} = 543$ K $\qquad w_{min} = C_p \Delta T = 1.0(543 - 300)$

$$= 243 \text{ kJ/kg}$$

25. B $T_2 = 300\left(\frac{100}{400}\right)^{0.2857} = 202 \text{ K} \qquad \frac{V_2^2}{2\times 1000} = 1.0(300-202)$

$\therefore V_2 = 443 \text{ m/s}$

26. A $\eta = \frac{h_1 - h_2}{h_1 - h_{2'}} \quad 0.8 = \frac{3870 - h_2}{3870 - 2462} \quad \therefore h_2 = 2744 \text{ and } u_2 = 2560 \text{ where}$

$s_{2'} = s_1 = 7.170 = 1.026 + 6.645x_2$ so that $x_2 = 0.9246$ and $h_{2'} = 2462$

27. B

28. A

29. C $P_2 = P_1(T_2/T_1)^{k/(k-1)} = 100(754/300)^{3.5} = 2520 \text{ kPa}$

30. D $q = C_v\Delta T = 0.717(1600 - 754) = 607 \text{ kJ/kg}$

31. B $v_1/v_2 = (T_2/T_1)^{1/(k-1)} = (754/300)^{2.5} = 10$

32. D $q = \Delta h = C_p\Delta T = 1.0(1000 - 500) = 500 \text{ kJ/kg}$

33. C $P_2 / P_1 = (T_2/T_1)^{k/(k-1)} = (500/300)^{3.5} = 6$

34. D $w_C/w_T = \Delta h_C/\Delta h_T = C_p(\Delta T)_C/C_p(\Delta T)_T = \frac{500-300}{1000-600} = 0.5$

35. A $q = h_3 - h_2 = 3625 - 251 = 3374 \text{ kJ/kg}$

36. A $s_3 = s_4 = 6.903 = 0.832 + 7.7077x_4 \qquad \therefore x_4 = 0.858 \text{ and } h_4 = 2274$

$w_T = h_3 - h_4 = 3625 - 2274 = 1351 \text{ kJ/kg}$

37. B $s_3 = 6.903 \qquad P_3 = 4 \qquad \therefore h_4 = 3307$

$q_{\text{in}} = h_3 - h_2 + h_5 - h_4 = 3625 - 251 + 3674 - 3307 = 3741$

38. D $\dot{W}_P = \dot{m}\Delta P/\rho = (2400/60)(10\,000 - 20)/1000 = 400 \text{ kW}$

39. D $0.622P_v/P_a = 0.015 \qquad \text{and} \qquad 100 = P_v + P_a$

$\therefore P_v = 2.355 \qquad \therefore \phi = \frac{2.355}{3.169} = 0.743$

40. B At $\phi = 90\%$ and $T = 5°\text{C},\ \omega = 0.0048$

At $T = 25°\text{C}$ and $\omega = 0.0048,\ \phi = 25\%$

Final Exam No. 2

1. D
2. C
3. B
4. C
5. B $P = \rho g \Delta h = 1000 \times 9.8 \times 20 = 196\,000$ Pa
6. B $Q = m\Delta h \qquad 320 = 1(h_2 - 3051) \qquad \therefore h_2 = 3371$
   $P_2 = 1$ MPa $\qquad \therefore T_2 = 450°\text{C}$
7. B $W = P\Delta v = 1000 \times 1(0.3304 - 0.2579) = 72.5$ kJ
8. D $m_s C_{p,s} \Delta T_s = m_w C_{p,w} \Delta T_w$
   $200 \times 0.8(T_2 - 0) = 42 \times 4.18(80 - T_2) \qquad \therefore T_2 = 42°\text{C}$
9. C Interpolate: $T = \dfrac{3000 - 2950}{3119 - 2950} \times 100 + 400 = 430°\text{C}$
10. C $\dfrac{0.7 - 0.6548}{0.7726 - 0.6548}(3273 - 3067) + 3067 = 3146$ kJ/kg
11. D $m = \dfrac{V}{v} = 0.1 / 0.892 = 0.112$ kg
    $Q = m\Delta u = 0.112(3130 - 2529) = 67.3$ kJ
12. B $C_p = \Delta h / \Delta T = (3214 - 2961)/100 = 2.53$ kJ/kg·K (central differences)
13. D From last row: $U_1 = 10$. From 1st row: $40 - W = 30 - 10 \quad \therefore W = 20$ kJ
14. D $q = \Delta h = C_p \Delta T = 2.254 \times 300 = 676$ kJ/kg
15. A $Q = mC_p \Delta T = 1 \times 1.0(100 - 600) = -500$ kJ
16. C $\dfrac{v_3}{v_2} = \dfrac{T_3}{T_2} = \dfrac{373}{873} \qquad W_{1\text{-}2} = mRT \ln \dfrac{v_2}{v_3} = 1 \times 0.287 \times 873 \ln \dfrac{873}{373} = 213$ kJ
17. C $Q_{1\text{-}2} = W_{1\text{-}2} = 213$ kJ

18. C $h_2 = h_1 = 71.69 = 24.26 + 211.05x_2 \qquad \therefore x_2 = 0.225$

19. A $\dot{W} = \dot{m}\Delta h = \dot{m}C_p\Delta T = 1 \times 1.0(200 - 20) = 180 \text{ kW}$

20. A $q = \Delta h = C_p\Delta T \qquad 200 = 1.0\Delta T \qquad \therefore \Delta T = 200°\text{C}$

21. D $0 = h_2 - h_1 + \dfrac{V_2^2 - V_1^2}{2} \qquad 1.0 \times \Delta T = \dfrac{10^2 - 400^2}{2 \times 1000} \qquad \therefore \Delta T = -80°\text{C}$

22. B

23. C

24. D

25. D $\eta = \dfrac{7}{100} = 0.07 \qquad \eta_{\text{max}} = 1 - \dfrac{T_L}{T_H} = 1 - \dfrac{283}{303} = 0.066$

$\therefore$ Violation of second law

26. C

27. C $\eta_{\text{max}} = 1 - \dfrac{T_L}{T_H} = 1 - \dfrac{293}{873} = 0.664 = \dfrac{\dot{W}}{\dot{Q}_{\text{in}}} = \dfrac{\dot{W}}{100} \qquad \therefore \dot{W} = 66.4 \text{ kW}$

28. D $Q = mC_i\Delta T = 10 \times 0.45(300 - 0) = 1350 \text{ kJ}$

$\Delta S = 10 \times 0.45 \ \ln\dfrac{273}{573} + \dfrac{1350}{273} = 1.61 \text{ kJ/K}$

29. A $T_2 = T_1(P_2/P_1)^{(k-1)/k} = 293(100/2000)^{0.2857} = 124 \text{ K} \qquad \text{or} \qquad -149°\text{C}$

30. B $q_{2\text{-}3} = \Delta h = C_p\Delta T = 1.0(2000 - 950) = 1050 \text{ kJ/kg}$

31. C $v_1/v_2 = (T_2/T_1)^{1/(k-1)} = (950/300)^{2.5} = 17.8$

32. C $\dfrac{v_3}{v_2} = \dfrac{T_3}{T_2} = \dfrac{2000}{950} = 2.11 \qquad \dfrac{v_4}{v_3} = \dfrac{v_1}{v_2} \times \dfrac{v_2}{v_3} = \dfrac{17.8}{2.11} = 8.46$

$T_4 = T_3(v_3/v_4)^{k-1} = 2000 / 8.46^{0.4} = 851 \text{ K}$

33. B $q = h_3 - h_2 = 3538 - 335 = 3203 \text{ kJ/kg}$

34. A $w_P = \Delta P/\rho = (20\,000 - 40)/1000 = 20 \text{ kJ/kg}$

35. D $w_T = h_3 - h_4 = 3538 - 2637 = 901$ kJ/kg

36. C $\dot{m}_s \Delta h_s = \dot{m}_w \Delta h_w \quad 50(2637 - 335) = \dot{m}_w \times 4.18 \times 15 \quad \therefore \dot{m}_w = 1836$ kg/s

37. C From the psychrometric chart, $\phi = 60\%$

38. D $\dot{Q} = \dot{m}\Delta h = 10(50 - 29.5) = 205$ kJ using $\omega_1 = \omega_2$ (psychrometric chart)

39. A Theoretical air: $C_2H_6 + 3.5(O_2 + 3.76N_2) \Rightarrow 2CO_2 + 3H_2O + 13.16N_2$

$(5 - 3.5)/3.5 = 0.429$ or 42.9% excess air

40. D $AF = \dfrac{m_{\text{air}}}{m_{\text{fuel}}} = \dfrac{5 \times 4.76 \times 29}{1 \times 30} = 23.0$

INDEX

O

P

Q

R

S

T

U

V

W

Z